Nonlinear Systems - Recent Developments and Advances
http://dx.doi.org/10.5772/intechopen.100717
Edited by Bo Yang and Dušan Stipanović

Contributors
Bilal Bilalov, Sabina Sadigova, Zaur Kasumov, Sabrina Streipert, Kibru Teka, Pu Li, Abebe Geletu, Markus Vogl, Marc Delphin Monsia, Jean Akande, Kolawolé Kêgnidé Damien Adjaï, Marcellin Nonti, Dongran Song, Ziqun Li, Jian Yang, Mi Dong, Liansheng Huang, Xiaojiao Chen, Minyu Chen, Xing He, Erdal Şehirli, Noureddine Bouteraa, Habib Djourdem, Serge Gervais Ngueuteu Mbouna, Sujatha Vijayaraghavan

First published in London, United Kingdom, 2023 by IntechOpen
IntechOpen is the global imprint of INTECHOPEN LIMITED, registered in England and Wales, registration number: 11086078, 5 Princes Gate Court, London, SW7 2QJ, United Kingdom

British Library Cataloguing-in-Publication Data
A catalogue record for this book is available from the British Library

Additional hard and PDF copies can be obtained from orders@intechopen.com

Nonlinear Systems - Recent Developments and Advances
Edited by Bo Yang and Dušan Stipanović
p. cm.
Print ISBN 978-1-80356-926-0
Online ISBN 978-1-80356-927-7
eBook (PDF) ISBN 978-1-80356-928-4

Meet the editors

Prof. Bo Yang obtained his Ph.D. in electrical engineering from the University of Liverpool, UK in 2015 (jointly sponsored by the University and by China's Scholarship Council). He is now a professor in the Faculty of Electrical Power Engineering, at Kunming University of Science and Technology, Kunming. Prof. Yang's main research areas include the optimization and control of renewable energy and energy storage systems, and the application of artificial intelligence in smart grids. He has headed over 20 important research projects, including two projects for China's National Natural Science Foundation. Prof. Yang was declared a highly cited scholar (electrical engineering) by Elsevier China in 2021. He has published more than 170 peer-reviewed SCI/EI papers and a book on nonlinear robust control of renewable energy systems.

Prof. Dušan Stipanović received his BS degree in electrical engineering (specializing in controls) from the University of Belgrade, Serbia, in 1994, and the MSEE and Ph.D. degrees in electrical engineering from Santa Clara University, California, in 1996 and 2000, respectively. Prof. Stipanović was an adjunct lecturer and research associate in the Department of Electrical Engineering at Santa Clara University (1998-2001), and a research associate in Professor Claire Tomlin's Hybrid Systems Laboratory in the Department of Aeronautics and Astronautics at Stanford University (2001-2004). In 2004 he joined the University of Illinois at Urbana-Champaign where he is now a Professor in the Controls Group of the Coordinated Science Laboratory and Department of Industrial and Enterprise Systems Engineering.

Contents

Preface

In mathematics and science, a nonlinear system is one in which the change of the output is not proportional to the change of input. Because most systems in nature are nonlinear, nonlinear problems have aroused the interest of engineers, biologists, physicists, mathematicians, and many other scientists. The most prominent difference between a nonlinear system and a linear system is that a nonlinear system may lead to chaos, unpredictability, or non-intuitive results.

In general, the behavior of a nonlinear system can be mathematically described as a set of nonlinear simultaneous equations, in which the unknown number (or the unknown functions in the differential equations) appears as a polynomial variable higher than the first degree, or as a parameter of a polynomial function of a non-first degree. Generally speaking, the behavior of nonlinear systems is described mathematically by a set of nonlinear simultaneous equations, which contain non-first-degree polynomials composed of unknowns. In other words, a nonlinear equation cannot be written as a linear combination of its unknowns. A nonlinear differential equation refers to a term in which the power of the unknown function and its derivative function is not equal to one. When determining whether an equation is linear or nonlinear, only the part of the unknown number (or unknown function) needs to be considered, without checking whether there is a known nonlinear term in the equation. For example, in the differential equation, if the order of all unknown functions and unknown derivatives is one, it can still be regarded as a linear differential equation even if there is a nonlinear function composed of a known variable.

Since nonlinear equations are quite difficult to solve, we often need to approximate a nonlinear system with linear equations (linear approximation). This approximation is very accurate for the input values (variables) in a certain range, but after linear approximation, many interesting phenomena, such as solitary waves, chaos, and singularities cannot be explained. These strange phenomena often make the behavior of nonlinear systems seem counterintuitive, unpredictable, or even chaotic. Therefore, it is of great theoretical significance to analyze and solve nonlinear differential equations.

At the same time, economic advances and the continuous innovation and development of electronic computer technology have led to the rise of automatic control technology which is gradually being applied to aviation detection, engineering production, mechanical equipment research and development and management, weapons manufacturing and other areas, where it is of inestimable value. Automatic control technology has also begun to expand into the realm of social life, in areas such as biological manufacturing, medical research, and environmental management. Automatic control systems may be linear or nonlinear. A linear system has regularity, but it is nonlinear systems that are widely found in society. Compared with a linear system, a nonlinear system cannot meet the superposition, and it is unbalanced. Research into nonlinear control systems and related control strategies is of important practical significance and value for engineering applications.

This book investigates promising and in-depth research on nonlinear systems and related control strategies. Chapter 1 describes the extension of differential equations to different underlying time domains, so-called dynamic equations on time scales, whose calculus unifies the continuous and discrete calculus and extends it to any non-empty closed subset of real numbers. Dynamic equations on time scales allow the modeling of processes that are neither fully discrete nor fully continuous. Chapter 2 deals with a class of nonlinear fractional differential equations involving the Caputo fractional derivative with nonlocal boundary conditions. By applying the Leray–Schauder nonlinear alternative and Banach contraction principle, the existence and uniqueness of this problem are established. In Chapter 3, the dynamical behavior of the incommensurate fractional-order FitzHugh–Nagumo model of neurons is explored in detail from local stability analysis. The FitzHugh–Nagumo model is a mathematical simplification of the Hodgkin–Huxley model, which proves that the fractional-order FitzHugh–Nagumo model can be simulated by a simple electrical circuit where the capacitor and the inductor are replaced by corresponding fractional-order electrical elements. The local stability of this model is then studied using the theorem on the stability of an incommensurate fractional-order system combined with Cauchy's argument principle. Finally, the dynamic behaviors of the model are investigated. In Chapter 4, an m-th order elliptic equation is considered in Sobolev spaces generated by the norm of a grand Lebesgue space. Subspaces are determined in which the shift operator is continuous, and local solvability (in the strong sense) is established in these subspaces. Interior and up-to-boundary Schauder-type estimates are established with respect to these Sobolev spaces for m-th order elliptic operators, the trace of functions and trace operator are determined, the boundedness of trace operator and the extension theorem is proved, and the properties of the Riesz potential regarding these Sobolev spaces are studied. In Chapter 5, a practical introduction to a generalized nonlinear analysis framework tailored for time-series data is provided, enabling the safe quantification of underlying evolutionary dynamics, which describe the referring empirical data-generating process. The reader can incorporate the proposed analysis framework, conduct the analyses and reconstructions using the correct specifications, and learn about misleading propositions or parameter choices. Chapter 6 proposes random matrix theory (RMT) to handle this problem, which begins by modeling spatial-temporal datasets as sequences, each of whose terms is in the form of a random matrix. Fundamental RMT principles are briefly discussed, such as asymptotic spectrum laws, transforms, convergence rate, and free probability, in order to extract high-dimensional statistics from the random matrix as the indicators. The statistical properties of these indicators are discussed for a better understanding of the system, and potential applications are suggested. Chapter 7 studies the structural properties and convergence approach of chance-constrained optimization of boundary-value elliptic partial differential equation systems (CCPDEs). Real-world systems, such as physical and living systems, are generally subject to vibrations that can affect their long-term integrity and safety. Thus, the determination of the law that governs the evolution of the oscillatory quantity has become a major topic in modern engineering design. Chapter 8 reviews recent developments and advances in the theory of isochronous oscillations of nonlinear systems. Chapter 9 focuses on conventional and proposed control configuration selection methods for nonlinear systems. The proposed input-output pair method for nonlinear benchmark processes is also calculated. In Chapter 10, feedback linearization techniques, including input-state and input-output linearization methods, are described. The input-output linearization method is then used for the output voltage control of an interleaved boost converter.

Chapter 11 proposes an advanced nonlinear model predictive control solution, including multi-step prediction models, for the yaw control system of a horizontal-axis wind turbine. A notable feature of the proposed solution is the use of a finite control set under constraints for the potentially demanded yaw rate; the optimal control demand for the yaw system is conveniently solved using an exhaustive search method based on a sequential diagram.

In summary, this book aims to provide advanced research on nonlinear systems and control schemes for researchers and engineers working in relevant fields.

Bo Yang
Faculty of Electric Power Engineering,
Kunming University of Science and Technology,
Kunming, China

Dušan Stipanović
Coordinated Science Laboratory and ISE Department,
University of Illinois at Urbana-Champaign,
Urbana, Illinois, USA

Section 1

Analysis and Solution of Nonlinear Differential Equations

Chapter 1

Dynamic Equations on Time Scales

Sabrina Streipert

Abstract

An extension of differential equations to different underlying time domains are so called dynamic equations on time scales. Time scales calculus unifies the continuous and discrete calculus and extends it to any nonempty closed subset of the real numbers. Choosing the time scale to be the real numbers, a dynamic equation on time scales collapses to a differential equation, while the integer time scale transforms a dynamic equation into a difference equation. Dynamic equations on time scales allow the modeling of processes that are neither fully discrete nor fully continuous. This chapter provides a brief introduction to time scales and its applications by incorporating a selective collection of existing results.

Keywords: time scales, existence, uniqueness, linear, applications

1. Introduction

The modeling of processes using differential equations is a well-established method in multiple branches of sciences. Dependent on the model assumptions, the form of the differential equation can range from a comparably simple ordinary differential equation to more advanced formulations using nonlinear, higher order, and partial differential equations. Reasons to consider difference equations include computational benefits and, even more fundamental, a discrete modeling perspective. For example, when describing a zero-coupon bond where the invested amount at time t, M_t, receives interest r at the end of each year but remains unchanged during each year, the recursive model $M_{t+1} = (1 + r)M_t$ captures the change of the investment from time t to time $t + 1$. Difference equations are also a common tool to describe processes on a macro scale in time, for example, when describing non-overlapping generations. Even though the number of individuals may vary throughout the generation period, one may only be interested in the individuals at the beginning of each generation time, i.e., the size of each cohort. There are however processes that cannot be described accurately using differential or difference equations. For example, when modeling species that are reproducing continuously during certain months of the year before laying eggs right before hibernating. Another example are plant populations that grow continuously during some months of the year and plant their seeds prior to dying out. In [1], Robert May gives examples of insects that exhibit such hybrid continuous–discrete behavior.

Instead of introducing a set of simplifying assumptions and possibly discontinuous model parameters that impact the model analysis, dynamic equations on time scales can provide a simple alternative to describe such processes. Time scales calculus was introduced by Stefan Hilger in 1988 [2]. It unifies the continuous and discrete calculus and extends it to any nonempty closed subset of the real numbers called a time scale, denoted by $\mathbb{T}$. By introducing differentiation and integration on $\mathbb{T}$, the classical theory of differential equations can be extended to time scales, which allows the modeling of processes that are not changing continuously nor solely discretely in time. These so-called dynamic equations are essentially the time scales analogue of differential and difference equations and have gained increasing interest due to their potential in applications. Choosing the time scale to be the real numbers, a dynamic equation transforms into a differential equation and by choosing the time scale to be the integers, a corresponding difference equation is obtained. Thus, instead of studying differential equations and difference equations separately, time scales provides also a tool to investigate both by analyzing the corresponding dynamic equation. This is specifically interesting since certain difference equations exhibit significantly different behavior as their continuous analogues, see for example the "logistic map" and the "logistic differential equation". By analyzing a dynamic equation on time scales, the effect of the underlying time domain onto the behavior of solutions may be revealed.

2. Time scales fundamentals

In this subsection, the basic definitions of time scales calculus are introduced based on the introductory book [3].

Definition 1. A time scale, denoted by $\mathbb{T}$, is a nonempty closed subset of $\mathbb{R}$.

Examples of a time scale are $\mathbb{R}, \mathbb{Z}, h\mathbb{Z}, q^{\mathbb{N}_0} = \{1, q, q^2, q^3, \dots\} (q > 1), [a, b] \cup \{c, d\}$ where $a < b$ and $a, b, c, d \in \mathbb{R}$, and the Cantor set. It therefore contains the popular cases of the continuous, the discrete, and the quantum calculus.

Operators that aid the description of a time scale are the "forward jump operator", denoted by $\sigma(t)$, the "backward jump operator", denoted by $\rho(t)$, and the "graininess function", denoted by $\mu(t)$. These operators are defined for $t \in \mathbb{T}$ as

$$\sigma(t) := \inf\{s \in \mathbb{T} : s > t\}, \quad \rho(t) := \sup\{s \in \mathbb{T} : s < t\}, \quad \mu(t) := \sigma(t) - t. \tag{1}$$

Since $\mathbb{T}$ is closed, $\sigma, \rho : \mathbb{T} \to \mathbb{T}$ and $\mu : \mathbb{T} \to [0, \infty)$. **Table 1** provides values of the corresponding operators for different examples of time scales.

$\mathbb{T}$	$\sigma(t)$	$\rho(t)$	$\mu(t)$
$\mathbb{R}$	t	t	0
$\mathbb{Z}$	$t+1$	$t-1$	1
$q^{\mathbb{N}_0}$	qt	$\frac{t}{q}$	$t(q-1)$

Table 1.
The description of the time scales functions σ, ρ, μ for the examples of $\mathbb{R}$, $\mathbb{Z}$, and $q^{\mathbb{N}_0} (q > 1)$.

Using these operators, any $t \in \mathbb{T}$ can be classified as:

- *right-scattered (left-scattered)*, if $\sigma(t) > t$ ($\rho(t) < t$), and
- *right-dense (left-dense)*, if $\sigma(t) = t$ ($\rho(t) = t$).

We say that a point $t \in \mathbb{T}$ is isolated, if it is right- and left-scattered. We say that a point $t \in \mathbb{T}$ is dense, if it is right- and left-dense. Note that for $\mathbb{T} = \mathbb{R}$, every point is dense and, for $\mathbb{T} = \mathbb{Z}$, every point is isolated.

Example 2.1. El Nin̄o events can be described using a time scale. El Nin̄o events between 2002 and 2017 have been observed in the time intervals 2002–2003, 2004–2005, 2006–2007, 2009–2010, and 2014–2016 [4], which suggests the corresponding time scale (**Figure 1**, **Table 2**)

$$\mathbb{T} = \cup_{i=0}^{5} [a_i, a_{i+1}]$$

with $(a_0, a_1, a_2, \ldots, a_5) = (2002, 2004, 2006, 2009, 2014, 2015)$.

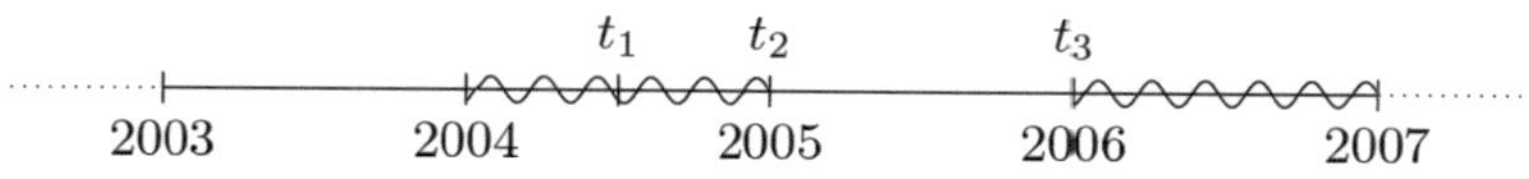

Figure 1.
Part of the time line containing points in the time scale $\mathbb{T}$. Curly lines identify intervals within $\mathbb{T}$. Here, $t_1 \in (2004, 2005)$, t_2 is the last point in the interval $[2004, 2005]$, and $t_3 = 2006$ is the first point in $[2006, 2007]$.

$t \in \mathbb{T}$	$\sigma(t)$	$\mu(t)$	$\rho(t)$
$t_1 \in (2004, 2005)$	$\sigma(t_1) = t_1$	$\mu(t_1) = 0$	$\rho(t_1) = t_1$
$t_2 = 2005$	$\sigma(t_2) = t_3$	$\mu(t_2) = 1$	$\rho(t_2) = t_2$
$t_3 = 2006$	$\sigma(t_3) = t_3$	$\mu(t_3) = 0$	$\rho(t_3) = t_2$

Table 2.
The functions σ, ρ, μ for the time points $t_1, t_2, t_3 \in \mathbb{T}$ based on ***Figure 1***.

The following notation is commonly used for $t \in \mathbb{T}$,

$$\sigma^n(t) = \underbrace{(\sigma \circ \sigma \circ \ldots \circ \sigma)}_{n-\text{times}}(t), \qquad \rho^n(t) = \underbrace{(\rho \circ \rho \circ \ldots \circ \rho)}_{n-\text{times}}(t).$$

2.1 Functions on time scales

We can now consider scalar functions on time scales, that is, $f : \mathbb{T} \to \mathbb{R}$, and discuss their properties. We define the subset $\mathbb{T}^\kappa$ as follows: If $\mathbb{T}$ has a left-scattered maximum $m \in \mathbb{T}$, then $\mathbb{T}^\kappa = \mathbb{T} \backslash \{m\}$, else $\mathbb{T}^\kappa = \mathbb{T}$.

Definition 2. f: $\mathbb{T} \to \mathbb{R}$ is called regressive, if, for all $t \in \mathbb{T}^\kappa$,

$$1 + \mu(t) f(t) \neq 0$$

and is called positively regressive, if, for all $t \in \mathbb{T}^\kappa$,

$$1 + \mu(t) f(t) > 0.$$

The following are properties of $f : \mathbb{T} \to \mathbb{R}$ that later identify integrability.

Definition 3. $f : \mathbb{T} \to \mathbb{R}$ is called regulated provided its right-sided limit exists (as a finite value) for all right-dense points and its left-sided limit exists (as a finite value) for all left-dense points.

Even though every regulated function on a compact interval is bounded, in general, $\max_{a \leq t \leq b} f(t)$ and $\min_{a \leq t \leq b} f(t)$ do not need to exist for regulated $f : \mathbb{T} \to \mathbb{R}$.

Definition 4. $f : \mathbb{T} \to \mathbb{R}$ is called rd-continuous if f is continuous at all right-dense points and its left-sided limit exists (as a finite value) for all left-dense points. The set of rd-continuous functions is denoted by $C_{rd} = C_{rd}(\mathbb{T}) = C_{rd}(\mathbb{T}, \mathbb{R})$.

Note that, if $f : \mathbb{T} \to \mathbb{R}$ is continuous, then f is rd-continuous. If f is rd-continuous, then f is regulated.

The set of rd-continuous and regressive (positively regressive) functions is denoted by $\mathcal{R} = \mathcal{R}(\mathbb{T}) = \mathcal{R}(\mathbb{T}, \mathbb{R})$ ($\mathcal{R}^+ = \mathcal{R}^+(\mathbb{T}) = \mathcal{R}^+(\mathbb{T}, \mathbb{R})$).

Beside the classical addition and subtraction of functions, time scales calculus introduces the so-called "circle plus", denoted by $\oplus$, and "circle minus", denoted by $\ominus$. These operations are defined for $f, g : \mathbb{T} \to \mathbb{R}$ as follows

$$(f \oplus g)(t) = f(t) + g(t) + (\mu f g)(t)$$

$$\text{and, for } g \in \mathcal{R}, \qquad (f \ominus g)(t) = \frac{f(t) - g(t)}{1 + (\mu g)(t)}.$$

A useful property is that if $f, g \in \mathcal{R}$ ($\mathcal{R}^+$), then $f \oplus g, f \ominus g \in \mathcal{R}$ ($\mathcal{R}^+$) implying that the (positively) regressive property is being carried over. Furthermore, $(\mathcal{R}, \oplus)$ forms an Abelian group with the inverse elements of $f \in \mathcal{R}$ given by $\ominus f$.

For $\mathbb{T} = \mathbb{R}$, the operators $\oplus$ and $\ominus$ correspond to the classical addition and subtraction.

2.2 Differentiation

Definition 5. Let $f : \mathbb{T} \to \mathbb{R}$ and $t \in \mathbb{T}^\kappa$. If there exists $f^\Delta(t) \in \mathbb{R}$ such that for all $\varepsilon > 0$, there exists $\delta > 0$ such that

$$|f(\sigma(t)) - f(s) - f^\Delta(t)(\sigma(t) - s)| \leq \varepsilon |\sigma(t) - s| \quad \textit{for all} \quad s \in (t - \delta, t + \delta) \cap \mathbb{T},$$

then we call $f^\Delta(t)$ the delta (or Hilger) derivative of f at $t \in \mathbb{T}^\kappa$.

If $f^\Delta(t)$ exists for all $t \in \mathbb{T}^\kappa$, we say that f is delta differentiable (or short: differentiable) and the function $f^\Delta : \mathbb{T} \to \mathbb{R}$ is called delta derivative of f on $\mathbb{T}^\kappa$.

If f is differentiable at $t \in \mathbb{T}^\kappa$, then

$$f(\sigma(t)) = f(t) + \mu(t) f^\Delta(t).$$

The following notations are used equivalently

$$(f^\sigma)(t) = (f \circ \sigma)(t) = f(\sigma(t)).$$

The definition of a delta derivative can be extended to consider higher order derivatives. We say that f is twice delta differentiable with the second (delta) derivative $f^{\Delta\Delta}$, if f^Δ is (delta) differentiable on $\mathbb{T}^{\kappa^2} = (\mathbb{T}^\kappa)^\kappa$.

Note that the definition of delta derivatives focuses on the change forward in time. A corresponding definition that focuses on the change backward in time is referred to as nabla derivative, see for example [5].

Theorem 2.2. [See [3, Theorem 1.16]] Let $f: \mathbb{T} \to \mathbb{R}$ and $t \in \mathbb{T}^\kappa$. Then, the following holds:

i. If t is right-dense, then

$$f^\Delta(t) = \lim_{s \to t} \frac{f(t) - f(s)}{t - s},$$

provided that the limit exists (as a finite number).

ii. If f is continuous at the right-scattered point t, then

$$f^\Delta(t) = \frac{f(\sigma(t)) - f(t)}{\mu(t)}.$$

Applying Theorem 2.2 for the case of $\mathbb{T} = \mathbb{R}$, shows that the delta derivative is consistent with the classical derivative, that is, $f^\Delta(t) = f(t)$ for $t \in \mathbb{T} = \mathbb{R}$. For $\mathbb{T} = \mathbb{Z}$, the delta derivative collapses to the forward Euler operator, widely accepted as the discrete analogue of a derivative, that is, $f^\Delta(t) = f(t+1) - f(t)$ if $\mathbb{T} = \mathbb{Z}$ (see **Table 3**).

$\mathbb{T}$	$\mathbb{T} = \mathbb{R}$	$\mathbb{T} = \mathbb{Z}$	$\mathbb{T} = q^{\mathbb{N}_0}$
$f^\Delta(t)$	$f'(t)$	$\Delta f(t)$	$\frac{f(qt)-f(t)}{t(q-1)}$

Table 3.
Derivatives for the examples of $\mathbb{T} = \mathbb{R}$, $\mathbb{T} = \mathbb{Z}$, *and* $\mathbb{T} = q^{\mathbb{N}_0}$ $(q > 1)$. *Note that* $\Delta f(t) = f(t+1) - f(t)$ *is the forward Euler operator.*

As in the continuous case, the differential operator is linear, that is, for $\alpha, \beta \in \mathbb{R}$, $t \in \mathbb{T}^\kappa$, and for (delta) differentiable functions $f, g: \mathbb{T} \to \mathbb{R}$,

$$(\alpha f + \beta g)^\Delta(t) = \alpha f^\Delta(t) + \beta g^\Delta(t).$$

The analogues of the product and the quotient rule on time scales take on slightly different forms. For (delta) differentiable functions $f, g: \mathbb{T} \to \mathbb{R}$, and $t \in \mathbb{T}^\kappa$,

$$(fg)^\Delta(t) = f^\Delta(t) g^\sigma(t) + f(t) g^\Delta(t) = f^\Delta(t) g(t) + f^\sigma(t) g^\Delta(t)$$

and, for $g(t), g^\sigma(t) \neq 0$,

$$\left(\frac{f}{g}\right)^\Delta(t) = \frac{f^\Delta(t) g(t) - f(t) g^\Delta(t)}{g(t) g^\sigma(t)}.$$

For $\mathbb{T} = \mathbb{R}$, we have $f^\sigma = f$ and $g^\sigma = g$ so that the classical product and quotient rule are retrieved. In the case of $\mathbb{T} = \mathbb{Z}$, we have the correspondent rules consistent with [6], namely

$$\Delta(fg)(t) = (\Delta f(t))g(t+1) + f(t)(\Delta g(t)) = (\Delta f(t))g(t) + f(t+1)(\Delta g(t)).$$

If $g(t), g(t+1) \neq 0$, then

$$\Delta\left(\frac{f(t)}{g(t)}\right) = \left(\frac{f}{g}\right)^{\Delta}(t) = \frac{(\Delta f(t))g(t) - (\Delta g(t))f(t)}{g(t)g(t+1)}.$$

The modifications in the product and quotient rule highlight that some of the well established differentiation rules only carry over to time scales calculus after some adjustments. In fact, the product rule on time scales reveals that the useful property of power functions $f(t) = t^n$ for $n \in \mathbb{N}_0$ is no longer the simple reduction of the power by one, because

$$\left(t^2\right)^{\Delta} = (t \cdot t)^{\Delta} = t + \sigma(t),$$

which may not be delta differentiable. This indicates already that the series representation of functions requires further thought.

Also, considering the chain rule, we note that for $\mathbb{T} = \mathbb{Z}$,

$$\Delta(f \circ f)(t) = f^{\sigma}(t)f^{\Delta}(t) + f(t)f^{\Delta}(t) = f^{\Delta}(t)(f(t) + f^{\sigma}(t)) \neq 2f(t)f^{\Delta}(t),$$

for $f^{\sigma}(t) \neq f(t)$. Thus, the powerful chain rule, often utilized in solving differential equations via a variable transformation, does not apply on time scales. In an attempt to generalize the chain rule for functions on time scales a few identities have been formulated. The next theorem provides such an expression based on works in [7, 8]. Other formulations can be found in [3].

Theorem 2.3. (See [3, Theorem 1.90]). Let $f : \mathbb{R} \to \mathbb{R}$ be continuously differentiable and suppose $g : \mathbb{T} \to \mathbb{R}$ is (delta) differentiable. Then $f \circ g : \mathbb{T} \to \mathbb{R}$ is (delta) differentiable and

$$(f \circ g)^{\Delta}(t) = \left\{\int_0^1 f'\left(g(t) + h\mu(t)g^{\Delta}(t)\right) \mathrm{d}h\right\} g^{\Delta}(t).$$

An interesting observation is that the operators, Δ and σ, do generally not commute, that is, $\left(f^{\Delta}\right)^{\sigma} \neq \left(f^{\sigma}\right)^{\Delta}$. Take for example $\mathbb{T} = q^{\mathbb{N}_0}$ with $q > 1$, then

$$\left(f^{\Delta}\right)^{\sigma}(t) = \frac{f(q^2t) - f(qt)}{\mu(qt)} \neq \frac{f(q^2t) - f(qt)}{\mu(t)} = \left(f^{\sigma}\right)^{\Delta}(t),$$

since $\mu(qt) = qt(q-1) \neq t(q-1) = \mu(t)$.

2.3 Integration

Definition 6. A continuous function $f : \mathbb{T} \to \mathbb{R}$ is called pre-differentiable with (region of differentiation) D, provided that $D \subset \mathbb{T}^{\kappa}$, $\mathbb{T}^{\kappa} \backslash D$ is countable and contains no right-scattered elements of $\mathbb{T}$, and f is (delta) differentiable at each $t \in D$.

Theorem 2.4. (See [3, Theorem 1.70]). Let $f : \mathbb{T} \to \mathbb{R}$ be regulated. Then there exists a function $F : \mathbb{T} \to \mathbb{R}$ which is pre-differentiable with region of differentiation D such that

$$F^{\Delta}(t) = f(t) \qquad \textit{for all } t \in D.$$

The function F is called an pre-antiderivative of $f(t)$.

If $F^{\Delta}(t) = f(t)$ for all $t \in \mathbb{T}^{\kappa}$, then F is called antiderivative of f.

We define the indefinite integral of a regulated function f by $\int f(t)\,\Delta t = F(t) + C$, where $C \in \mathbb{R}$ is an arbitrary integration constant and F is a pre-antiderivative of f. The Cauchy integral is defined by $\int_a^b f(t)\,\Delta t = F(b) - F(a)$ for all $a, b \in \mathbb{T}$.

Theorem 2.5. (See [3, Theorem 1.74]). Every rd-continuous function f has an antiderivative. In particular, if $t_0 \in \mathbb{T}$, then F defined by

$$F(t) := \int_{t_0}^{t} f(s)\,\Delta s \qquad \textit{for all } t \in \mathbb{T}$$

is an antiderivative of f.

For $\mathbb{T} = \mathbb{R}$, the integral is consistent with the Rieman integral (see **Table 4**).

$\mathbb{T}$	$\mathbb{T} = \mathbb{R}$	$\mathbb{T} = \mathbb{Z}$	$\mathbb{T} = q^{\mathbb{N}_0}$	$\mathbb{T}_{\mathcal{I}}$
$\int_s^t f(\tau)\,\Delta\tau$	$\int_s^t f(\tau)\,d\tau$	$\sum_{\tau=s}^{t-1} f(\tau)$	$\sum_{n=0}^{k} sq^k(q-1)f(q^k s)$	$\sum_{\tau \in [a,b) \cap \mathbb{T}} \mu(\tau) f(\tau)$

Table 4.
Integrals for the examples of $\mathbb{T} = \mathbb{R}$, $\mathbb{T} = \mathbb{Z}$, and $\mathbb{T} = q^{\mathbb{N}_0}$ $(q > 1)$, and isolated time scales $\mathbb{T}_{\mathcal{I}}$, for which all points in $\mathbb{T}_{\mathcal{I}}$ are assumed to be isolated. In all cases, $s, t \in \mathbb{T}$ and $s < t$. In the case of $\mathbb{T} = q^{\mathbb{N}_0}$, we assume $t = q^k s$.

The integral operator is linear so that for $f, g \in C_{rd}$ and $a < b$, $a, b \in \mathbb{T}$, and $\alpha, \beta \in \mathbb{R}$,

$$\int_a^b (\alpha f + \beta g)(s)\,\Delta s = \alpha \int_a^b f(s)\,\Delta s + \beta \int_a^b g(s)\,\Delta s.$$

With the definition of integration on time scales, we have the machinery to introduce a series representation for time scales functions. In [9], see also [3], a time scales analogue of polynomials that allows a corresponding Taylor series expression was introduced using the recursive formulation

$$g_0(t,s) = h_0(t,s) \equiv 1 \qquad \text{for all } t, s \in \mathbb{T},$$

and, for every $k \in \mathbb{N}_0$,

$$g_{k+1}(t,s) = \int_s^t g_k(\sigma(\tau), s)\,\Delta\tau \qquad \text{for all } s, t \in \mathbb{T},$$

and

$$h_{k+1}(t,s) = \int_s^t h_k(\tau, s)\,\Delta\tau \qquad \text{for all } s, t \in \mathbb{T}.$$

Now, $h_k^{\Delta}(t,s) = h_k(t,s)$ and $g_k^{\Delta}(t,s) = g_k(\sigma(t), s)$ for $k \in \mathbb{N}$ and t, s T^{κ}. Two Taylor series representations can be formulated for a time scales function f, one that uses the time scales polynomials g_k and one that uses the polynomials h_k, see Section 1.6 in [3] for more details.

3. Linear dynamic equations

This chapter provides a brief introduction to first order dynamic equations and provides a selected summary of [3], extended by applications. A first order dynamic equation is of the form

$$y^{\Delta}(t) = f(t, y, y^{\sigma}), \tag{2}$$

for $y : \mathbb{T} \to \mathbb{R}^n$ and $f : \mathbb{T} \times \mathbb{R}^n \times \mathbb{R}^n \to \mathbb{R}^n$ with $n \in \mathbb{N}_1 = \{1, 2, 3, \dots\}$. A first order initial value problem (short: IVP) is then given by (2) with an initial condition $y(t_0) = y_0 \in \mathbb{R}^n$ for $t_0 \in \mathbb{T}$. A function $y : \mathbb{T} \to \mathbb{R}^n$ is called a *solution* of (2) if y satisfies the equation for all $t \in \mathbb{T}^{\kappa}$.

We call (2) linear if

$$f(t, y, y^{\sigma}) = f_1(t)y + f_2(t), \quad \text{or } f(t, y, y^{\sigma}) = f_1(t)y^{\sigma} + f_2(t),$$

where $f_1, f_2 : \mathbb{T} \to \mathbb{R}^n$. We say the linear dynamic equation is homogeneous, if $f_2 \equiv 0$.

3.1 Scalar case

We first focus on the scalar case of (2), that is, $f : \mathbb{T} \to \mathbb{R}$. Based on the above definition of linearity, there are two forms a linear, homogeneous, first order dynamic equation can have:

$$y^{\Delta} = p(t)y, \tag{3}$$

$$y^{\Delta} = p(t)y^{\sigma}, \quad \text{for } p : \mathbb{T} \to \mathbb{R} \tag{4}$$

Note that for $\mathbb{T} = \mathbb{R}$, $y^{\sigma} = y$ and therefore $y' = p(t)y^{\sigma} = p(t)y$ so that both, (3) and (4), are the time scales analogues of $y' = p(t)y$.

If $p \in \mathcal{R}$, then (3) is called *regressive* and if $-p \in \mathcal{R}$, then (4) is called *regressive*.

The unique solution to (3) with initial condition $y(t_0) = 1$ for some $t_0 \in \mathbb{T}$ is denoted by $y(t) = e_p(t, t_0)$ and is called the time scales exponential function. The unique solution to (4) with initial condition $y(t_0) = 1$ is $y(t) = e_{\ominus(-p)}(t, t_0)$.

Table 5 contains the time scales analogues of the exponential function for the dense time scale $\mathbb{T} = \mathbb{R}$, the discrete time scale $\mathbb{T} = \mathbb{Z}$, and the quantum time scale $\mathbb{T} = q^{\mathbb{N}_0}$.

$\mathbb{T}$	**Dynamic Eq. (3)**	$e_p(t, t_0)$
$\mathbb{R}$	$y' = p(t)y$	$\exp\left\{\int_{t_0}^{t} p(s)\,\mathrm{d}s\right\}$
$\mathbb{Z}$	$\Delta y = p(t)y$	$\prod_{i=t_0}^{t-1}(1 + p(i))$
$q^{\mathbb{N}_0}$	$y^{\Delta} = p(t)y$	$\prod_{s\in[t_0,t)\cap\mathbb{T}}(1 + s(q-1)p(s))$

Table 5.
The exponential function for the continuous, discrete, and quantum time scale ($q > 1$), assuming $p \in \mathcal{R}$.

The **Table 5** reveals a crucial aspect of the time scales exponential function, namely that the positivity property, known for the traditional exponential function, does not uphold on time scales. Take for example, $\mathbb{T} = \mathbb{Z}$ and $p = -3$, then $p \in \mathcal{R}$ as $1 + p = -2 \neq 0$, but $e_p(t, 0) = (-2)^t$ which is negative for odd values of t. If however $p \in \mathcal{R}^+$, then $e_p(t, t_0) > 0$, restoring the positivity property. Note that if $\mathbb{T} = \mathbb{R}$, then any function $p \in \mathcal{R}^+$ since $1 + \mu(t)p(t) = 1 > 0$.

Some of the properties of the time scales exponential function are consistent with the convenient properties in the continuous case. If $p, q \in \mathcal{R}$ and $t, s \in \mathbb{T}$, then

i. $e_0(t, s) = 1$, $e_p(t, t) = 1$,

ii. $e_{p \oplus q}(t, s) = e_p(t, s)e_q(t, s)$,

iii. $e_{\ominus p}(t, s) = e_p(s, t) = \frac{1}{e_p(t, s)}$,

iv. $e_p(t, r)e_p(r, s) = e_p(t, s)$,

v. $e_p(\sigma(t), s) = (1 + \mu(t)p(t))e_p(t, s)$.

Theorem 3.1. [See [3, Theorem 2.39]] If $p \in \mathcal{R}$ and $a, b, c \in \mathbb{T}$, then

$$\int_a^b p(t)e_p(t, c)\,\Delta t = e_p(b, c) - e_p(a, c)$$

$$\int_a^b p(t)e_p(c, \sigma(t))\,\Delta t = e_p(c, a) - e_p(c, b).$$

As an application of linear, homogeneous, first order dynamic equations, one may consider the Malthusian growth model. In "An essay on the principle of population" from 1798, Thomas Robert Malthus proposed an exponential law of population growth with the corresponding differential equation

$$P' = rP, \quad P(t_0) = P_0,$$

where P is the population at time t, r is the inherent growth rate, and P_0 is the initial population level at time $t_0 \in \mathbb{R}$. This linear, homogeneous, first order differential equation has the solution $P(t) = e^{r(t-t_0)}P_0$. Assuming a positive initial population level $P_0 > 0$, it follows that for a positive growth rate $r > 0$, the population increases exponentially. If instead $r < 0$ and $P_0 > 0$, then the population goes extinct as $\lim_{t\to\infty} e^{r(t-t_0)}P_0 = 0$. Despite its simplicity and the unrealistic behavior of unbounded population levels for $r > 0$, the Malthusian model can sometimes serve short-term predictions.

Let us now consider the corresponding time scales model (3) with initial condition $P(t_0) = P_0 > 0$ and inherent growth rate $r > 0$, that is, $P^\Delta = rP$ with $P(t_0) = P_0$ for $t_0 \in \mathbb{T}$. The respective solution is then $P(t) = e_r(t, t_0)P_0$, which is unbounded for $r, P_0 > 0$, see **Figure 2**. Thus, for $r, P_0 > 0$, the behavior of the solution is consistent with the solution in the continuous case. However, for $r < 0$, the population does not have to go extinct but can result in biologically unmeaningful behavior as solutions can become negative.

Using the time scales exponential function that solves a linear, homogeneous, first order dynamic equation, we can use the variation of constants formula to obtain the solution to a linear, nonhomogeneous, first order dynamic equation.

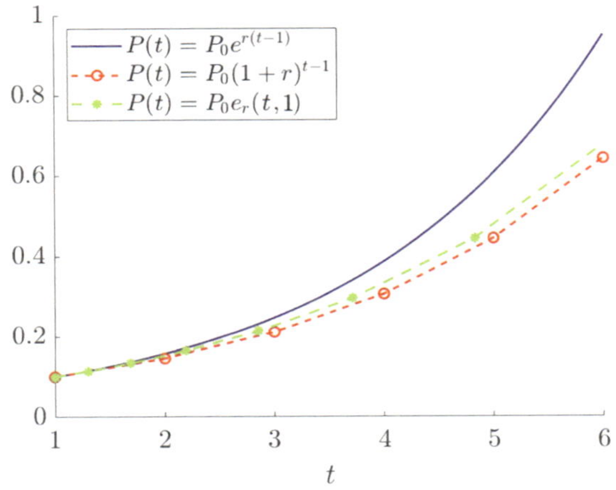

Figure 2.
The behavior of the solution to $P^{\Delta} = rP$ *with* $P(t_o) = P_o$ *where* $r = 0.45$, $t_o = 1$, *and* $P_o = 0.1$, *for* $\mathbb{T} = \mathbb{R}$, $\mathbb{T} = \mathbb{Z}$ *and* $\mathbb{T} = 1.3^{\mathbb{N}_0}$. *The solid line represents the solution in the continuous case, the open circle represents the solution in the discrete case, and the stars represent the solution in the quantum calculus case with* $q=1.3$.

Theorem 3.2. [See [3, Theorems 2.74 & 2.77]] Suppose $p \in \mathcal{R}, f \in C_{rd},\ t_0 \in \mathbb{T}$ and $y_0 \in \mathbb{R}$ then the unique solution to

$$y^{\Delta} = p(t)y + f(t), \qquad y(t_0) = y_0$$

is given by

$$y(t) = e_p(t,t_0)y_0 + \int_{t_0}^{t} e_p(t,\sigma(s))f(s)\,\Delta s.$$

Furthermore, the unique solution to

$$y^{\Delta} = -p(t)y^{\sigma} + f(t), \qquad y(t_0) = y_0$$

is given by

$$y(t) = e_{\ominus p}(t,t_0)y_0 + \int_{t_0}^{t} e_{\ominus p}(t,s)f(s)\,\Delta s.$$

Example 3.3. Suppose that the life span of a certain species is one time unit. Suppose that just before the species dies out, eggs are laid that are hatch after one time unit. The species is therefore only alive on $\mathbb{T} = \cup_{k=0}^{\infty}[2k, 2k+1]$, see also [3, Example 1.39] and [10]. Suppose further that during the specie's, life cycle, the species reduces due to external factors with rate $d \in (0,1)$ and at the end of the life cycle $t = 2k+1$, the individuals alive in $(2k, 2k+1)$ lay eggs that result in the reproduction rate $r > 0$. The corresponding dynamic equation for the species $N(t)$ at time t, is then

$$N^{\Delta}(t) = p(t)N(t), \qquad \textit{with} \quad p(t) = \begin{cases} -d & t \in [2k, 2k+1) \\ r & t = 2k+1 \end{cases}$$

and initial condition $N(0) = N_0$. We note that even though $p(t)$ is discontinuous at $t = 2k+1$, $p(t) \in \mathcal{R}$. Theorem 3.2 gives the population at time $t \in [2m, 2m+1]$ as

$$N(t) = N_0 e_p(t, t_0) = N_0 e_p(t, 2m) \prod_{k=0}^{m-1} \left(e_p(2k+1, 2k) e_p(2k+2, 2k+1)\right)$$
$$= N_0 \exp\left\{\int_{2m}^{t} -d\, \mathrm{d}s\right\} \left\{\prod_{k=0}^{m-1} \exp\left\{\int_{2k}^{2k+1} -d\, \mathrm{d}s\right\}(1+r)\right\} = N_0 e^{-d(t-m)}(1+r)^m.$$

Example 3.4. Newton's law of cooling suggests that the temperature of an object at time t, $T(t)$, changes dependent on the temperature of its surrounding, T_m. Then, $T'(t) = -\kappa(T - T_m)$, where κ is the heat transfer coefficient. Suppose that an object with initial temperature T_0 is cooled in a lab environment. Due to safety regulations, once the lab assistant leaves the work space, the object can only be exposed to an environment that preserves the current temperature of the object. The cooling of the object can be modeled using time scales with the underlying time domain to be the working hours of the lab assistant. Assume that the lab assistant's working hours, and therefore the time scale, is of the form $\mathbb{T} = \cup_{i=0}^{\infty}[a_i, b_i] \cup [c_i, d_i]$, where the interval $[a_i, b_i]$ are the working hours prior to lunch, and $[c_i, d_i]$ are the working hours of the lab assistant after lunch of day i. One way of modeling this scenario on time scales is

$$T^\Delta = -p(t)(T - T_m), \qquad p(t) = \begin{cases} \kappa & t \in [a_i, b_i) \cup [c_i, d_i) \\ 0 & t \in \{b_i, d_i\} \end{cases}$$

with initial temperature $T(t_0) = T_0$ for $t_0 \in \mathbb{T}$. Since $p(t)$ is rd-continuous and regressive, the theorems above can be applied despite the discontinuity of $p(t)$.

Example 3.5. The following example is from [11], where a Keynesian-Cross model with lagged income is considered. Here, the aggregated income y changes according to

$$y^\Delta = \delta[d^\sigma(t) - y], \qquad t \geq t_0 \in \mathbb{T},$$

where $d(t)$ is the aggregated demand at time t and $\delta \in (0, 1)$ is the "adjustment speed". Since $d(t)$ can be expressed as the addition of aggregated consumption (c), aggregated investment (I), and governmental spending (G), we have $d(t) = c(t) + I + G$ for $I, G \in (0, \infty)$. Under the assumption that aggregated consumption is itself linear in the aggregated income, we have $c(t) = a + by(t)$ with $a, b > 0$ so that the model reads as

$$y^\Delta = \delta[a + by^\sigma + I + G - y].$$

Under the assumption that $p(t) := 1 - \delta b\mu(t) \neq 0$, we can apply $y^\sigma = y + \mu y^\Delta$, and express the dynamic equation as

$$y^\Delta = \frac{\delta(a + I + G)}{p(t)} + \frac{\delta(b-1)}{p(t)} y.$$

which is a linear, non-homogeneous, first order dynamic equation. It is left as an exercise to apply the techniques of this subsection to derive an explicit solution to this dynamic equation.

Example 3.6. Let us consider a time scales analogue of the popular logistic growth model $y' = ry\left(1 - \frac{y}{K}\right)$, namely,

$$y^\Delta = ry^\sigma\left(1 - \frac{y}{K}\right), \qquad y(t_0) = y_0, \tag{5}$$

with growth rate $r > 0$, and carrying capacity $K > 0$, and initial population size $y(t_0) > 0$ at time $t_0 \in \mathbb{T}$. Even though this is an example of a nonlinear dynamic equation of first order, we can apply the substitution $z = \frac{1}{y}$ for $y \neq 0$, to obtain the linear dynamic equation

$$z^\Delta = \frac{-y^\Delta}{yy^\sigma} = -rz + \frac{r}{K}, \qquad z(t_0) = \frac{1}{y_0}.$$

For $-r \in \mathcal{R}$, the solution is then given by Theorem 3.2. Using also Theorem 3.1 and resubstituting yields

$$y(t) = \frac{y_0 K}{e_{-r}(t, t_0)(K - y_0) + y_0}. \tag{6}$$

It can be easily checked that $y(t_0) = y_0$ and that y solves (5), see also [12].

Note that for $\mathbb{T} = \mathbb{R}$, (5) collapses to the Verhulst model $y' = ry\left(1 - \frac{y}{K}\right)$ and the solution (6) reads in this case as

$$y(t) = \frac{y_0 K}{e^{-r(t-t_0)}(K - y_0) + y_0},$$

which coincides with the classical solution.

3.2 Linear systems

Let us now consider (2) with $f: \mathbb{T} \times \mathbb{R}^n \times \mathbb{R}^n \to \mathbb{R}^n$ for $n \in \mathbb{N} = \{1, 2, 3, \ldots\}$. In order to extend the solution methods for linear first order dynamic equations that were introduced in the previous section for scalar functions, the definitions of rd-continuity and delta differentiability have to be first extended to matrix valued functions $A : \mathbb{T} \to \mathbb{R}^{m \times n}$. This adjustment is mostly proposed element-wise. More precisely, A is rd-continuous on $\mathbb{T}$ if a_{ij} is rd-continuous on $\mathbb{T}$ for all $1 \leq i \leq n$, $1 \leq j \leq m$. The class of all such rd-continuous $m \times n$-matrix-valued functions on $\mathbb{T}$ is then denoted by $C_{rd}(\mathbb{T}, \mathbb{R}^{m \times n})$. Similarly, we say that A is delta differentiable (or short: differentiable), if a_{ij} is delta differentiable for all $1 \leq i \leq n$, $1 \leq j \leq m$. Similar to the scalar case, the following identity holds for any matrix-valued (delta) differentiable function A,

$$A^\sigma(t) = A(t) + \mu(t) A^\Delta(t).$$

The property of regressive is however not defined elementwise. Instead, we say that $A \in \mathbb{R}^{n \times n}$ is regressive if $I_n + \mu(t)A(t)$ is invertible for all $t \in \mathbb{T}^\kappa$, where $I_n \in \mathbb{R}^{n \times n}$ is the identity matrix. The class of rd-continuous and regressive functions is denoted by $\mathcal{R}(\mathbb{T}, \mathbb{R}^{n \times n})$ (or short $\mathcal{R}$).

Note that even if all entries of A are regressive, A does not have to be regressive. Take for example $\mathbb{T} = \mathbb{Z}$ with

$$A = \begin{bmatrix} a_{11} & a_{12} \\ a_{21} & a_{22} \end{bmatrix} = \begin{bmatrix} 0 & -2 \\ -2 & 3 \end{bmatrix}.$$

Then all entries are regressive as $1 + a_{ij} \neq 0$ for all $1 \leq i, j \leq 2$ but $\det(I + A) = 0$.
As for the scalar case, differentiation is linear, that is,

$$(\alpha A + \beta B)^{\Delta}(t) = \alpha A^{\Delta}(t) + \beta B^{\Delta}(t)$$

for differentiable $m \times n$-matrix-valued functions A, B, and $\alpha, \beta \in \mathbb{R}$.
We consider

$$y^{\Delta} = A(t)y \tag{7}$$

to be the system analogue of (3). If A is $n \times n$ matrix valued function, then, the unique solution to (7) with $y(t_0) = I_n$, where I_n is the $n \times n$ identity matrix, is denoted by $y(t) = e_A(t, t_0)$. If $A \in \mathbb{R}^{n \times n}$ and $\mathbb{T} = \mathbb{R}$ then $e_A(t, t_0) = e^{A(t-t_0)}$, and if $\mathbb{T} = \mathbb{Z}$, then $e_A(t, t_0) = (I + A)^{t-t_0}$. The analogue of (4) in higher dimensions is

$$y^{\Delta} = -A^*(t)y^{\sigma},$$

where $A^*(t)$ is the conjugate transpose of $A(t)$.

Theorem 3.7. (See [3, Theorems 5.24 & 5.27]). Let $A \in \mathcal{R}(\mathbb{T}, \mathbb{R}^{n\times n}, \mathbb{R}^{n\times n})$ and suppose that $f\colon \mathbb{T} \to \mathbb{R}^n$ is rd-continuous. Let $t_0 \in \mathbb{T}$ and $y_0 \in \mathbb{R}^n$. Then, the initial value problem

$$y^{\Delta} = A(t)y + f(t), \qquad y(t_0) = y_0$$

is given by

$$y(t) = e_A(t, t_0)y_0 + \int_{t_0}^{t} e_A(t, \sigma(\tau))f(\tau)\,\Delta\tau.$$

The unique solution to

$$y^{\Delta} = -A^*(t)y^{\sigma} + f(t), \qquad y(t_0) = y_0$$

is given by

$$y(t) = e_{\ominus A^*}(t, t_0)y_0 + \int_{t_0}^{t} e_{\ominus A^*}(t, \tau)f(\tau)\,\Delta\tau.$$

Example 3.8. In [13], the authors consider the Cucker-Smale type model on an isolated $\mathbb{T}$ (i.e., every $t \in \mathbb{T}$ is isolated) with $\sup\mathbb{T} = \infty$ and $\sup\{\mu(t) : t \in \mathbb{T}\} < \infty$,

$$\begin{aligned} x_i^{\Delta} &= v_i \\ v_i^{\Delta} &= \frac{1}{N}\sum_{j=1}^{N} a_{ij}(v_j - v_i), \end{aligned} \tag{8}$$

where $a_{ij} \in \mathbb{R}_0^+ = [0, \infty)$ and $i \in \{1, 2, \ldots, N\}$ represents the impact of agent's j opinion onto the agent's i opinion. The variable x_i represents the state of agent i, and v_i is the consensus parameter of agent i. The original Cucker-Smale model, see [14], is a discrete time system discussing the flock behavior of birds, where v_i represents the velocity of bird i and x_i is its position. The weights a_{ij} quantify the way the birds influence each other.

Note that since $\mathbb{T}$ is isolated, we can equivalently write (8) as

$$x_i(\sigma(t)) = x_i(t) + \mu(t)v_i(t), \qquad v_i(\sigma(t)) = v_i(t) + \frac{\mu(t)}{N}\sum_{j=1}^{N} a_{ij}(v_j(t) - v_i(t)),$$

or in form of a system in $y = (x_1, x_2, \dots, x_N, v_1, v_2, \dots, v_N)^T$,

$$y^\Delta = By, \qquad B = \frac{1}{N}\begin{bmatrix} 0_N & NI_N \\ 0_N & A - D, \end{bmatrix}, \tag{9}$$

where $(A)_{ij} = a_{ij}$ for $i,j \in \{1, 2, \dots, N\}$, $D = diag(d_1, d_2, \dots, d_N)$ with $d_k = \sum_{j=1}^{N} a_{kj}$, 0_N is a matrix of dimension $N \times N$ with all entries being zero, and I_N is the identity matrix of dimension $N \times N$.

If $B \in \mathcal{R}$, then the solution to (9) with initial condition $y(t_0) = y_0$ is $y(t) = e_B(t,t_0)\, y_0$. In order for $B \in \mathcal{R}$, $NI_N + \mu(t)(A\text{-}D)$ must be invertible because

$$\tilde{B}(t) = I_{2N} + \mu(t)B = \begin{bmatrix} I_N & \mu(t)I_n \\ 0_N & C(t) \end{bmatrix}, \qquad C(t) = I_N + \mu(t)\frac{1}{N}(A - D),$$

and

$$\det\left(\tilde{B}(t)\right) = \det(I_{2N} + \mu(t)B) = \det(I_N)\det(C(t)).$$

We conclude this section by examples of nonlinear dynamic equations that can be transformed into a system of linear dynamic equations of first order, so that Theorem 3.7 provides its solution.

Example 3.9. Let $\mathbb{T}$ be again an isolated time scale, that is, every point in $\mathbb{T}$ is isolated and $inf\{\mu(t) : t \in \mathbb{T}\} > 0$. Consider

$$x^{\sigma^k} = \frac{Kx}{(1 - \mu(t)\alpha)K + \mu(t)\alpha x}, \tag{10}$$

with initial values $\vec{x}_0 = (x_0, x_1, \dots, x_{k-1}) \in (0, \infty)^k$, $K > 0$, and $-\alpha \in \mathbb{R}^+$. Eq. (10) is a delayed Beverton-Holt model and can be used to model mature individuals of a population, assuming that it takes k reproductive cycles for an individual to become mature, where the length of a reproductive cycle starting at t is $\mu(t)$. An application may be populations where the lengths between breeding cycles is temperature dependent. Model (10) has been considered in [15] (and, for $\mathbb{T} = \mathbb{Z}$, in [16]), where the authors applied the transformation $y := \frac{K}{x}$ for $x \neq 0$ to obtain

$$Y^\Delta = A(t)Y + \mathbf{b}(t) \quad \text{with} \quad A(t) = \frac{1}{\mu(t)}\begin{bmatrix} \mathbf{0}_{k-1} & I_{k-1} \\ -\mu\alpha & -\mathbf{s} \end{bmatrix}, \ \mathbf{b}(t) = \begin{pmatrix} \mathbf{0}_{k-1} \\ \alpha \end{pmatrix}, \tag{11}$$

where $\mathbf{s} = \left(\binom{k}{1}, \binom{k}{2}, \binom{k}{3}, \dots, \binom{k}{k-1}\right)$ and $\mathbf{0}_{k-1} \in \mathbb{R}^{k-1\times 1}$ is vector of zeros. Applying Theorem 3.7, to (11) yields the solution.

Example 3.10. In [17], the authors proposed the following nonlinear system of dynamic equations to model the spread of a contagious disease,

$$\begin{aligned} S^\Delta &= -\beta(t)S^\sigma I - \nu(t)S + \gamma(t)I + \nu(t)\kappa, \\ I^\Delta &= \beta(t)S^\sigma I - \gamma(t)I - \nu(t)I. \end{aligned}$$

In line with well-established epidemic models, the population was compartmentalized into susceptible S and infected I individuals. The model assumes that the disease is spread by contact with an infected individual with a transmission rate of $\beta > 0$. The recovery rate is assumed to be $\gamma > 0$ and recovered individuals rejoin the group of susceptible individuals. The death rate is $\nu(t)$ across the population and $\nu(t)\kappa$ newborns join the group of susceptibles.

By introducing a new variable $w := S + I$, $w^\Delta = -\nu(t)w + \nu(t)\kappa$. This first order, linear, nonhomogeneous dynamic equation can be solved using Theorem 3.2, assuming $-\nu(t) \in \mathcal{R}$. The solution is then $w(t) = e_{-\nu}(t, t_0)(I_0 + S_0 - \kappa) + \kappa$, so that, after recalling that $S = w(t) - I$, the dynamic equation in I can be expressed as

$$I^\Delta = \beta(t)(w^\sigma - I^\sigma)I - \gamma(t)I - \nu(t)I.$$

Although the dimension has been reduced to one, the dynamic equation is still nonlinear. Defining however $y = \frac{1}{I}$ for $I \neq 0$ yields again a linear dynamic equation, namely

$$y^\Delta = (-\beta(t)w^\sigma(t) + \gamma(t) + \nu(t))y^\sigma + \beta(t).$$

Applying Theorem 3.2 gives the solution

$$y(t) = e_{\ominus p}(t, t_0)y_0 + \int_{t_0}^{t} e_{\ominus p}(t, s)\beta(s)\,\Delta s,$$

where $p(t) = \beta(t)w(\sigma(t)) - (\gamma(t) + \nu(t))$ is assumed to be an element of $\mathcal{R}$. Resubstituting yields then the solution I and using $S = w - I$ yields S.

For more epidemic models on time scales that are systems of first order nonlinear dynamic equations, see [18–21]. While the dynamic Susceptible-Infected-Recovered epidemic model introduced in [18] can be solved explicitly via variable transformations, in most cases, including [19], explicit solutions to nonlinear dynamic equations are not available. In these cases, properties of solutions such as existence and uniqueness are of fundamental interest. The interested reader is referred to [22, Section 2] and [3, Section 8.2].

Author details

Sabrina Streipert
McMaster University, Hamilton, Ontario, Canada

*Address all correspondence to: streipes@mcmaster.ca

References

[1] May R. Simple mathematical models with very complicated dynamics. Nature. 1976;**261**:459-467

[2] Hilger S. Analysis on measure chains—A unified approach to continuous and discrete calculus. Research in Mathematics. 1990;**18**:18-56

[3] Bohner M, Peterson A. Dynamic Equations on Time Scales. Boston, MA: Birkhäuser Boston Inc.; 2001 ISBN 0-8176-4225-0. An introduction with applications

[4] Historical el nino/la nina episodes (1950–present), climate prediction center. Available from: https://origin.cpc.ncep.noaa.gov/products/analysis_monitoring/ensostuff/ONI_v5.php [Accessed: March 15, 2019]

[5] Anderson DR, Krueger RJ, Peterson AC. Delay dynamic equations with stability. Advances in Difference Equations. 2006;**2006**:19

[6] Kelley WG, Peterson AC. Difference Equations: An Introduction with Applications. Boston, MA: Academic Press, Inc.; 1991 ISBN 0-12-403325-3

[7] Pötzsche C. Chain rule and invariance principle on measure chains. Journal of Computational and Applied Mathematics. 2002;**141**(1):249-254 ISSN 0377-0427. Dynamic Equations on Time Scales

[8] Keller S. Asymptotisches Verhalten invarianter Faserbundel bei Diskretisierung und Mittelwertbildung im Rahmen der Analysis auf Zeitskalen. Augsburg: Universität Augsburg; 1999. Thesis (Ph.D.)

[9] Agarwal RP, Bohner M. Basic calculus on time scales and some of its applications. Results in Mathematics. 1999;**35**(1–2):3-22

[10] Christiansen FB, Fenchel TM. Theories of Populations in Biological Communities. Ecological Studies, Berlin Heidelberg: Springer; 2012. Available from: https://books.google.de/books?id=HAL8CAAAQBAJ

[11] Tisdell CC, Zaidi A. Basic qualitative and quantitative results for solutions to nonlinear, dynamic equations on time scales with an application to economic modelling. Nonlinear Analysis. 2008; **68**(11):3504-3524

[12] Bohner M, Warth H. The Beverton–Holt dynamic equation. Applicable Analysis. 2007;**86**(8):1007-1015

[13] Girejko E, Machado L, Malinowska AB, Martins N. On consensus in the Cucker-Smale type model on isolated time scales. Discrete & Continuous Dynamical System Series. 2018;**11**(1):77-89. ISSN 1937-1632

[14] Cucker F, Smale S. Emergent behavior in flocks. IEEE Transactions on Automatic Control. 2007;**52**(5): 852-862. DOI: 10.1109/TAC.2007.895842

[15] Bohner M, Cuchta T, Streipert S. Delay dynamic equations on isolated time scales and the relevance of one-periodic coefficients. Mathematicsl Methods in the Applied Sciences. 2022: 1-18. DOI: 10.1002/mma.8141

[16] Bohner M, Dannan FM, Streipert S. A nonautonomous Beverton-Holt equation of higher order. Journal of Mathematical Analysis and Applications. 2018;**457**(1):114-133

[17] Bohner M, Streipert S. An integrable SIS model on time scales. In: Bohner M, Siegmund S, Šimon Hilscher R, Stehlík P, editors. Difference Equations and Discrete Dynamical Systems with Applications. Cham: Springer International Publishing; 2020. pp. 187-200

[18] Bohner M, Streipert S, Torres DFM. Exact solution to a dynamic sir model. Nonlinear Analysis Hybrid Systems. 2019;**32**:228-238

[19] Ferreira RAC, Silva CM. A nonautonomous epidemic model on time scales. Journal of Difference Equations and Applications. 2018;**24**(8):1295-1317

[20] Sae-Jie W, Bunwong K, Moore E. The effect of time scales on sis epidemic model. WSEAS Transactions on Mathematics. 2010;**9**(10):757-767

[21] Yeni G. Modeling of HIV, SIR and SIS Epidemics on Time Scales and Oscillation Theory. ProQuest LLC, Ann Arbor, MI: Missouri University of Science and Technology; 2019. ISBN 978-1392-67226-6. Thesis (Ph.D.)

[22] Lakshmikantham V, Kaymakçalan B, Sivasundaram S. Dynamic Systems on Measure Chains, Volume 370 of Mathematics and its Applications. Dordrecht: Kluwer Academic Publishers; 1996

Chapter 2

Existence Results for Boundary Value Problem of Nonlinear Fractional Differential Equation

Noureddine Bouteraa and Habib Djourdem

Abstract

In this chapter, we investigate the existence and uniqueness of solutions for class of nonlinear fractional differential equations with nonlocal boundary conditions. The existence results are obtained by using Leray-Schauder nonlinear alternative and Banach contraction principle. An illustrative example is presented at the end to illustrated the validity of our results.

Keywords: fractional differential equations, existence, nonlocal boundary, fixed-point theorem

1. Introduction

In this chapter, we are interested in the existence of solutions for nonlinear fractional difference equations

$$^{c}D_{0^+}^{\alpha}u(t) - A^{c}D_{0^+}^{\beta}u(t) = f\left(t, u(t), ^{c}D_{0^+}^{\beta}u(t), ^{c}D_{0^+}^{\alpha}u(t)\right), \quad t \in J = [0, T], \tag{1}$$

subject to the three-point boundary conditions

$$\begin{cases} \lambda u(0) - \mu u(T) - \gamma u(\eta) = d, \\ \lambda u'(0) - \mu u'(T) - \gamma u'(\eta) = l, \end{cases} \tag{2}$$

where $T > 0$, $0 \leq \eta \leq T$, $\lambda \neq \mu + \gamma$, d, l, λ, μ, $\gamma \in \mathbb{R}$, $\beta + 1 < \alpha$, A is an $\mathbb{R}^{n \times n}$ matrix and $^{c}D_{0^+}^{\alpha}$, $^{c}D_{0^+}^{\beta}$ are the Caputo fractional derivatives of order $1 < \alpha \leq 2$, $0 < \beta \leq 1$, respectively.

The first definition of fractional derivative was introduced at the end of the nineteenth century by Liouville and Riemann, but the concept of non-integer derivative and integral, as a generalization of the traditional integer order differential and integral calculus, was mentioned already in 1695 by Leibniz and L'Hospital. In fact, fractional derivatives provide an excellent tool for the description of memory and hereditary properties of various materials and processes. The mathematical modeling of systems and processes in the fields of physics, chemistry, aerodynamics, electrodynamics of complex medium, polymer rheology, Bode's analysis of feedback amplifiers, capacitor theory, electrical circuits, electro-analytical chemistry, biology, control theory, fitting of experimental data, involves derivatives of fractional order. In consequence, the subject of fractional differential equations is gaining much importance and attention. For more details we refer the reader to [1–5] and the references cited therein.

Fractional differential equation theory have recieved increasing attention. This theory has been developed very quickly and attracted a considerable interest from researches in this field, which revealed the flexibility of fractional calculus theory in designing various mathematical models. The main methods conducted in this articles are by terms of fixed point techniques [6]. Boundary value problems for nonlinear differential equations arise in a variety of areas of applied mathematics, physics and variational problems of control theory. A point of central importance in the study of nonlinear boundary value problems is to understand how the properties of nonlinearity in a problem influence the nature of the solutions to the boundary value problems. The multi-point boundary conditions are important in various physical problems of applied science when the controllers at the end points of the interval (under consideration) dissipate or add energy according to the sensors located, at intermediate points, see [7, 8] and the references therein. We quote also that realistic problems arising from economics, optimal control, stochastic analysis can be modeled as differential inclusion. The study of fractional differential inclusions was initiated by EL-Sayad and Ibrahim [9]. Also, recently, several qualitative results for fractional differential inclusion were obtained in [10–13] and the references therein.

The techniques of nonlinear analysis, as the main method to deal with the problems of nonlinear differential equations (DEs), nonlinear fractional differential equations (FDEs), nonlinear partial differential equations (PDEs), nonlinear fractional partial differential equations (FPDEs), nonlinear stochastic fractional partial differential equations (SFPDEs), plays an essential role in the research of this field, such as establishing the existence, uniqueness and multiplicity of solutions (or positive solutions) and mild solutions for nonlinear of different kinds of FPDEs, FPDEs, SFPDEs, inclusion differential equations and inclusion fractional differential equations with various boundary conditions, by using different techniques (approaches). For more details, see [14–36] and the references therein. For example, iterative method is an important tool for solving linear and nonlinear boundary value problems. It has been used in the research areas of mathematics and several branches of science and other fields. However, many authors showed the existence of positive solutions for a class of boundary value problem at resonance case. Some recent devolopment for resonant case can be found in [37, 38]. Let us cited few papers. In [39], the authors studied the boundary value problems of the fractional order differential equation:

$$\begin{cases} D^{\alpha}_{0+}u(t) = f(t, u(t)) = 0, \quad t \in (0,1), \\ u(0) = 0, \ D^{\beta}_{0+}u(1) = aD^{\beta}_{0+}u(\eta), \end{cases}$$

where $1 < \alpha \leq 2, \ 0 < \eta < 1, \ 0 < a, \ \beta < 1, \ f \in C([0,1] \times \mathbb{R}^2, \mathbb{R})$ and $D^{\alpha}_{0+}, \ D^{\beta}_{0+}$ are the standard Riemann-Liouville fractional derivative of order α. They obtained the multiple positive solutions by the Leray-Schauder nonlinear alternative and the fixed point theorem on cones.

In 2017, Resapour et al. [40] investigated a Caputo fractional inclusion with integral boundary condition for the following problem

$$\begin{cases} {}^cD^\alpha u(t) \in F\left(t, u(t), {}^cD^\beta u(t), u'(t)\right), \\ u(0) + u'(0) + {}^cD^\beta u(0) = \int_0^\eta u(s)ds, \\ u(1) + u'(1) + {}^cD^\beta u(1) = \int_0^\nu u(s)ds, \end{cases}$$

where $1 < \alpha \leq 2$, $\eta, \nu, \beta \in (0,1)$, $F : [0,1] \times \mathbb{R} \times \mathbb{R} \times \mathbb{R} \to 2^{\mathbb{R}}$ is a compact valued multifunction and ${}^cD^\alpha$ denotes the Caputo fractional derivative of order α.

In 2017, Sheng and Jiang [41] studied the existence and uniqueness of the solutions for fractional damped dynamical systems

$$\begin{cases} {}^cD^\alpha_{0^+} u(t) - A\,{}^cD^\beta_{0^+} u(t) = f(t, u(t)), \quad t \in [0, T], \\ u(0) = u_0, \qquad u'(0) = u'_0, \end{cases}$$

where $0 < \beta \leq 1 < \alpha \leq 2$, $0 < T < \infty$, $u \in \mathbb{R}^n$, A is an $\mathbb{R}^{n \times n}$ matrix, $f : [0,1] \times \mathbb{R}^n \to \mathbb{R}^n$ jointly continuous function and ${}^cD^\alpha_{0^+}$, ${}^cD^\beta_{0^+}$ are the Caputo derivatives of order α, β, respectively.

In 2018, Abbes et al. [42] studied the existence and uniqueness of the solutions for fractional damped dynamical systems

$$\begin{cases} {}^cD^\alpha_{0^+} u(t) - A\,{}^cD^\beta_{0^+} u(t) = f\left(t, u(t), {}^cD^\beta_{0^+} u(t), {}^cD^\alpha_{0^+} u(t)\right), \quad t \in [0, T], \\ u(0) = u(T), \qquad u'(0) = u'(T), \end{cases}$$

where $0 < \beta \leq 1 < \alpha \leq 2$, $0 < T < \infty$, $u \in \mathbb{R}^n$, A is an $\mathbb{R}^{n \times n}$ matrix and $f : [0,1] \times \mathbb{R}^n \to \mathbb{R}^n$ jointly continuous.

In 2019, Tao Zhu [43] studied the existence and uniqueness of positive solutions of the following fractional differential equations

$$\begin{cases} D^\alpha_{0^+} u(t) - A\,{}^cD^\beta_{0^+} u(t) = f(t, u(t)), \quad t \in [0, T), \quad 0 < \beta < \alpha < 1, \\ u(0) = u_0. \end{cases}$$

Inspired and motivated by the works mentioned above, we establish the existence results for the nonlocal boundary value problem (1.1)–(1.2) by using Leray-Schauder nonlinear alternative and the Banach fixed point theorem. Note that our work generalized the three works cited above [41–43]. The chapter is organized as follows. In Section 2, we recall some preliminary facts that we need in the sequel. In Section 3, deals with main results and we give an example to illustrate our results.

2. Existence and uniqueness results for our problem

2.1 Preliminaries

Let as introduce notations, definitions and preliminary facts that will be need in the sequel. For more details, see for example [44–46].

Definition 2.1. The Caputo fractional derivative of order α for the function $u \in C^n([0, \infty), \mathbb{R})$ is defined by

$$ {}^cD_{0^+}^{\alpha}u(t) = \frac{1}{\Gamma(n-\alpha)} \int_0^t (t-s)^{n-\alpha-1} u^{(n)}(s)ds. $$

where $\Gamma(\cdot)$ is the Eleur gamma function and $\alpha > 0$, $n = [\alpha] + 1$, $[\alpha]$ denotes the integer part of the real number α.

Definition 2.2. The Riemann-Liouville fractional integral of order $\alpha > 0$ of a function $u : (0, \infty) \to \mathbb{R}$ is given by

$$ I_{0^+}^{\alpha}u(t) = \frac{1}{\Gamma(\alpha)} \int_0^t (t-s)^{\alpha-1} u(s)ds, \quad t > 0. $$

where $\Gamma(\cdot)$ is the Eleur gamma function, provided that the right side is pointwise defined on $(0, \infty)$.

Lemma 2.1. *Let* $u \in AC^n[0, T]$, $n \in \mathbb{N}$ *and* $u(\cdot) \in C[0, T]$. *Then, we have*

$$ {}^cD_{0^+}^{\beta}\left(I_{0^+}^{\alpha}u(t)\right) = I_{0^+}^{\alpha-\beta}u(t), $$

$$ I_{0^+}^{\alpha}\left({}^cD_{0^+}^{\alpha}u(t)\right) = u(t) - \sum_{k=0}^{n-1} \frac{t^k}{k!} u^{(k)}(0), \quad t > 0, \quad n-1 < \alpha < n, $$

Especially, when $1 < \alpha < 2$, *then we have*

$$ I_{0^+}^{\alpha}\left({}^cD_{0^+}^{\alpha}u(t)\right) = u(t) - u(0) - tu'(0). $$

Lemma 2.2. ([10]) *Let* $0 < \beta < 1 < \alpha < 2$, *then we have*

$$ I_{0^+}^{\alpha}\left({}^cD_{0^+}^{\beta}u(t)\right) = I_{0^+}^{\alpha-\beta}u(t) - \frac{u(0)t^{\alpha-\beta}}{\Gamma(\alpha-\beta+1)}. $$

2.2 Existence results

Let $C(J, \mathbb{R}^n)$ be the Banach space for all continuous function from J into $\mathbb{R}^n$ equipped with the norm

$$ \|u\|_{\infty} = \sup\{\|u(t)\| : t \in J\}, $$

where $\|\cdot\|$ denotes a suitable complete norm on $\mathbb{R}^n$. Denote $L^1(J, \mathbb{R}^n)$ the Banach space of the measurable functions $u : J \to \mathbb{R}^n$ that are Lebesgue integrable with norm

$$ \|u\|_{L^1} = \int_0^T \|u(t)\|dt. $$

Let $AC(J, \mathbb{R}^n)$ be the Banach space of absolutely continuous valued functions on J and set

$$ AC^n(J) = \left\{u : J \to \mathbb{R}^n : u, u', u'', \ldots, u^{(n-1)} \in C(J, \mathbb{R}^n)\right\} \ \textit{and} \ \ u^{(n-1)} \in AC(J, \mathbb{R}^n). $$

By

$$C^1(J) = \{u : J \to \mathbb{R}^n \ where \ \ u' \in C(J, \mathbb{R}^n)\},$$

we denote the Banach space equipped with the norm

$$\|u\|_1 = \max\left\{\|u\|_\infty, \|u'\|_\infty\right\}.$$

For the sake of brevity, we set

$$\begin{aligned}
&\delta = (\lambda - \mu - \gamma)\Gamma(\alpha - \beta + 1) + A\left(\mu T^{\alpha-\beta} + \gamma\eta^{\alpha-\beta}\right), \quad \Delta = \frac{\Gamma(\alpha - \beta + 1)}{\delta} \\
&\sigma = A(\alpha - \beta)\left(\mu T^{\alpha-\beta-1} + \gamma\eta^{\alpha-\beta-1}\right), \quad \Lambda = (\lambda - \mu - \gamma) - (\mu T + \gamma\eta)\left(\frac{\sigma}{\delta}\right), \\
&R_1 = \left(1 + \frac{\|A\|T^{\alpha-\beta}}{\Gamma(\alpha - \beta + 1)}\right)M_1 + TM_2 + \frac{\|A\|T^{\alpha-\beta}}{\Gamma(\alpha - \beta + 1)}, \\
&R_2 = M_2 + \frac{(\alpha - \beta)\|A\|T^{\alpha-\beta-1}}{\Gamma(\alpha - \beta + 1)}M_1 + \frac{\|A\|T^{\alpha-\beta-1}}{\Gamma(\alpha - \beta)} \\
&M_1 = \Delta\left\{\left[\Lambda^{-1}\left(\frac{\sigma}{\delta}\right)(\mu T + \gamma\eta) + 1\right]\Phi + \Lambda^{-1}(\mu T + \gamma\eta)\Theta\right\}
\end{aligned}$$

and

$$M_2 = \Lambda^{-1}\left\{\left(\frac{\sigma}{\delta}\right)\Phi + \Theta\right\},$$

with

$$\Phi = \frac{\|A\|}{\Gamma(\alpha - \beta + 1)}\left(\mu T^{\alpha-\beta} + \gamma\eta^{\alpha-\beta}\right) + \frac{(\mu T^\alpha + \gamma\eta^\alpha)\left(L_1\Gamma(2 - \beta) + T^{1-\beta}(L_3\|A\| + L_2)\right)}{\Gamma(\alpha + 1)\Gamma(2 - \beta)(1 - L_3)},$$

$$\Theta = \frac{\|A\|}{\Gamma(\alpha - \beta)}\left(\mu T^{\alpha-\beta-1} + \gamma\eta^{\alpha-\beta-1}\right) + \frac{\left(\mu T^{\alpha-1} + \gamma\eta^{\alpha-1}\right)\left(L_1\Gamma(2 - \beta) + T^{1-\beta}(L_3\|A\| + L_2)\right)}{\Gamma(\alpha)\Gamma(2 - \beta)(1 - L_3)}.$$

Lemma 2.3. *Let $y(\cdot) \in C(J, \mathbb{R}^n)$. The function $u(\cdot) \in C^1(J, \mathbb{R}^n)$ is a solution of the fractional differential problem*

$$\begin{cases} {}^cD^\alpha_{0^+}u(t) - A\,{}^cD^\beta_{0^+}u(t) = y(t), & t \in J = [0, T], \\ \lambda u(0) - \mu u(T) - \gamma u(\eta) = d, \\ \lambda u'(0) - \mu u'(T) - \gamma u'(\eta) = l, \end{cases} \tag{3}$$

if and only if, u is a solution of the fractional integral equation

$$\begin{aligned} u(t) = \left(1 - \frac{At^{\alpha-\beta}}{\Gamma(\alpha - \beta + 1)}\right)u(0) + tu'(0) + \frac{A}{\Gamma(\alpha - \beta)}\int_0^t (t - s)^{\alpha-\beta-1}u(s)ds \\ + \frac{1}{\Gamma(\alpha)}\int_0^t (t - s)^{\alpha-1}y(s)ds, \end{aligned} \tag{4}$$

with

$$u(0)=\Delta\Big\{\Big[\Lambda^{-1}\Big(\frac{\sigma}{\delta}\Big)(\mu T+\gamma\eta)+1\Big]\Big[AI_{0^+}^{\alpha-\beta}(\mu u(T)+\gamma u(\eta))+I_{0^+}^{\alpha}(\mu y(T)+\gamma y(\eta))+d\Big]$$
$$+\Lambda^{-1}(\mu T+\gamma\eta)\Big[AI_{0^+}^{\alpha-\beta-1}(\mu u(T)+\gamma u(\eta))+I_{0^+}^{\alpha-1}(\mu y(T)+\gamma y(\eta))+l\Big]\Big\}, \tag{5}$$

and

$$u'(0)=\Lambda^{-1}\Big\{\Big(\frac{\sigma}{\delta}\Big)\Big[AI_{0^+}^{\alpha-\beta}(\mu u(T)+\gamma u(\eta))+I_{0^+}^{\alpha}(\mu y(T)+\gamma y(\eta))+d\Big]$$
$$+\Big[AI_{0^+}^{\alpha-\beta-1}(\mu u(T)+\gamma u(\eta))+I_{0^+}^{\alpha-1}(\mu y(T)+\gamma y(\eta))\Big]+l\Big\}, \tag{6}$$

where

$$I_{0^+}^{\alpha}u(T)=\frac{1}{\Gamma(\alpha)}\int_0^T(T-s)^{\alpha-1}u(s)ds,$$

$$I_{0^+}^{\alpha}u(\eta)=\frac{1}{\Gamma(\alpha)}\int_0^\eta(\eta-s)^{\alpha-1}u(s)ds,$$

$$I_{0^+}^{\alpha-\beta}u(T)=\frac{1}{\Gamma(\alpha-\beta)}\int_0^T(T-s)^{\alpha-\beta-1}u(s)ds,$$

$$I_{0^+}^{\alpha-\beta}u(\eta)=\frac{1}{\Gamma(\alpha-\beta)}\int_0^\eta(\eta-s)^{\alpha-\beta-1}u(s)ds.$$

Proof. From Lemmas 2.1 and 2.2, we have

$$u(t)=u(0)+tu'(0)-\frac{At^{\alpha-\beta}}{\Gamma(\alpha-\beta+1)}u(0)+\frac{A}{\Gamma(\alpha-\beta)}\int_0^t(t-s)^{\alpha-\beta-1}u(s)ds$$
$$+\frac{1}{\Gamma(\alpha)}\int_0^t(t-s)^{\alpha-1}y(s)ds$$

Applying conditions (2), we obtain (5) and (6).

Conversely, assume that u satisfies the fractional integral (4), and using the facts that ${}^cD_{0^+}^{\alpha}$ is the left inverse of $I_{0^+}^{\alpha}$ and the fact that ${}^cD_{0^+}^{\alpha}C=0$, where C is a constant, we get

$${}^cD_{0^+}^{\alpha}u(t)-A\,{}^cD_{0^+}^{\beta}u(t)=f(t,u(t),u'(t)),\quad t\in J=[0,T].$$

Also, we can easily show that

$$\begin{cases}\lambda u(0)-\mu u(T)-\gamma u(\eta)=d,\\ \lambda u'(0)-\mu u'(T)-\gamma u'(\eta)=l.\end{cases}$$

The proof is complete.

To simplify the proofs in the forthcoming theorem, we etablish the bounds for the integrals and the bounds for the term arising in the sequel.

Lemma 2.4. *For $y(\cdot) \in C(J, \mathbb{R}^n)$, we have*

$$\left|I^{\alpha}_{0^+}y(\eta)\right| = \left|\int_0^{\eta} \frac{(\eta-\tau)^{\alpha-1}}{\Gamma(\alpha)} y(\tau)d\tau\right| \leq \frac{\eta^{\alpha}}{\Gamma(\alpha+1)}\|y\|.$$

Proof. Obviously,

$$\int_0^{\eta} \frac{(\eta-\tau)^{\alpha-1}}{\Gamma(\alpha)} y(\tau)d\tau = \left[-\frac{(\eta-\tau)^{\alpha}}{\alpha\Gamma(\alpha)}\right]_0^{\eta} = \frac{\eta^{c}}{\alpha\Gamma(\alpha)} = \frac{s^{\alpha}}{\Gamma(\alpha+1)}.$$

Hence

$$\left|\int_0^{\eta} \frac{(s-\tau)^{\alpha-1}}{\Gamma(\alpha)} y(\tau)d\tau\right| \leq \frac{\eta^{\alpha}}{\Gamma(\alpha+1)}\|y\|.$$

Lemma 2.5. *For $u(\cdot) \in C^1(J, \mathbb{R}^n)$ and $0 < \beta \leq 1$, we have*

$$\left\|{}^{c}D^{\beta}_{0^+}u(t)\right\|_{\infty} \leq \frac{T^{1-\beta}}{\Gamma(2-\beta)}\|u'\|_{\infty},$$

and, so

$$\left\|{}^{c}D^{\beta}_{0^+}u(t)\right\|_{\infty} \leq \frac{T^{1-\beta}}{\Gamma(2-\beta)}\|u\|_{1}.$$

Proof.

Clearly, when $\beta = 1$, the conclusion are true. So, consider the case $0 < \beta < 1$. By Definition 2.1, for each $u(\cdot) \in C^1(J, \mathbb{R}^n)$ and $t \in J$, we have

$$\begin{aligned}
\left|D^{\beta}_{0^+}u(t)\right| &= \frac{1}{\Gamma(1-\beta)}\left|\int_0^{t}(t-s)^{-\beta}u'(s)ds\right| \\
&\leq \|u'\|_{\infty}\frac{1}{\Gamma(1-\beta)}\int_0^{t}(t-s)^{-\beta}ds \\
&= \|u'\|_{\infty}\frac{t^{1-\beta}}{\Gamma(1-\beta)} \\
&\leq \frac{T^{1-\beta}}{\Gamma(1-\beta)}\|u'\|_{\infty} \\
&\leq \frac{T^{1-\beta}}{\Gamma(1-\beta)}\|u'\|_{1}.
\end{aligned}$$

We need to give the following hypothesis:

(H_1) there existe a constants $L_1,\ L_2 > 0$ and $0 < L_3 < 1$ such that

$$|f(t,u,v,w) - f(t,\overline{u},\overline{v},\overline{w})| \le L_1\|u-\overline{u}\| + L_2\|v-\overline{v}\| + L_3\|w-\overline{w}\|,$$

for any $u, v, \overline{u}, \overline{v}, \overline{w} \in \mathbb{R}^n$ and $t \in J$.

Now we are in a position to present the first main result of this paper. The existence results is based on Banach contraction principle.

Theorem 1.1. ([47] Banach's fixed point theorem). Let C be a non-empty closed subset of a Banach space E, then any contraction mapping T of C into itself has a unique fixed point.

Theorem 1.2. Assume that (H_1) holds. If

$$\max(R_1, R_2) < 1, \tag{7}$$

then the boundary value problem (1.1)–(1.2) has a unique solution on J.

Proof. We transform the problem (1.1)–(1.2) into fixed point problem. Let $N : C^1(J,\mathbb{R}^n) \to C^1(J,\mathbb{R}^n)$ the operator defined by

$$\begin{aligned}(Nu)(t) &= \left(1 - \frac{At^{\alpha-\beta}}{\Gamma(\alpha-\beta+1)}\right)B + tD + \frac{A}{\Gamma(\alpha-\beta)}\int_0^t (t-s)^{\alpha-\beta-1}u(s)ds \\ &+ \frac{1}{\Gamma(\alpha)}\int_0^t (t-s)^{\alpha-1}g(s)ds,\end{aligned} \tag{8}$$

with

$$\begin{aligned}B = \Delta\Big\{&\left[\Lambda^{-1}\left(\frac{\sigma}{\delta}\right)(\mu T + \gamma\eta) + 1\right]\left[AI_{0^+}^{\alpha-\beta}(\mu u(T) + \gamma u(\eta)) + I_{0^+}^{\alpha}(\mu y(T) + \gamma y(\eta)) + d\right] \\ &+ \Lambda^{-1}(\mu T + \gamma\eta)\left[AI_{0^+}^{\alpha-\beta-1}(\mu u(T) + \gamma u(\eta)) + I_{0^+}^{\alpha-1}(\mu y(T) + \gamma y(\eta)) + l\right]\Big\},\end{aligned}$$

and

$$\begin{aligned}D = \Lambda^{-1}\Big\{&\left(\frac{\sigma}{\delta}\right)\left[AI_{0^+}^{\alpha-\beta}(\mu u(T) + \gamma u(\eta)) + I_{0^+}^{\alpha}(\mu g(T) + \gamma g(\eta)) + d\right] \\ &+ \left[AI_{0^+}^{\alpha-\beta-1}(\mu u(T) + \gamma u(\eta)) + I_{0^+}^{\alpha-1}(\mu g(T) + \gamma g(\eta))\right] + l\Big\},\end{aligned}$$

where $g \in C(J,\mathbb{R}^n)$ be such that

$$g(t) = f\left(t, u(t), {}^cD_{0^+}^{\beta}u(t), g(t) + A\,{}^cD_{0^+}^{\beta}u(t)\right)$$

For every $u \in C^1(J,\mathbb{R}^n)$ and any $t \in J$, we have

$$\begin{aligned}(Nu)(t) &= D - \frac{(\alpha-\beta)At^{\alpha-\beta-1}}{\Gamma(\alpha-\beta+1)}B + \frac{A}{\Gamma(\alpha-\beta-1)}\int_0^t (t-s)^{\alpha-\beta-2}u(s)ds \\ &+ \frac{1}{\Gamma(\alpha-1)}\int_0^t (t-s)^{\alpha-2}g(s)ds.\end{aligned} \tag{9}$$

Clearly, the fixed points of operator N are solutions of problem (1.1)–(1.2).
It is clear that $(Nu) \in C(J, \mathbb{R}^n)$, consequently, N is well defined.
Let $u, v \in C(J, \mathbb{R}^n)$. Then for $t \in J$, we have

$$\begin{aligned}\|(Nu)(t) - (Nv)(t)\| \leq & \left(1 + \frac{\|A\|T^{\alpha-\beta}}{\Gamma(\alpha-\beta+1)}\right)\|B - B_1\| + T\|D - D_1\| \\ & + \frac{\|A\|}{\Gamma(\alpha-\beta)}\int_0^T (T-s)^{\alpha-\beta-1}\|u(s) - v(s)\|ds \\ & + \frac{1}{\Gamma(\alpha)}\int_0^T (T-s)^{\alpha-1}\|g(s) - h(s)\,|ds,\end{aligned}$$

with

$$\begin{aligned}B_1 = \Delta\Big\{ & \left[\Lambda^{-1}\left(\frac{\sigma}{\delta}\right)(\mu T + \gamma\eta) + 1\right]\left[AI_{0^+}^{\alpha-\beta}(\mu v(T) + \gamma v(\eta)) + I_{0^+}^{\alpha}(\mu h(T) + \gamma h(\eta)) + d\right] \\ & + \Lambda^{-1}(\mu T + \gamma\eta)\left[AI_{0^+}^{\alpha-\beta-1}(\mu v(T) + \gamma v(\eta)) + I_{0^+}^{\alpha-1}(\mu h(T) + \gamma h(\eta)) + l\right]\Big\},\end{aligned}$$

and

$$\begin{aligned}D_1 = \Lambda^{-1}\Big\{ & \left(\frac{\sigma}{\delta}\right)\left[AI_{0^+}^{\alpha-\beta}(\mu v(T) + \gamma v(\eta)) + I_{0^+}^{\alpha}(\mu h(T) + \gamma h(\eta)) + d\right] \\ & + \left[AI_{0^+}^{\alpha-\beta-1}(\mu v(T) + \gamma v(\eta)) + I_{0^+}^{\alpha-1}(\mu h(T) + \gamma h(\eta))\right] + l\Big\},\end{aligned}$$

From (H), for any $t \in J$, we have

$$\begin{aligned}\|g(t) - h(t)\| = & L_1\|u(t) - v(t)\| + L_2\left\|{}^cD_{0^+}^{\beta}u(t) - {}^cD_{0^+}^{\beta}v(t)\right\| \\ & + L_3\left\|g(t) + A\,{}^cD_{0^+}^{\beta}u(t) - h(t) - A\,{}^cD_{0^+}^{\beta}v(t)\right\| \\ \leq & L_1\|u(t) - v(t)\| + L_2\left\|{}^cD_{0^+}^{\beta}u(t) - {}^cD_{0^+}^{\beta}v(t)\right\| \\ & + L_3\|g(t) - h(t)\| + L_3\|A\|\left\|{}^cD_{0^+}^{\beta}u(t) - {}^cD_{0^+}^{\beta}v(t)\right\| \\ \leq & L_1\|u(t) - v(t)\| + L_3\|g(t) - h(t)\| + (L_3\|A\| + L_2)\left\|{}^cD_{0^+}^{\beta}(u(t) - v(t))\right\|.\end{aligned}$$

Thus

$$\begin{aligned}\|g(t) - h(t)\| \leq & \frac{L_1}{1 - L_3}\|u(t) - v(t)\| + \frac{L_3\|A\| + L_2}{1 - L_3}\left\|{}^cD_{0^+}^{\beta}(u(t) - v(t))\right\| \\ \leq & \frac{L_1}{1 - L_3}\|u - v\|_{\infty} + \frac{L_3\|A\| + L_2}{1 - L_3}\left\|{}^cD_{0^+}^{\beta}(u - v)\right\|_{\infty}.\end{aligned}$$

Then, according to the Lemma 3.2, we get

$$\begin{aligned}\|g(t) - h(t)\| \leq & \frac{L_1}{1 - L_3}\|u - v\|_1 + \frac{T^{1-\beta}(L_3\|A\| + L_2)}{\Gamma(2-\beta)(1 - L_3)}\|u - v\|_1 \\ = & \frac{L_1\Gamma(2-\beta) + T^{1-\beta}(L_3\|A\| + L_2)}{\Gamma(2-\beta)(1 - L_3)}\|u - v\|_1.\end{aligned} \tag{10}$$

By employing (10) and Lemma 3.1, we get

$$\begin{aligned}\|B_1 - B_2\| &\leq \Delta\left\{\left[\Lambda^{-1}\left(\frac{\sigma}{\delta}\right)(\mu T + \gamma\eta) + 1\right]\Phi + \Lambda^{-1}(\mu T + \gamma\eta)\Theta\right\}\|u - v\|_1 \\ &= M_1\|u - v\|_1.\end{aligned}$$

and

$$\begin{aligned}\|D_1 - D_2\| &\leq \Lambda^{-1}\left\{\left(\frac{\sigma}{\delta}\right)\Phi + \Theta\right\}\|u - v\|_1 \\ &= M_2\|u - v\|_1,\end{aligned}$$

where Φ and Θ defined above.
Thus, for $t \in J$, we have

$$\begin{aligned}\|(Nu)(t) - (Nv)(t)\| &\leq \left[\left(1 + \frac{\|A\|T^{\alpha-\beta}}{\Gamma(\alpha-\beta+1)}\right)M_1 + TM_2 + \frac{\|A\|T^{\alpha-\beta}}{\Gamma(\alpha-\beta+1)}\right. \\ &\quad \left. + \frac{T^{\alpha}L_1\Gamma(2-\beta) + T^{1-\beta+\alpha}(L_3\|A\| + L_2)}{\Gamma(\alpha+1)\Gamma(2-\beta)(1-L_2)}\right]\|u - v\|_1 \\ &= R_1\|u - v\|_1.\end{aligned}$$

Also

$$\begin{aligned}\|(Nu)(t) - (Nv)(t)\| &\leq \|D_2 - D_1\| + \frac{(\alpha-\beta)\|A\|T^{\alpha-\beta-1}}{\Gamma(\alpha-\beta+1)}\|B_1 - B_2\| \\ &\quad + \frac{\|A\|}{\Gamma(\alpha-\beta-1)}\int_0^T (T-s)^{\alpha-\beta-2}\|u(s) - v(s)\|ds \\ &\quad + \frac{1}{\Gamma(\alpha-1)}\int_0^T (T-s)^{\alpha-2}\|g(s) - h(s)\|ds.\end{aligned}$$

By employing (10) and Lemma 3.2, we get

$$\begin{aligned}\|(Nu)(t) - (Nv)(t)\| &\leq \left[M_2 + \frac{(\alpha-\beta)\|A\|T^{\alpha-\beta-1}}{\Gamma(\alpha-\beta+1)}M_1 + \frac{\|A\|T^{\alpha-\beta-1}}{\Gamma(\alpha-\beta)}\right. \\ &\quad \left. + \frac{T^{\alpha-1}L_1\Gamma(2-\beta) + T^{\alpha-\beta}(L_3\|A\| + L_2)}{\Gamma(\alpha)\Gamma(2-\beta)(1-L_2)}\right]\|u - v\|_1 \\ &= R_2\|u - v\|_1.\end{aligned}$$

Therefore

$$\|(Nu)(t) - (Nv)(t)\| \leq \max\{R_1, R_2\}\|u - v\|_1.$$

Thus, by (10) the operator N is a contraction. Hence it follows by Banach's contraction principle that the boundary value problem (1)–(12) has a unique solution on J.

Now we are in a position to present the second main result of this paper. The existence results is based on Leray-Schauder nonlinear alternative.

Theorem 1.3. ([6] Nonlinear alternative for single valued maps). Let E be a Banach space, C a closed, convex subset of E and U an open subset of C with $0 \in U$. Suppose that $F : \overline{U} \to C$ is a continuous and compact (that is $F(\overline{U})$ is relatively compact subset of C) map. Then either

i. F has a fixed point in $\overline{U}$, or

ii. there is a $u \in \partial U$ (the boundary of U in C) and $\lambda \in (0,1)$ with $u = \lambda F(u)$.

Set

$$l_1 = M_3 + TM_4 + \frac{\|A\|T^{\alpha-\beta}}{\Gamma(\alpha-\beta+1)}M_3 + TM_4 + \frac{\|A\|rT^{\alpha-\beta}}{\Gamma(\alpha-\beta+1)} + \frac{T^{\alpha}}{\Gamma(\alpha+1)}M,$$

and

$$l_2 = M_4 + \frac{(\alpha-\beta)\|A\|T^{\alpha-\beta-1}}{\Gamma(\alpha-\beta+1)}M_3 + \frac{\|A\|rT^{\alpha-\beta-1}}{\Gamma(\alpha-\beta)} + \frac{T^{\alpha}M}{\Gamma(\alpha)}.$$

Theorem 1.4. Assume that (H_1) holds and there exists a positive constant $M > 0$ such that $\max\{l_1, l_2\} = l < M$. Then the boundary value problem (1.1)–(1.2) has at least one solution on J.

Proof. Let N be the operator defined in (8).

N is continuous. Let (u_n) be a sequence such that $u_n \to u$ in $C(J, \mathbb{R}^n)$. Then for $t \in J$, we have

$$\begin{aligned}\|(Nu)(t) - (Nu_n)(t)\| &\le \left(1 + \frac{\|A\|T^{\alpha-\beta}}{\Gamma(\alpha-\beta+1)}\right)\|B_1 - B_{n2}\| + T\|D_1 - D_{n2}\| \\ &+ \frac{\|A\|}{\Gamma(\alpha-\beta)}\int_0^T (T-s)^{\alpha-\beta-1}\|u(s) - u_n(s)\|ds \\ &+ \frac{1}{\Gamma(\alpha)}\int_0^T (T-s)^{\alpha-1}\|g(s) - g_n(s)\|ds,\end{aligned}$$

where $B_{n2}, D_{n2} \in \mathbb{R}^n$, with

$$\begin{aligned}B_{n2} = \Delta\Big\{&\left[\Lambda^{-1}\left(\frac{\sigma}{\delta}\right)(\mu T + \gamma\eta) + 1\right]\left[AI_{0^+}^{\alpha-\beta}(\mu u_n(T) + \gamma u_n(\eta)) + I_{0^+}^{\alpha}(\mu g_n(T) + \gamma g_n(\eta)) + d\right] \\ &+ \Lambda^{-1}(\mu T + \gamma\eta)\left[AI_{0^+}^{\alpha-\beta-1}(\mu u_n(T) + \gamma u_n(\eta)) + I_{0^+}^{\alpha-1}(\mu g_n(T) + \gamma g_n(\eta)) + l\right]\Big\}, \\ D_{n2} = \Lambda^{-1}\Big\{&\left(\frac{\sigma}{\delta}\right)\left[AI_{0^+}^{\alpha-\beta}(\mu u_n(T) + \gamma u_n(\eta)) + I_{0^+}^{\alpha}(\mu g_n(T) + \gamma g_n(\eta)) + d\right] \\ &+ AI_{0^+}^{\alpha-\beta-1}(\mu u_n(T) + \gamma u_n(\eta)) + I_{0^+}^{\alpha-1}(\mu g_n(T) + \gamma g_n(\eta)) + l\Big\},\end{aligned}$$

and

$$g_n(t) = f\left(t, u_n(t), {}^cD_{0^+}^{\beta}u_n(t), g_n(t) + A\,{}^cD_{0^+}^{\beta}u_n(t)\right).$$

From (H), for any $t \in J$, we have

$$\begin{aligned}\left\|g(t) - g_n(t)\right\| &= L_1\|u(t) - u_n(t)\| + L_3\left\|g(t) + A^c D^\beta_{0^+} u(t) - g_n(t) - A^c D^\beta_{0^+} u_n(t)\right\| \\ &\quad + L_2\left\|{}^c D^\beta_{0^+} u(t) - {}^c D^\beta_{0^+} u_n(t)\right\| \\ &\le L_1\|u(t) - u_n(t)\| + L_2\left\|{}^c D^\beta_{0^+} u(t) - {}^c D^\beta_{0^+} u_n(t)\right\| \\ &\quad + L_3\left\|g(t) - g_n(t)\right\| + L_3\|A\|\left\|{}^c D^\beta_{0^+} u(t) - {}^c D^\beta_{0^+} u_n(t)\right\| \\ &\le L_1\|u(t) - u_n(t)\| + L_3\left\|g(t) - g_n(t)\right\| + (L_3\|A\| + L_2)\left\|{}^c D^\beta_{0^+}(u(t) - u_n(t))\right\|.\end{aligned}$$

Thus

$$\begin{aligned}\left\|g(t) - g_n(t)\right\| &\le \frac{L_1}{1 - L_3}\|u(t) - u_n(t)\| + \frac{L_3\|A\| + L_2}{1 - L_3}\left\|{}^c D^\beta_{0^+}(u(t) - u_n(t))\right\| \\ &\le \frac{L_1}{1 - L_3}\|u - u_n\|_\infty + \frac{L_3\|A\| + L_2}{1 - L_3}\left\|{}^c D^\beta_{0^+}(u - u_n)\right\|_\infty.\end{aligned}$$

Then, according to the Lemma 3.2, we get

$$\begin{aligned}\left\|g(t) - g_n(t)\right\| &\le \frac{L_1}{1 - L_3}\|u - u_n\|_1 + \frac{T^{1-\beta}(L_3\|A\| + L_2)}{(1 - L_3)\Gamma(2 - \beta)}\|u - u_n\|_1 \\ &= \frac{L_1\Gamma(2 - \beta) + T^{1-\beta}(L_3\|A\| + L_2)}{(1 - L_3)\Gamma(2 - \beta)}\|u - u_n\|_1.\end{aligned}$$

By employing (10) and Lemma 3.1, we get

$$\begin{aligned}\|B_1 - B_{n2}\| &\le \Delta\left\{\left[\Lambda^{-1}\left(\frac{\sigma}{\delta}\right)(\mu T + \gamma\eta) + 1\right]\Phi + \Lambda^{-1}(\mu T + \gamma\eta)\Theta\right\}\|u - u_n\|_1, \\ &= M_1\|u - u_n\|_1.\end{aligned}$$

and

$$\begin{aligned}\|D_1 - D_{n2}\| &\le \Lambda^{-1}\left\{\left(\frac{\sigma}{\delta}\right)\Phi + \Theta\right\}\|u - u_n\|_1 \\ &= M_2\|u - u_n\|_1,\end{aligned}$$

Thus, for $t \in J$, we have

$$\begin{aligned}\|(Nu)(t) - (Nu_n)(t)\| &\le \left[\left(1 + \frac{\|A\|T^{\alpha-\beta}}{\Gamma(\alpha - \beta + 1)}\right)M_1 + TM_2 + \frac{\|A\|T^{\alpha-\beta}}{\Gamma(\alpha - \beta + 1)}\right. \\ &\quad \left. + \frac{T^\alpha L_1\Gamma(2 - \beta) + T^{1-\beta+\alpha}(L_3\|A\| + L_2)}{\Gamma(\alpha + 1)\Gamma(2 - \beta)(1 - L_2)}\right]\|u - u_n\|_1 \\ &= R_1\|u - u_n\|_1.\end{aligned}$$

Also

$$\|(Nu)(t) - (Nu_n)(t)\| \leq \|D_{n2} - D_1\| + \frac{(\alpha-\beta)\|A\|T^{\alpha-\beta-1}}{\Gamma(\alpha-\beta+1)}\|B_1 - B_{n2}\|$$
$$+\frac{\|A\|}{\Gamma(\alpha-\beta-1)}\int_0^T (T-s)^{\alpha-\beta-2}\|u(s) - u_n(s)\|ds$$
$$+\frac{1}{\Gamma(\alpha-1)}\int_0^T (T-s)^{\alpha-2}\|g(s) - g_n(s)\|ds.$$

By employing (10), we get

$$\|(Nu)(t) - (Nu_n)(t)\| \leq \left[M_2 + \frac{(\alpha-\beta)\|A\|T^{\alpha-\beta-1}}{\Gamma(\alpha-\beta+1)}M_1 + \frac{\|A\|T^{\alpha-\beta-1}}{\Gamma(\alpha-\beta)}\right.$$
$$\left. + \frac{T^{\alpha-1}L_1\Gamma(2-\beta)(L_3\|A\|+L_2)T^{\alpha-\beta-1}}{\Gamma(\alpha-\beta+1)}M_1 + \frac{\|A|\,T^{\alpha-\beta-1}}{\Gamma(\alpha-\beta)}\right]\|u-u_n\|_1.$$
$$= R_2\|u-u_n\|_1.$$

Thus $\|Nu - Nu_n\|_1 \to 0$ as $n \to \infty$, which implies that the operator N is continuous.

Now, we show N maps bounded sets into bounded sets in $C(J, \mathbb{R}^n)$. For a positive number r, let $B_r = \{u \in C^1(J, \mathbb{R}^n) : \|u\|_1 \leq r\}$ be a bounded set in $C(J, \mathbb{R}^n)$. Then we have

$$\|g(t)\| \leq \left\| f\left(t, u(t), g(t) + A^c D^\beta_{0^+}u(t), D^\beta_{0^+}u(t)\right) - f(t,0,0,0)\right\| + \|f(t,0,0,0)\|$$
$$\leq L_1\|u(t)\| + L_3\left\|g(t) + A^c D^\beta_{0^+}u(t)\right\| + L_2\left\|D^\beta_{0^+}u(t)\right\| + \|f(t,0,0,0)\|$$
$$\leq L_1\|u\|_\infty + L_3\|g(t)\| + (L_3\|A\| + L_2)\left\|D^\beta_{0^+}u\right\|_\infty + f^*,$$

where $\sup_{t\in J}|f(t,0,0,0)| = f^* < \infty$. Thus

$$\|g(t)\| \leq \frac{L_1}{1-L_3}\|u\|_\infty + \frac{L_3\|A\| + L_2}{1-L_3}\left\|D^\beta_{0^+}u\right\|_\infty + \frac{f^*}{1-L_3}.$$

Then, By Lemma 3.2, we have

$$\|g(t)\| \leq \frac{L_1}{1-L_3}\|u\|_\infty + \frac{(L_3\|A\|+L_2)T^{1-\beta}}{(1-L_3)\Gamma(2-\beta)}\|u'\|_\infty + \frac{f^*}{1-L_3}$$
$$\leq \frac{L_1}{1-L_3}\|u\|_1 + \frac{(L_3\|A\|+L_2)T^{1-\beta}}{(1-L_3)\Gamma(2-\beta)}\|u\|_1 + \frac{f^*}{1-L_3} \tag{11}$$
$$\leq \frac{L_1 r}{1-L_3} + \frac{(L_3\|A\|+L_2)rT^{1-\beta}}{(1-L_3)\Gamma(2-\beta)} + \frac{f^*}{1-L_3} = M,$$

which implies that

$$\|B\| \leq r\|A\|\Delta\Big[\Big(\Lambda^{-1}\Big(\frac{\sigma}{\delta}\Big)(\mu T+\gamma\eta)+1\Big)\Big(\frac{\mu T^{\alpha-\beta}+\gamma\eta^{\alpha-\beta}}{\Gamma(\alpha-\beta+1)}\Big)$$

$$+\Lambda^{-1}(\mu T+\gamma\eta)\Big(\frac{\mu T^{\alpha-\beta-1}+\gamma\eta^{\alpha-\beta-1}}{\Gamma(\alpha-\beta)}\Big)\Big]$$

$$+M\Delta\Big[\Big(\Lambda^{-1}\Big(\frac{\sigma}{\delta}\Big)(\mu T+\gamma\eta)+1\Big)\Big(\frac{\mu T^{\alpha}+\gamma\eta^{\alpha}}{\Gamma(\alpha+1)}\Big)+\Lambda^{-1}(\mu T+\gamma\eta)\Big(\frac{\mu T^{\alpha-1}+\gamma\eta^{\alpha-1}}{\Gamma(\alpha)}\Big)\Big]$$

$$+\Delta\Lambda^{-1}(\mu T+\gamma\eta)\Big[l+d\Big(\Big(\frac{\sigma}{\delta}\Big)+1\Big)\Big]=M_3,$$

and

$$\|D\| \leq r\|A\|\Big[\Lambda^{-1}\Big(\frac{\sigma}{\delta}\Big)\frac{\mu T^{\alpha-\beta}+\gamma\eta^{\alpha-\beta}}{\Gamma(\alpha-\beta+1)}+\frac{\mu T^{\alpha-\beta-1}+\gamma\eta^{\alpha-\beta-1}}{\Gamma(\alpha-\beta)}\Big]$$

$$+M\Lambda^{-1}\Big[\Big(\frac{\sigma}{\delta}\Big)\Big(\frac{\mu T^{\alpha}+\gamma\eta^{\alpha}}{\Gamma(\alpha+1)}\Big)+\Big(\frac{\mu T^{\alpha-1}+\gamma\eta^{\alpha-1}}{\Gamma(\alpha)}\Big)\Big]+\Lambda^{-1}\Big[\Big(\frac{\sigma}{\delta}\Big)d+l\Big]=M_4.$$

Thus (8) implies

$$\|(Nu)(t)\| \leq M_3+\frac{\|A\|T^{\alpha-\beta}}{\Gamma(\alpha-\beta+1)}M_3+TM_4+\frac{\|A\|rT^{\alpha-\beta}}{\Gamma(\alpha-\beta+1)}+\frac{T^{\alpha}}{\Gamma(\alpha+1)}M=l_1,$$

and

$$\|(Nu)(t)\| \leq M_4+\frac{(\alpha-\beta)\|A\|T^{\alpha-\beta-1}}{\Gamma(\alpha-\beta+1)}M_3+\frac{\|A\|rT^{\alpha-\beta-1}}{\Gamma(\alpha-\beta)}+\frac{T^{\alpha}M}{\Gamma(\alpha)}=l_2.$$

Therefore

$$\|(Nu)\|_1 \leq \max\{l_1,l_2\}=l. \tag{12}$$

Now, we show that N maps bounded sets into equicontinuous sets of $C^1(J,\mathbb{R}^n)$. Let $t_1,t_2\in[0,1]$ with $t_1<t_2$ and $u\in B_r$ is bounded sets of $C^1(J,\mathbb{R}^n)$. Then

$$\|(Nu)(t_2)-(Nu)(t_1)\| \leq M_4(t_2-t_1)+\Big(1+\frac{\|A\|M_3}{\Gamma(\alpha-\beta+1)}\Big)\Big(t_2^{\alpha-\beta}-t_1^{\alpha-\beta}\Big)$$

$$+\frac{\|A\|r}{\Gamma(\alpha-\beta)}\int_{t_1}^{t_2}(t_2-s)^{\alpha-\beta-1}ds+\frac{\|A\|r}{\Gamma(\alpha-\beta)}\int_0^{t_1}\Big[(t_2-s)^{\alpha-\beta-1}-(t_1-s)^{\alpha-\beta-1}\Big]ds$$

$$+\frac{M_1}{\Gamma(\alpha)}\Big[\int_{t_1}^{t_2}(t_2-s)^{\alpha-1}ds+\int_0^{t_1}\Big[(t_2-s)^{\alpha-1}-(t_1-s)^{\alpha-1}\Big]ds\Big]$$

Obviously, the right-hand side of the above inequality tends to zero as $t_2\to t_1$. Similarly, we have

$$\|(Nu)(t_2) - (Nu)(t_1)\| \le \frac{(\alpha-\beta)\|A\|M_3}{\Gamma(\alpha-\beta+1)}\left(t_2^{\alpha-\beta-1} - t_1^{\alpha-\beta-1}\right)$$

$$+\frac{\|A\|r}{\Gamma(\alpha-\beta-1)}\int_{t_1}^{t_2}(t_2-s)^{\alpha-\beta-2}ds + \frac{\|A\|r}{\Gamma(\alpha-\beta-2)}\int_0^{t_1}\left[(t_2-s)^{\alpha-\beta-2} - (t_1-s)^{\alpha-\beta-2}\right]ds$$

$$+\frac{M}{\Gamma(\alpha-1)}\left[\int_{t_1}^{t_2}(t_2-s)^{\alpha-2}ds + \int_0^{t_1}\left[(t_2-s)^{\alpha-2} - (t_1-s)^{\alpha-2}\right]ds\right]$$

Again, it is seen that the right-hand side of the above inequality tends to zero as $t_2 \to t_1$. Thus, $\|(Nu)(t_2) - (Nu)(t_1)\| \to 0$, as $t_2 \to t_1$. This shows that the operator N is completely continuous, by the Ascoli-Arzela theorem. Thus, the operator N satisfies all the conditions of Theorem 3.4, and hence by its conclusion, either condition (i) or condition (ii) holds. We show that the condition (ii) is not possible.

Let $U = \{u \in C^1(J, \mathbb{R}^n) : \|u\| < M\}$ with $\max\{l_1, l_2\} = l < M$. In view of condition $l < M$ and by (12), we have

$$\|Nu\| \le \max\{l_1, l_2\} < M.$$

Now, suppose there exists $u \in \partial U$ and $\lambda \in (0,1)$ such that $u = \lambda Nu$. Then for such a choice of u and the constant λ, we have

$$M = \|u\| = \lambda\|Nu\| < \max\{l_1, l_2\} < M,$$

which is a contradiction. Consequently, by the Leray-Schauder alternative, we deduce that F has a fixed point $u \in \overline{U}$ which is a solution of the boundary value problem (1)–(12). The proof is completed.

We construct an example to illustrate the applicability of the results presented.

Example 2.1. *Consider the following fractional differential equation*

$$^cD_{0^+}^{\alpha}u(t) - A\,^cD_{0^+}^{\beta}u(t) = f\left(t, u(t), {}^cD_{0^+}^{\beta}u(t), {}^cD_{0^+}^{\alpha}u(t)\right), \quad t \in J = [0,1], \tag{13}$$

subject to the three-point boundary conditions

$$\begin{cases} u(0) - u(1) - u\left(\frac{1}{2}\right) = 1, \\ u'(0) - u'(1) - u'\left(\frac{1}{2}\right) = 1, \end{cases} \tag{14}$$

where $\alpha = 2, \quad \beta = 1, \quad \lambda = \mu = d = l = 1, \quad A = \begin{pmatrix} 2 & 1 \\ 0 & 2 \end{pmatrix}$ and

$$f_i(t, u, v, w) = \frac{c_i t}{8}\, arctan\,(|u| + |v| + |w|), \; i = 1, \; 2,$$

such that $f = (f_1, f_2)$ with $0 < c_i < 1$, $i = 1, 2$.

For every $u_i, \; v_i \in \mathbb{R}^2, \; i = 1, \; 2, \; 3,$ we have

$$|f_i(t, u_1, u_2, u_3) - f_i(t, v_1, v_2, v_3)| \leq \frac{c_i}{8}(|u_1 - v_1| + |u_2 - v_2| + |u_3 - v_3|),\ i = 1,\ 2,$$

where $L_1 = L_2 = L_3 = \frac{c_i}{8}$ for appropriate choice of constants $c_i,\ i = 1,\ 2.$ we check the condition of Theorem 2.2. Clearly, assumption (H_1) holds. A simple computations of $R_1,\ R_2,\ l_1$ and l_2 shows tha the second condition of Theorems 3.3 and 3.5 is satisfied. Thus the conclusion of Theorems 3.3 and 3.5 applies, and hence the problem (13)–(14) has a unique solution and at least one solution on $[0, 1]$.

3. Conclusions

This chapter concerns the boundary value problem of a class of fractional differential equations involving the Riemann-Liouville fractional derivative with nonlocal boundary conditions. By using Leray-Schauder nonlinear alternative and the Banach fixed point theorem, we shows the existence and uniqueness of positive solutions of our problem. In addition, an example is provided to demonstrate the effectiveness of the main results. The results of the present chapter are significantly contribute to the existing literature on the topic.

Acknowledgements

The authors want to thank the anonymous referee for the thorough reading of the manuscript and several suggestions that help us improve the presentation of the chapter.

Conflict of interest

The authors declare no conflict of interest.

Author details

Noureddine Bouteraa[1,2*†] and Habib Djourdem[1,3*†]

1 Laboratory of Fundamental and Applied Mathematics of Oran (LMFAO), University of Oran, Ahmed Benbella, Algeria

2 Oran Graduate School of Economics, Bir El Djir, Algeria

3 University of Ahmed Zabbana, Relizane, Algeria

*Address all correspondence to: bouteraa-27@hotmail.fr and djourdem.habib7@gmail.com

† These authors contributed equally.

References

[1] Kac V, Cheung P. Quantum Calculus. New York: Springer; 2002

[2] Lakshmikantham V, Vatsala AS. General uniqueness and monotone iterative technique for fractional differential equations. Applied Mathematics Letters. 2008;**21**(8): 828-834

[3] Miller S, Ross B. An Introduction to the Fractional Calculus and Fractional Differential Equations. New York: John Wiley and Sons, Inc.; 1993

[4] Rudin W. Functional analysis. In: International Series in Pure and Applied Mathematics. 2nd ed. New York: McGraw Hill; 1991

[5] Samko SG, Kilbas AA, Marichev OI. Fractional Integrals and Derivatives: Theory and Applications. Yverdon: Gordon & Breach; 1993

[6] Deimling K. Functional Analysis. Berlin: Springer; 1985

[7] Jarad F, Abdeljaw T, Baleanu D. On the generalized fractional derivatiives and their Caputo modification. Journal of Nonlinear Sciences and Applications. 2017;**10**(5):2607-2619

[8] Tian Y. Positive solutions to m-point boundary value problem of fractional differential equation. Acta Mathematicae Applicatae Sinica, English Series. 2013; **29**:661-672

[9] EL-Sayed AMA, EL-Haddad FM. Existence of integrable solutions for a functional integral inclusion. Differential Equations & Control Processes. 2017;**3**: 15-25

[10] Bouteraa N, Benaicha S. Existence of solutions for nonlocal boundary value problem for Caputo nonlinear fractional differential inclusion. Journal of Mathematical Sciences and Modelling. 2018;**1**(1):45-55

[11] Bouteraa N, Benaicha S. Existence results for fractional differential inclusion with nonlocal boundary conditions. Rivista di Matematica della Università di Parma. 2020;**11**: 181-206

[12] Cernia A. Existence of solutions for a certain boundary value problem associated to a fourth order differential inclusion. International Journal of Analysis and Applications. 2017;**14**:27-33

[13] Ntouyas SK, Etemad S, Tariboon J, Sutsutad W. Boundary value problems for Riemann-Liouville nonlinear fractional diffrential inclusions with nonlocal Hadamard fractional integral conditions. Mediterranean Journal of Mathematics. 2015;**2015**:16

[14] Bouteraa N, Benaicha S. Triple positive solutions of higher-order nonlinear boundary value problems. Journal of Computer Science and Computational Mathematics. June 2017; **7**(2):25-31

[15] Bouteraa N, Benaicha S. Existence of solutions for three-point boundary value problem for nonlinear fractional equations. Analele Universitatii din Oradea. Fascicola Matematica. Tom 2017;**XXIV**(2):109-119

[16] Benaicha S, Bouteraa N. Existence of solutions for three-point boundary value problem for nonlinear fractional differential equations. Bulltin of the Transilvania University of Brasov, Series III: Mathematics, Informtics, Physics. 2017;**10**(2):59

[17] Bouteraa N, Benaicha S. Existence of solutions for third-order three-point boundary value problem. Mathematica. 2018;**60**(1):21-31

[18] Bouteraa N, Benaicha S. The uniqueness of positive solution for higher-order nonlinear fractional differential equation with nonlocal boundary conditions. Advances in the Theory of Nonlinear and Its Application. 2018;**2**(2):74-84

[19] Bouteraa N, Benaicha S, Djourdem H. Positive solutions for nonlinear fractional differential equation with nonlocal boundary conditions. Universal Journal of Mathematics and Applications. 2018;**1**(1):39-45

[20] Bouteraa N, Benaicha S. The uniqueness of positive solution for nonlinear fractional differential equation with nonlocal boundary conditions. Analele Universitatii din Oradea. Fascicola Matematica. Tom 2018; **XXV**(2):53-65

[21] Bouteraa N, Benaicha S, Djourdem H, Benatia N. Positive solutions of nonlinear fourth-order two-point boundary value problem with a parameter. Romanian Journal of Mathematics and Computer Science. 2018;**8**(1):17-30

[22] Bouteraa N, Benaicha S. Positive periodic solutions for a class of fourth-order nonlinear differential equations. Siberian Journal of Numerical Mathematics. 2019;**22**(1):1-14

[23] Bouteraa N. Existence of solutions for some nonlinear boundary value problems [thesis]. Ahmed Benbella, Algeria: University of Oran; 2018

[24] Bouteraa N, Benaicha S, Djourdem H. On the existence and multiplicity of positive radial solutions for nonlinear elliptic equation on bounded annular domains via fixed point index. Maltepe Journal of Mathematics. 2019;**I**(1):30-47

[25] Bouteraa N, Benaicha S, Djourdem H. Positive solutions for systems of fourth-order two-point boundary value problems with parameter. Journal of Mathematical Sciences and Modeling. 2019;**2**(1):30-38

[26] Bouteraa N, Benaicha S. Existence and multiplicity of positive radial solutions to the Dirichlet problem for the nonlinear elliptic equations on annular domains. Studia Universitatis Babeș-Bolyai Mathematica. 2020;**65**(1):109-125

[27] Benaicha S, Bouteraa N, Djourdem H. Triple positive solutions for a class of boundary value problems with integral boundary conditions. Bulletin of Transilvania University of Brasov, Series III: Mathematics, Informatics, Physics. 2020;**13**(1):51-68

[28] Bouteraa N, Djourdem H, Benaicha S. Existence of solution for a system of coupled fractional boundary value problem. Proceedings of International Mathematical Sciences. 2020;**II**(1):48-59

[29] Bouteraa N, Benaicha S. Existence results for second-order nonlinear differential inclusion with nonlocal boundary conditions. Numerical Analysis and Applications. 2021;**14**(1):30-39

[30] Bouteraa N, Inc M, Akinlar MA, Almohsen B. Mild solutions of fractional PDE with noise. Mathematical Methods in the Applied Sciences. 2021:1-15

[31] Bouteraa N, Benaicha S. A study of existence and multiplicity of positive solutions for nonlinear fractional differential equations with nonlocal boundary conditions. Studia

Universitatis Babeș-Bolyai Mathematica. 2021;**66**(2):361-380

[32] Djourdem H, Benaicha S, Bouteraa N. Existence and iteration of monotone positive solution for a fourth-order nonlinear boundary value problem. Fundamental Journal of Mathematics and Applications. 2018;**1**(2):205-211

[33] Djourdem H, Benaicha S, Bouteraa N. Two positive solutions for a fourth-order three-point BVP with sign-changing green's function. Communications in Advanced Mathematical Sciences. 2019;**II**(1):60-68

[34] Djourdem H, Bouteraa N. Mild solution for a stochastic partial differential equation with noise. WSEAS Transactions on Systems. 2020;**19**:246-256

[35] Ghorbanian R, Hedayati V, Postolache M, Rezapour SH. On a fractional differential inclusion via a new integral boundary condition. Journal of Inequalities and Applications. 2014; **2014**:20

[36] Inc M, Bouteraa N, Akinlar MA, Chu YM, Weber GW, Almohsen B. New positive solutions of nonlinear elliptic PDEs. Applied Sciences. 2020;**10**:4863. DOI: 10.3390/app10144863

[37] Bouteraa N, Benaicha S. Nonlinear boundary value problems for higher-order ordinary differential equation at resonance. Romanian Journal of Mathematic and Computer Science. 2018;**8**(2):83-91

[38] Bouteraa N, Benaicha S. A class of third-order boundary value problem with integral condition at resonance. Maltepe Journal of Mathematics. 2020; **II**(2):43-54

[39] Lin X, Zhao Z, Guan Y. Iterative technology in a singular fractional boundary value problem with q-difference. Applied Mathematics. 2016;**7**:91-97

[40] Rezapour SH, Hedayati V. On a Caputo fractional differential inclusion with integral boundary condition for convex-compact and nonconvex-compact valued multifunctions. Kragujevac Journal of Mathematics. 2017;**41**:143-158

[41] Sheng S, Jiang J. Existence and uniqueness of the solutions for fractional damped dynamical systems. Advances in Difference Equations. 2017;**2017**:16

[42] Abbas S, Benchohra M, Bouriah S, Nieto JJ. Periodic solution for nonlinear fractional differential systems. Differential Equations Applications. 2018;**10**(3):299-316

[43] Zhu T. Existence and uniqueness of positive solutions for fractional differential equations. Boundary Value Problems. 2019;**22**:11

[44] Annaby MH, Mansour ZS. q-Fractional calculus and equations. In: Lecture Notes in Mathematics. Vol. 2056. Berlin: Springer-Verlag; 2012

[45] Agrawal O. Some generalized fractional calculus operators and their applications in integral equations. Fractional Calculus and Applied Analysis. 2012;**15**:700-711

[46] Kilbas AA, Srivastava HM, Trijull JJ. Theory and Applications of Fractional Differential Equations. Amsterdam: Elsevier Science B.V.; 2006

[47] Ahmad B, Ntouyas SK, Alsaedi A. Coupled systems of fractional differential inclusions with coupled boundary conditions. Electronic Journal of Differential Equations. 2019; **2019**(69):1-21

Chapter 3

Fractional Calculus-Based Generalization of the FitzHugh-Nagumo Model: Biophysical Justification, Dynamical Analysis and Neurocomputational Implications

Serge Gervais Ngueuteu Mbouna

Abstract

In this chapter, the dynamical behavior of the incommensurate fractional-order FitzHugh-Nagumo model of neuron is explored in details from local stability analysis. First of all, considering that the FitzHugh-Nagumo model is a mathematical simplification of the Hodgkin-Huxley model, the considered model is derived from the fractional-order Hodgkin-Huxley model obtained taking advantage of the powerfulness of fractional derivatives in modeling certain biophysical phenomena as the dielectrics losses in cell membranes, and the anomalous diffusion of particles in ion channels. Then, it is shown that the fractional-order FitzHugh-Nagumo model can be simulated by a simple electrical circuit where the capacitor and the inductor are replaced by corresponding fractional-order electrical elements. Then, the local stability of the model is studied using the Theorem on the stability of incommensurate fractional-order systems combined with the Cauchy's argument Principle. At last, the dynamical behavior of the model are investigated, which confirms the results of local stability analysis. It is found that the simple model can exhibit, among others, complex mixed mode oscillations, phasic spiking, first spike latency, and spike timing adaptation. As the dynamical richness of a neuron expands its computational capacity, it is thus obvious that the fractional-order FitzHugh-Nagumo model is more computationally efficient than its integer-order counterpart.

Keywords: fractional-order FitzHugh-Nagumo model, fractional derivatives, slow-fast dynamics, mixed mode oscillations, first spike latency

1. Introduction

When excited sufficiently, the neuron, which is the primary unit of brain and nerves electrical activity, generates an action potential, also known as neuronal spike, which is a rapid increase then decrease in the neuronal voltage (see for example [1]). This action potential is at the basis of many mechanisms of communication between neurons and therefore is fundamental to understanding signal processing in the brain and nerves activity [2, 3]. Indeed, action potentials can propagate in essentially constant shape away from the cell body along nerves axons and toward synaptic connections with other cells [3]. Neurons exhibit many different spike-based behaviors including regular periodic and chaotic spiking (train of spikes) and bursting (alternation between a quiescent state and repetitive spiking) [3]. Bursting activity via action potentials plays a crucial role in neuronal communication, including robust transmission of signals [4, 5]. Another interesting spike-based behavior is the mixed mode oscillations (MMOs) pattern, which is an alternation between oscillations of distinct large and small amplitudes [6], where large amplitude oscillations are spikes. Certain MMOs patterns are considered as a type of bursting patterns where the period of small amplitude oscillations is considered as the quiescent phase of the bursting pattern.

Since the pioneering work of Hodgkin and Huxley [1], many relevant conductance-based models and simplified phenomenological models have been developed to describe the brain and nerves microscopic dynamical functions. Studying these computational models with the help of the tools of nonlinear dynamics and singular perturbation theory has shed light on the dynamical processes that support the generation of spiking, bursting and MMOs. Indeed, all these models are nonlinear and almost all of them are singularly perturbed systems which are systems involving multiple time scales. The simplest singularly perturbed systems evolve on on two time scales and are therefore named slow-fast systems. Spike as a form of relaxation, bursting and MMOs are the dynamical signatures of the slow-fast property of a system. The complex slow-fast dynamical behaviors that are bursting and MMOs have been intensively investigated in three-dimensional slow-fast systems and have been found previously to occur only in slow-fast systems involving at least three variables because the successive trigger and extinction of spikes suppose the multiple time scale trip of the system trajectory on a complex well organized high-dimensional phase space. A long time afterwards, it was found that noise can induce MMOs in simple two-variable systems as the van der Pol oscillator with constant forcing [7] and FitzHugh-Nagumo-like oscillators [8, 9]. More recently, we found that MMOs can emerge in two-variable systems due to fractional derivation, while studying the fractional-order van der Pol oscillator with constant forcing where the forcing value was set near the Hopf bifurcation point [10]. Subsequently, this result was confirmed by Abdelouahab *et al.* while studying Hopf-like bifurcation and bifurcating oscillatory states in a fractional-order FitzHugh-Nagumo model [11].

Recent studies have shown that the fractional derivation increases the dynamical richness of neuronal models. For example, in Ref. [12], Teka *et al.* have implemented the fractional dynamics on each gating variable of the Hodgkin-Huxley model and they found that the obtained fractional-order versions of the Hodgkin-Huxley model exhibit not only spiking behavior as their integer-order counterpart, but in addition, square wave bursting, MMOs, pseudo-plateau potentials, and phasic spiking. In Ref. [13], Shi and Wang considered a fractional-order Morris-Lecar model and found that this new model exhibits a rich variety of bursting patterns that appear in some common neuron models with properly chosen parameters but do not exist in the corresponding integer-order Morris-Lecar model. In Ref. [14], Mondal *et al.* considered a fractional-order FitzHugh-Rinzel model whose integer-order counterpart is an elliptic burster and found that it exhibits a wide range of neuronal responses including regular spiking, fast spiking, bursting, and MMOs.

In Ref. [15], Teka *et al.* studied the fractional-order Izhikevich model and found that the model produces a wide range of neuronal spike responses, including regular spiking, fast spiking, intrinsic bursting, MMOs, regular bursting and chattering, by adjusting only the fractional derivatives order. The dynamical richness of these fractional-order systems with at least three variables would justify the occurrence of MMOs in the aforementioned two-variable fractional-order slow-fast systems.

In this chapter, we explore further the dynamical behaviors of fractional-order two-variable slow-fast systems by considering an incommensurate fractional-order FitzHugh-Nagumo (FHN) model derived on the basis of biophysical concepts. Commensurate fractional-order FHN models that are widely considered in previous works are just mathematical generalizations that are not supported by any biophysical justification (see for example Ref. [11]). Exploring the behavior of the incommensurate fractional-order FHN model, we have found that depending on the orders of fractional derivatives the model can exhibit two types of MMOs. In the first case, they are identical to classical folded nodes-induced MMOs observed in integer-order systems, also known as canard generated MMOs [16]. In the second case, obtained for lower derivatives orders, the small oscillations of the MMOs pattern start with very low amplitude which then grows slowly before the oscillations enter the spiking phase. This last type of MMOs is identical to singular Hopf bifurcation-induced MMOs observed in integer-order systems [16]. In the second case, the active MMOs phase is sometimes, depending on initial conditions, preceded by a quiescent state which is known in the context of neuroscience as first spike latency. And in addition, the MMOs regime shows a prominent spike timing adaptation as the order of fractional derivatives decrease. The rest of the paper is organized as follows. In Section 2, the incommensurate fractional-order FHN model is derived from the fractional-order Hodgkin-Huxley model obtained using biophysical concepts. In Section 3, it is shown that the fractional-order FHN model can be simulated using an electrical circuit that is similar to the circuit proposed by Nagumo *et al.* [17] with the only difference that the capacitor and the inductor are replaced by corresponding fractional-order electrical elements. Section 4 is devoted to the study of local stability of the fractional-order FHN model, which allows to derive the conditions of occurrence of Hopf-like bifurcation with respect to fractional derivatives orders. In Section 5, the dynamical behavior of the fractional-order FHN model is revealed in order to confirm the results of local stability analysis carried out in the preceding section. A particular attention is granted to MMOs and first spike latency as they are new dynamical features due to fractional derivatives. The chapter ends with a conclusion where the results are summarized.

2. Model and biophysical justification

The FHN model

$$\begin{aligned} \varepsilon\frac{dx}{dt} &= y + x - \frac{1}{3}x^3 + I, \\ \frac{dy}{dt} &= -x - \delta y + \gamma, \end{aligned} \tag{1}$$

is a simple representative of a class of excitable-oscillatory systems including the Hodgkin-Huxley model of the squid giant axon. The derivation of the FHN model as a simple nerve membrane model is based on phase space methods applied to the

Hodgking-Huxley model. Indeed, the phase diagram of the FHN model and a properly chosen projection from the 4-dimensional Hodgking-Huxley phase space onto a plane are similar, which underlines the relationship between the two models.

As the FHN model is just a mathematical representation, to derive its generalization with fractional derivatives, we consider as starting point the generalization of the Hodgking-Huxley model which was derived using biophysical concepts. The Hodgking-Huxley model is given by the following set of differential equations:

$$\begin{aligned} C_m \frac{dv}{dt} &= I - \bar{g}_K n^4 (v - v_K) - \bar{g}_{Na} m^3 h (v - v_{Na}) - \bar{g}_l (v - v_l), \\ \frac{dn}{dt} &= a_n (1 - n) - b_n n, \\ \frac{dm}{dt} &= a_m (1 - m) - b_m m, \\ \frac{dh}{dt} &= a_h (1 - h) - b_h h, \end{aligned} \tag{2}$$

where $v = E_m - E_r$ is the displacement of the membrane potential E_m from its resting value E_r; n, m, and h are the potassium current activation, sodium current activation, and sodium current inactivation gating variables, respectively. C_m is the membrane capacity per unit area, I is an applied stimulus current. $v_K = E_K - E_r$, $v_{Na} = E_{Na} - E_r$, and $v_l = E_l - E_r$, where E_K and E_{Na} are the equilibrium potentials for the sodium and potassium ions, and E_l is the potential at which the leakage current due to chloride and other ions is zero. $\bar{g}_K n^4 = g_K, \bar{g}_{Na} m^3 h = g_{Na}$, where g_K, g_{Na} and $\bar{g}_l$ are ionic conductances. $a_n, b_n, a_m, b_m, a_h, b_h$ are rate constants which vary with membrane voltage. However, Hodgking and Huxley claimed that "The only major reservation which must be made about [the first equation in Eq. (2): $I = C_m \frac{dv}{dt} + I_{Na} + I_K + I_l$] is that it takes no account of dielectric loss in the membrane" [1]. Now, to account for dielectric losses in a capacitor, Curie proposed the following empiric relation between a DC voltage U_0 applied at $t = 0$ and the current i it will produce [18]:

$$i(t) = \frac{U_0}{h_1 t^\alpha}, \tag{3}$$

where h_1 is a constant related to the capacitance of the capacitor and the kind of dielectric, $0 < \alpha < 1$, α is related to the losses of the capacitor. The lower the losses, the closer to 1 is α. Westerlund and Ekstam [19] showed that for a general input voltage $v(t)$, Eq. (3) turns into:

$$i(t) = C_\alpha \frac{d^\alpha v(t)}{dt^\alpha}, \tag{4}$$

where $C_\alpha = \Gamma(1 - \alpha)/h_1$, $\frac{d^\alpha v(t)}{dt^\alpha}$ is the fractional (α-order) time derivative of $v(t)$, and $\Gamma(\cdot)$ is the Gamma function. On the other hand, Cole claimed that alternating current resistance and capacity measurements over a wide frequency range show that biological materials may be considered electrically equivalent to a circuit including a polarization element considered as a resistance and a capacity in series [20]. When a constant current i is started through this element at time $t = 0$, the potential difference across the element may be found to be:

$$e(t) = \frac{z_1 i t^{\alpha}}{\Gamma(1+\alpha)}, \tag{5}$$

where $z(\omega) = z_1(j\omega)^{-\alpha}$ is the impedance of the element, with $j^2 = -1$. The phase of this element: $\varphi = \alpha\pi/2$ allowed to provide the value of α for diverse biological materials. For example, it was found that α is ranged from 0.62 to 0.71 for frog sciatic nerve. This justifies why the power-law dynamics can occur in the membrane electrical activity. Note that Eqs. (3) and (5) are equivalent. Fractional derivatives with power-law kernel have also been used to account for the multiple timescale dynamics in neuroscience i.e. for processes which cannot be dynamically characterized with a single time constant. For example, Lundstrom *et al.* [21] showed that single rat neocortical pyramidal neurons adapt with a time scale that depends on the time scale of changes in stimulus statistics and that this multiple time scale adaptation is consistent with fractional-order differentiation, such that the neuron firing rate is a fractional derivative of slowly varying stimulus parameters. Subsequently, Teka *et al.* [22] developed a fractional-order leaky integrate-and-fire model which can reproduce upward and downward spike adaptations found experimentally by Lundstrom *et al.* [21]. In Ref. [22], the fractional derivation operator was applied on the membrane potential. So, the spike timing adaptation would be a dynamical manifestation of dielectric losses in the membrane. So, in order to account for the dielectric losses in the membrane, the total membrane current should be written as follows:

$$I = C_m^{\alpha} \frac{d^{\alpha} v}{dt^{\alpha}} + I_{\mathrm{Na}} + I_{\mathrm{K}} + I_l, \tag{6}$$

where C_m^{α} is a constant related to the capacitance of the membrane. Taking into account the ionic conductances and the gating variables, Eq. (6) is rewritten as follows:

$$C_m^{\alpha} \frac{d^{\alpha} v}{dt^{\alpha}} = I - \overline{g}_{\mathrm{K}} n^4 (v - v_{\mathrm{K}}) - \overline{g}_{\mathrm{Na}} m^3 h (v - v_{\mathrm{Na}}) - \overline{g}_l (v - v_l). \tag{7}$$

On the other hand, it has been shown that the dynamics of the gating variables would better be described using fractional derivatives instead of first-order derivatives [12]. Indeed, the gating dynamics is more complicated than what is traditionally assumed. Let us recall that ion channels are complex membrane proteins which provide ion-conducting, nanoscale pores in the biological membranes [23]. The gating dynamics is the spontaneous conformational dynamics of these proteins resulting in stochastic transitions between the conducting and non-conducting states also known as open and closed states of the pore, respectively. It was considered that the transitions between the open state and the closed state is a Markovian stochastic process that can be characterized by exponential residence time distributions of open and closed time intervals. However, experimental investigations had revealed that the distributions of the residence time intervals of closed states are typically not exponential. For several ion channels, these residence time distributions can be fitted by a stretched exponential function [24], or by a power-law function [25]. In Ref. [23], the time derivative of the Mittag-Leffler function which interpolates between the stretched exponential (at small time) and the asymptotic long time power-law function, was considered to show that the closed state of the ion channel can be modeled as an anomalous diffusion process over a large number of traps, described by a time-fractional diffusion equation. Accordingly, the dynamics of gating variables in the Hodgking-Huxley equation should be described by fractional-order differential equations as shown by Teka *et al.* [12], i.e.:

$$\frac{d^{\beta}\mathbf{z}}{dt^{\beta}} = a_{\mathbf{z}}(1-\mathbf{z}) - b_{\mathbf{z}}\mathbf{z}, \tag{8}$$

where $\mathbf{z} = [n, m, h]$, and β is a parameter related to the power-law index of anomalous diffusion. As shown in Ref. [23], the diffusion process over the substates of the closed state is a subdiffusive phenomenon, which corresponds to $0 < \beta < 1$.

Combining Eqs. (7) and (8), one obtains the incommensurate fractional-order Hodgking-Huxley model. Now, the variables x and y of the FHN model have been considered to describe the membrane voltage and the coarse-grained action of the gating variables, respectively. So, the variables x and y correspond to v in Eq. (7) and $\mathbf{z}$ in Eq. (8), respectively. Thus, the incommensurate fractional-order Hodgking-Huxley model given by Eqs. (7) and (8) is reduced to the following incommensurate fractional-order FHN model:

$$\begin{aligned} \varepsilon D_t^{\alpha} x &= y + x - \frac{1}{3}x^3 + I, \\ D_t^{\beta} y &= -x - \delta y + \gamma, \end{aligned} \tag{9}$$

where $D_t^{\alpha} x = \frac{d^{\alpha}x}{dt^{\alpha}}$ is the fractional-order (α-order) time derivative of $x(t)$ defined in the sense of Caputo [26] as follows:

$$D_t^{\alpha} x(t) = \frac{1}{\Gamma(1-\alpha)} \int_0^t \frac{x^{(1)}(\tau)}{(t-\tau)^{\alpha}} d\tau, \tag{10}$$

where $\alpha \in \mathbb{R}$ and $0 < \alpha < 1$, and $\Gamma(\cdot)$ is the Gamma function. The Caputo's definition of fractional derivative is suitable for initial value problems. Note that $\lim_{\alpha \to 1^-} D_t^{\alpha} x(t) = x^{(1)}(t) = \frac{dx(t)}{dt}$ [27] and then for $\alpha \to 1$ and $\beta \to 1$, the set of eqs. (9) is reduced to the classical FHN model given by Eq. (1). For $I = \delta = 0$, Eq. (9) amounts to the incommensurate fractional-order van der Pol oscillator with constant forcing studied by us in Ref. [10]. If we apply the transformation $y \to -y$ and set $\beta = \alpha$ in Eq. (9), we uncover another version of the fractional-order FHN model considered by Abdelouahab *et al.* [11].

3. Equivalent electrical circuit

An example of electrical circuit capable to reproduce the dynamical behavior of the fractional-order FHN model is apparently the same as the one considered by Nagmo *et al.* [17] as shown in **Figure 1**, where L is a fractional-order inductor, C is a fractional-order capacitor, and the other electronic elements are classic. Since, there is not yet a conventional manner to represent fractional-order electronic elements, we prefer to represent them as classical elements. TD is a tunnel diode whose voltage-current characteristic is given by $f(e) = i_0 - \left((e-e_0) - (e-e_0)^3/3K^2\right)/\rho$, where $i_0 = f(e_0)$. The fractional capacitor and the fractional inductor are characterized by

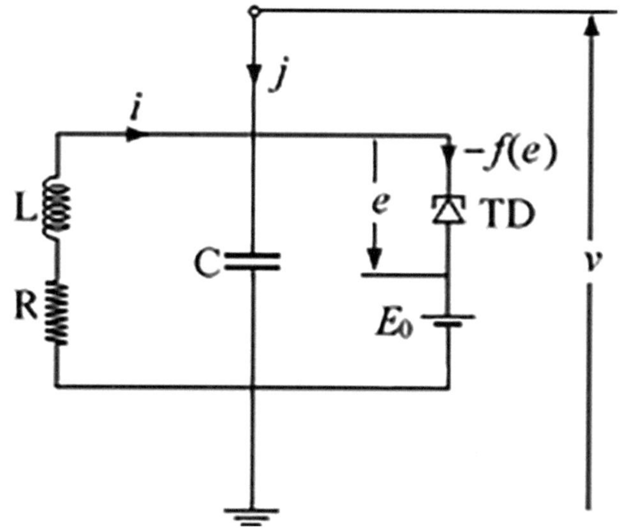

Figure 1.
An electrical circuit simulating the fractional-order FHN model.

$i_C = C^\alpha \frac{d^\alpha v}{d\tau^\alpha}$, and $v_L = L^\beta \frac{d^\beta i}{d\tau^\beta}$, respectively, where C^α and L^β are parameters related to their capacitance and inductance, with $0 < \alpha < 1$, $0 < \beta < 1$. Some of these coefficients can be found in Refs. [19, 28] for real capacitors and in Ref. [29] for real inductors. Applying the Kirchoff's law, it comes out that the circuit of **Figure 1** is described by the following set of fractional-order differential equations:

$$\begin{aligned} C^\alpha \frac{d^\alpha v}{d\tau^\alpha} &= i + i_0 - \frac{1}{\rho}\left((E_0 - v - e_0) - \frac{1}{3K^2}(E_0 - v - e_0)^3 \right) + j, \\ L^\beta \frac{d^\beta i}{d\tau^\beta} &= -Ri - v. \end{aligned} \tag{11}$$

Let $\tau_L = L/\rho$ and $\tau_C = \rho C$, the time constants related to the dynamics of the inductor and capacitor, respectively. Let us introduce the following dimensionless variables: $t = \tau/\tau_L$, $x = (v + e_0 - E_0)/K$, and $y = \rho(i + i_0)/K$, and use the fractional differential operator D_t^α. Then, Eq. (11) can be rewritten as follows:

$$\begin{aligned} \varepsilon \tau_L^{1-\alpha} D_t^\alpha x &= y + x - \frac{1}{3}x^3 + I, \\ \tau_L^{1-\beta} D_t^\beta y &= -x - \delta y + \gamma, \end{aligned} \tag{12}$$

where $\varepsilon = \tau_C/\tau_L = \rho^2 C/L$, $I = \rho j/K$, $\delta = R/\rho$, $\gamma = (Ri_0 + e_0 - E_0)/K$. The difference of scales between τ_L and τ_C is at the basis of the slow-fast dynamics that results in relaxation oscillations in the FHN system behavior. Indeed, $\tau_C \ll \tau_L$, then $\varepsilon = \tau_C/\tau_L \ll 1$ acts as a time scales ratio between x and y. Without harming any generality, we will consider $\tau_L = 1$, since the only effect of this parameter is to reinforce the relaxation that is expressed yet in ε. Then, Eq. (12) can be rewritten without τ_L, which amounts to the fractional-order FHN model given by Eq. (9). So, the fractional-order FHN model can be simulated with the electrical circuit in **Figure 1** where the capacitor and the inductor are fractional-order electrical elements known as fractances.

4. Local stability analysis and Hopf-like bifurcation

The equilibrium points $E(x_*, y_*)$ of Eq. (9) are solutions of the following set of algebraic equations:

$$\begin{aligned} x_*^3 - 3\left(1 - \frac{1}{\delta}\right)x_* - 3\left(I + \frac{\gamma}{\delta}\right) = 0, \\ y_* = \frac{-x_* + \gamma}{\delta}. \end{aligned} \tag{13}$$

We will consider the case where this equation admits only one solution, i.e. for $-4(1 - 1/\delta)^3 + 9(I + \gamma/\delta)^2 > 0$, according to the Cardan's method. The fractional dynamics does not affect neither the number of equilibrium points nor their positions, but it may change their stability [26]. So, it is proper to study the stability of E in this particular context and conclude about the effect of the fractional derivatives. To do so, we will first consider the local stability of the classical integer-order FHN model. Let $\{\lambda\}$ be the eigenvalues spectrum of the Jacobian matrix $\mathbf{J}_E$ of Eq. (9) evaluated at equilibrium point $E(x_*, y_*)$. The corresponding eigenvalues are conjugate complex numbers given by:

$$\lambda_{1,2} = \frac{-(\varepsilon\delta - 1 + x_*^2) \pm j\sqrt{\Delta}}{2\varepsilon}, \tag{14}$$

where $j^2 = -1$, and $\Delta = 4\varepsilon[1 - \delta(1 - x_*^2)] - (\varepsilon\delta - 1 + x_*^2)^2 > 0$. Then, E is a focus, a key ingredient for the occurrence of Hopf bifurcation. According to Eq. (14), E is stable (for the integer-order system) if:

$$\varepsilon\delta - 1 + x_*^2 > 0. \tag{15}$$

Thus, Hopf bifurcations occur at $x_* = x_H^+ = \sqrt{\varepsilon\delta - 1}$ and $x_* = x_H^- = -\sqrt{\varepsilon\delta - 1}$ which correspond via Eq. (13) to two values of the stimulus current, namely I_H^+ and I_H^-, respectively. **Figure 2** shows the bifurcation diagram computed using Matcont Matlab toolbox, around I_H^+ for a set of parameters chosen very close to the one used by FitzHugh in its pioneering work [30], namely $\varepsilon = 0.1$, $\delta = 0.8$, $\gamma = 0.7$, that will be used all through the paper. **Figure 2** shows that the Hopf bifurcation obtained for this set of parameters is subcritical.

Secondly, we consider the fractional-order FHN model with different rational orders $\alpha = m/m'$ and $\beta = n/n'$, with $m, n, m', n' \in \mathbb{N}$. Let M be the less common multiple of m' and n'. According to the Theorem on the stability of incommensurate fractional-order systems [26], the equilibrium point E is asymptotically stable if all the roots λ_k of the following equation:

$$\det(\mathbf{diag}[\lambda^{M\alpha} \quad \lambda^{M\beta}] - \mathbf{J}_E) = 0, \tag{16}$$

satisfy the following condition:

$$|\arg(\lambda_k)| > \frac{\pi}{2M}, \quad \forall\ k, \tag{17}$$

where $\mathbf{J}_E$ is the Jacobian matrix of the system evaluated at E. The stability condition given by Eq. (17) is equivalent to the following:

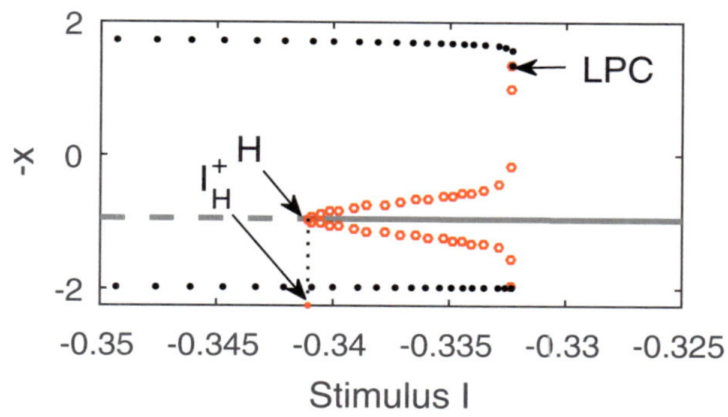

Figure 2.
One parameter bifurcation diagram of the FHN system with respect to the stimulus current I, for $\varepsilon = 0.1$, $\delta = 0.8$, $\gamma = 0.7$. The solid (resp. dashed) gray line depicts stable (resp. unstable) equilibrium point; the black solid (resp. red hollow) circle markers depict the extremums of stable (resp. unstable) periodic orbits. H and LPC label Hopf bifurcation and fold bifurcation of periodic orbits also known as saddle-node bifurcation of limit cycles and limit point bifurcation for cycles, respectively.

$$\lambda_k \notin \mathrm{D} = \left\{ z \in \mathbb{C}/z = re^{j\theta},\quad 0 \le r < R \to \infty,\quad |\theta| \le \frac{\pi}{2M} \right\},\ \forall\ k. \tag{18}$$

And Eq. (17) is equivalent to the following characteristic equation:

$$P(\lambda) = \lambda^{M(\alpha+\beta)} + \delta\lambda^{M\alpha} - \frac{1 - x_*^2}{\varepsilon}\lambda^{M\beta} + \frac{1 - \delta(1 - x_*^2)}{\varepsilon} = 0. \tag{19}$$

Let $\mathrm{B_D}$ be the boundary of D. According to the Cauchy's argument Principle [31, 32], if Eq. (19) has no root on $\mathrm{B_D}$ the closed curve $P(\mathrm{B_D})$ encircles the origin N times, where N is the number of roots of Eq. (19) inside the domain D. Accordingly, the stability condition given by Eq. (18) requires that $N = 0$. Therefore, the stability condition can be resumed in the following theorem [32]:

Theorem 1: The equilibrium point E of the fractional-order FHN model is stable if $P(\mathrm{B_D})$ neither encircles nor gets through the origin O in the complex plan.

Drawing one's inspiration from the method of exploitation of Theorem 1 proposed in Ref. [32] and improved in Ref. [10], one can derive the following stability condition for the equilibrium point E of the incommensurate fractional-order FHN model:

$$\zeta^{\alpha+\beta}\cos\frac{(\alpha+\beta)\pi}{2} + \delta\zeta^{\alpha}\cos\frac{\alpha\pi}{2} - \frac{1 - x_*^2}{\varepsilon}\zeta^{\beta}\cos\frac{\beta\pi}{2} + \frac{1 - \delta(1 - x_*^2)}{\varepsilon} > 0, \tag{20}$$

where ζ is solution of the following equation:

$$\zeta^{\alpha+\beta}\sin\frac{(\alpha+\beta)\pi}{2} + \delta\zeta^{\alpha}\sin\frac{\alpha\pi}{2} - \frac{1 - x_*^2}{\varepsilon}\zeta^{\beta}\sin\frac{\beta\pi}{2} = 0. \tag{21}$$

where $\zeta = r^M$, with $r \in \mathbb{R}_+$. For a given value of I, Eq. (13) is solved and the solution x_* is introduced into Eq. (21) which is solved its turn for a given set (α, β), and the solution ζ is introduced into Eq. (20) to check the stability condition. Let us recall that in the case of the integer-order system, the stability changes via Hopf bifurcations. Now, considering its definition, a Hopf bifurcation cannot occur in a fractional-order system which cannot have exact periodic solutions on a finite time interval [33]. However, S-asymptotically T-periodic functions, can occur as solutions of a fractional-order autonomous system with fixed bounded lower terminal, instead of normal T-periodic solutions [34]. Then, the concept of Hopf-like bifurcation has been introduced to characterize the stability change of an equilibrium point giving rise to S-asymptotically T-periodic solutions [11]. As the stimulus current I is the bifurcation parameter, the numerical simulation of the set of Eqs. (13), (20) and (21) shows that the stability of the equilibrium point E of the fractional-order FHN model switches via Hopf-like bifurcations at two critical points that we will refer to as I_{H}^{+} and I_{H}^{-}, whose values depend on α and β . It is worthwhile pointing out that for $\beta = \alpha \rightarrow 1$ these bifurcation points merge with those obtained above in the case of integer-order FHN model.

In the case, where the fractional dynamics appears only in the membrane potential (i.e., for $0 < \alpha < 1$ and $\beta \rightarrow 1$), and in the case where the fractional dynamics appears only in the gating variables (i.e., for $\alpha \rightarrow 1$ and $0 < \beta < 1$), the stability conditions can be derived easily from Eqs. (20) and (21). These two limiting cases are depicted in **Figures 3** and **4**, where we can see how the positions of the bifurcations points I_{H}^{+} and

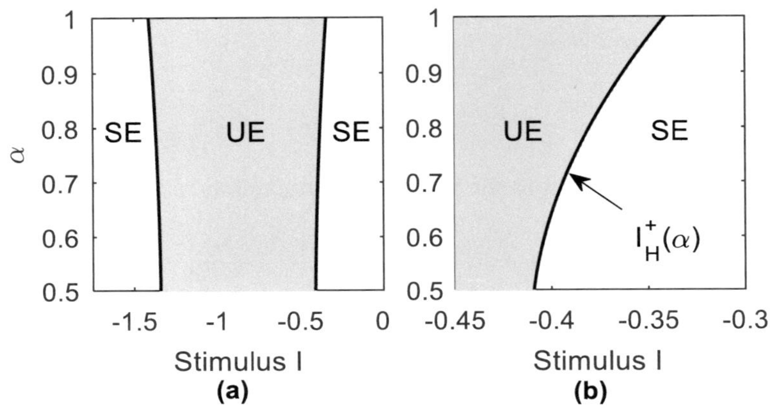

Figure 3.
Stability chart of the equilibrium point E *in* (I, α) *space for* $\beta \rightarrow 1$*: (b) is a zoom of (a). The gray colored area is the oscillatory region, which corresponds to unstable* E*. The black solid lines depict the Hopf-like bifurcations or stability boundaries* $I_{\mathrm{H}}^{+}(\alpha)$ *and* $I_{\mathrm{H}}^{-}(\alpha)$*. UE (resp. SE): Unstable (resp. stable) equilibrium point* E*.*

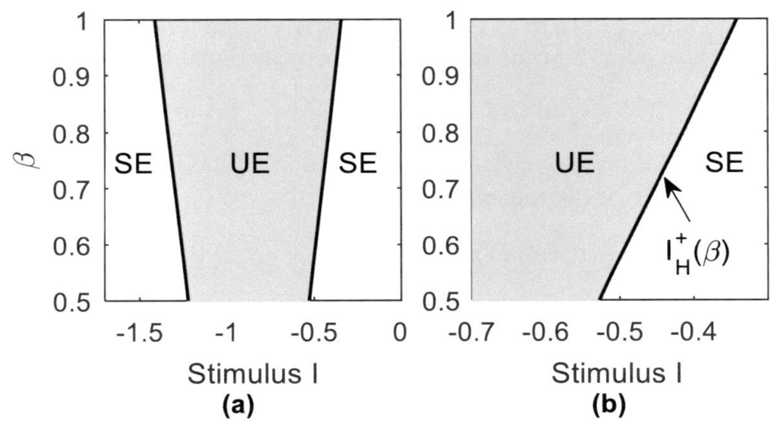

Figure 4.
Stability chart of the equilibrium point E *in* (I, β) *space for* $\alpha \rightarrow 1$*: (b) is a zoom of (a). The gray colored area is the oscillatory region, which corresponds to unstable* E*. The black solid lines depict the Hopf-like bifurcations or stability boundaries* $I_{\mathrm{H}}^{+}(\beta)$ *and* $I_{\mathrm{H}}^{-}(\beta)$*. UE (resp. SE): Unstable (resp. stable) equilibrium point* E*.*

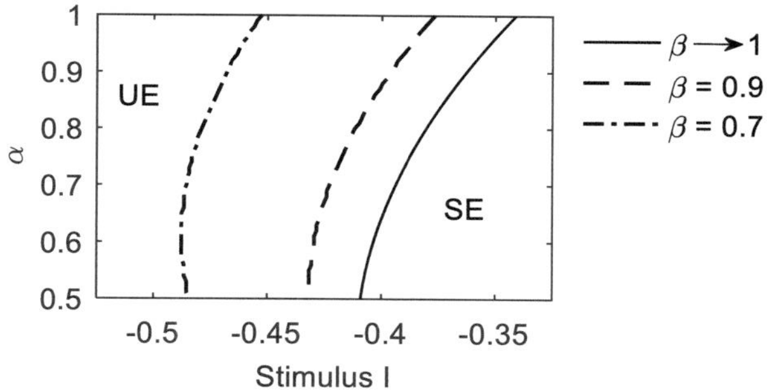

Figure 5.
Effect of α and β on the stability boundary (Hopf-like bifurcation) I_H^+.

I_H^- vary with respect to the fractional derivatives orders α and β. **Figure 5** shows the effects of the fractional-order derivatives on the stability boundaries of **Figure 4(b)**. **Figure 5** corresponds to the general case where the fractional dynamics appear in both the membrane potential and gating variables, that is, for $0 < \alpha < 1$ and $0 < \beta < 1$. Overall, as shown in **Figures 3–5**, as the derivatives orders α and β decrease, the region corresponding to unstable equilibrium shrinks, involving the expansion of stability regions. Thus, as expected, the fractional-order derivation enhances stability in the dynamics of the FHN system.

In what comes next, keeping in mind the above local stability analysis, we will examine the oscillatory behaviors of the fractional-order FHN system.

5. Dynamical behavior and neurocomputational implications

The incommensurate fractional-order FHN model given by Eq. (9) is solved numerically thanks to the Adams-Bashforth-Moulton predictor-corrector scheme [35], with the set of parameters used for **Figure 2**, that is, $\varepsilon = 0.1$, $\delta = 0.8$, $\gamma = 0.7$. Unless otherwise indicated, the initial conditions are set as $x(0) = 0$ and $y(0) = 0$. The fractional derivatives orders α and β, and the input stimulus current I are assumed to be control parameters.

Let us recall that, for the chosen set of parameters, the dynamics of the classical integer-order FHN model (Eq. (1) or Eq. (9) for $\alpha \to 1$ and $\beta \to 1$) converges either to the equilibrium point (resting state) or to a limit cycle with relaxation oscillations (spiking state), depending on the strength of the stimulus I. Since, the transition between these two regimes occurs via a subcritical Hopf bifurcation, there is an interval of I where the two attractors coexist.

In the case, where the fractional dynamics appears only in the membrane potential (i.e. for $0 < \alpha < 1$ and $\beta \to 1$), a narrow region of a new regime, namely phasic spiking, appears between the regions of existence of resting state and spiking state. The phasic spiking pattern is made up of a spiking phase transient to resting. The size of its region of existence in the parameter space increases with decreasing value of α. **Figure 6** shows an illustration of these three dynamical regimes. Note that the region of existence of spiking state extends beyond the stability threshold, which means that the subcritical Hopf-like bifurcation persists. Keeping the value of $\alpha \in]0, 1[$, the value of β is reduced a bit, say from 1 to 0.9. The bifurcation scenario changes a lot. When the system loses stability, MMOs take place (see **Figure 7**) and when the value of I decreases further, there is a transition to spiking state. For high values of α, at the

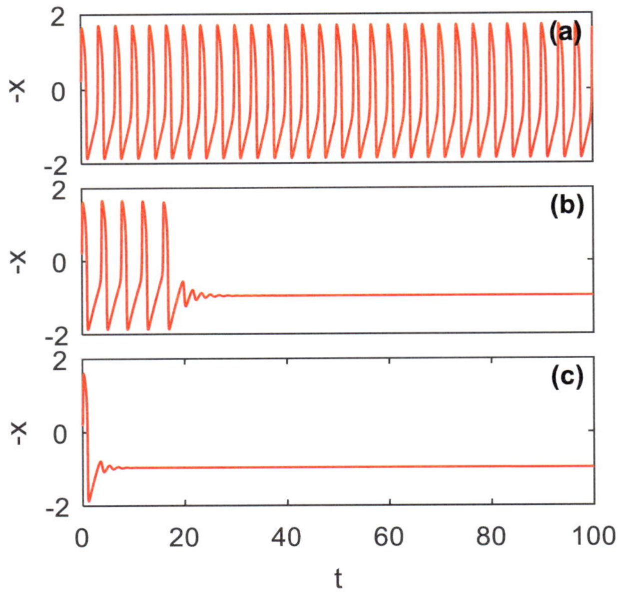

Figure 6.
Time series $-x(t)$ showing the dynamical regimes exhibited by the FHN model with fractional-order dynamics only in the membrane potential (i.e. $\beta \to 1$) for $\alpha = 0.8$, and different values of the stimulus I: (a) $I = -0.400$: Spiking; (b) $I = -0.345$: Phasic spiking; (c) $I = -0.330$: Resting.

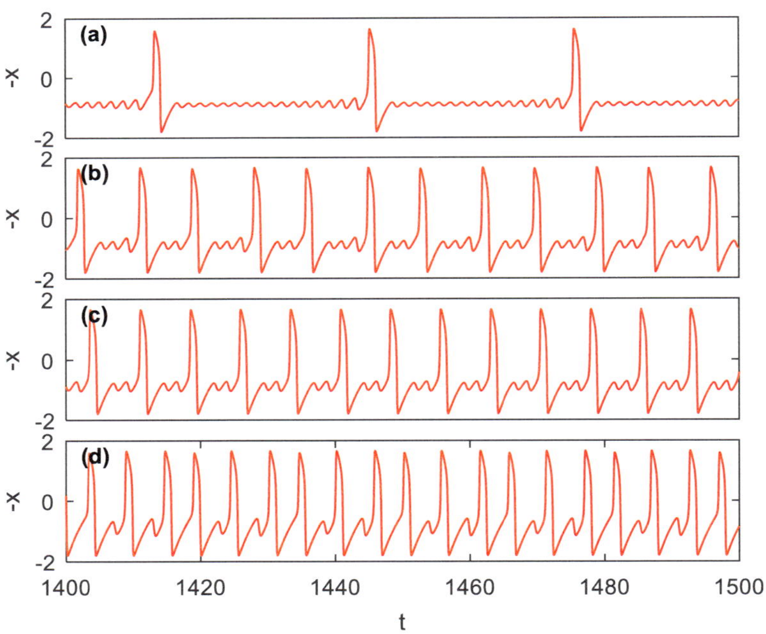

Figure 7.
Time series $-x(t)$ showing some MMOs patterns exhibited by the fractional-order FHN model for $\alpha = 0.8$, $\beta = 0.9$, and different values of I: (a) 1^{15} pattern for $I = -0.4146$; (b) $1^3 1^2$ pattern for $I = -0.4290$; (c) 1^2 pattern for $I = -0.4320$; (d) $1^1 2^1$ pattern for $I = -0.4472$.

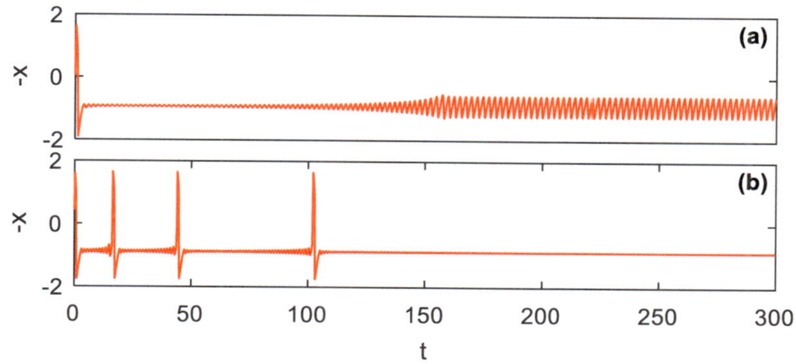

Figure 8.
Time series $-x(t)$ showing some dynamical regimes exhibited by the fractional-order FHN model: (a) small amplitude oscillations for $\alpha = 0.985$, $\beta = 0.9$, and $I = -0.3824$; (b) phasic spiking for $\alpha = 0.6$, $\beta = 0.9$, and $I = -0.427$.

transition between resting state and MMOs, a narrow window of small amplitude oscillations (see **Figure 8**) appears. A very narrow region of another type of phasic spiking appears, for relatively small values of α, at the transition from resting state to MMOs. This new type of phasic spiking is made up of transient MMOs to resting (see **Figure 8**). The lifetime and the number of spikes of this transient MMOs regime increases as the strength of the stimulus current goes away from the bifurcation point.

Regular MMOs have been often referred to as L^s patterns, where L and s are the number of large amplitudes and the number small amplitudes in one pattern, respectively. For example, **Figure 7** shows MMOs with 1^{15} pattern in (a), and 1^2 pattern in (c). When the value of the control parameter I varies, MMOs with complex patterns develop at the transition between two MMOs states with regular patterns; for example, **Figure 7** shows MMOs with $1^3 1^2$ pattern in (b); and $1^1 2^1$ pattern in (d). The $1^3 1^2$ MMOs develops between 1^3 MMOs and 1^2 MMOs, and $1^1 2^1$ MMOs develops between 1^1 MMOs and 2^1 MMOs. **Figure 7** also shows that when the value of I is close to the bifurcation point $I_{\mathrm{H}}^{+} \approx -0.41185$, we have 1^s MMOs – the value of s decreases for decreasing value of I. As the value of I decreases further and the value of s reaches 1, one obtains complex MMOs patterns with many large amplitude oscillations, which finally leads to L^1 MMOs (where L increases with decreasing value of I). The MMOs patterns shown in **Figure 7** are identical to classical folded nodes-induced MMOs observed in integer-order systems, also known as canard generated MMOs [16].

In order to illustrate the bifurcation scenarios described above, we map all the dynamical regimes in the parameter space as shown in **Figure 9**. As mentioned above, the Hopf-like bifurcation obtained there is subcritical as shown in **Figure 9(a)**. Indeed, there is a parameter region where oscillatory states coexist with resting state. However, **Figure 9(b)** show that when the value of β decreases, the limit between oscillatory states and resting state is marked clearly by the stability threshold obtained by local stability analysis.

We now examine the case where the fractional dynamics appears only in the gating variables, that is, for $\alpha \to 1$ and $0 < \beta < 1$. The resting state exists for high values of I. As the value I decreases the equilibrium state loses stability and MMOs take place. At the transition between resting state and MMOs, a very narrow window of small amplitude oscillations appears. As the value of I decreases further, there is a transition from MMOs to spiking state. Keeping $0 < \beta < 1$, the value of α is reduced a bit, say from 1 to 0.9. The bifurcation scenario does not change significantly. Finally, we map

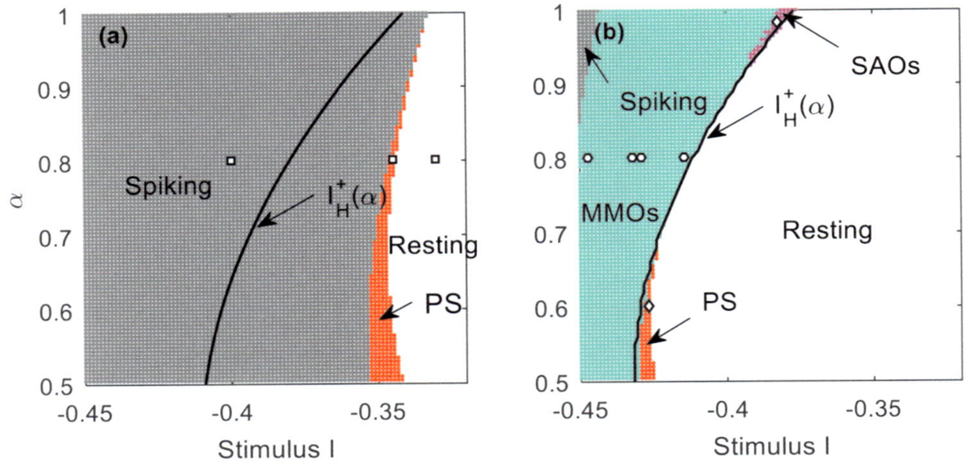

Figure 9.
Dynamical regimes maps in (I, α) space superimposed on the stability boundaries $I_H^+(\alpha)$ obtained analytically: (a) $\beta \to 1$; (b) $\beta = 0.9$. Gray colored region: Spiking; cyan colored region: MMOs; magenta colored region: Small amplitude oscillations (SAOs); red colored region: Phasic spiking (PS), white region: Resting. The square markers in (a), circle markers and diamond marker in (b) show the parameters used for ***Figures 6–8****, respectively.*

all the dynamical behaviors in the parameter space as shown in **Figure 10**. One can notice that the result of the global dynamics analysis obtained numerically agrees very well with the local stability analysis result. This figure shows that the domain of existence of MMOs widens as the value of β decreases. The major difference between the cases depicted in the subsets (a) and (b) of **Figure 10** is that the region of existence of small amplitude oscillations shrinks with decreasing value of α.

These phase diagrams in **Figures 9** and **10** show the rich variety of dynamical behaviors of the fractional-order FHN neuron model.

For low values of β (say, lesser than about 0.7) one observes another type of MMOs for which small oscillations start with very low amplitude which then grows slowly before the oscillations enter the spiking phase (see **Figure 11**). This last type of MMOs is identical

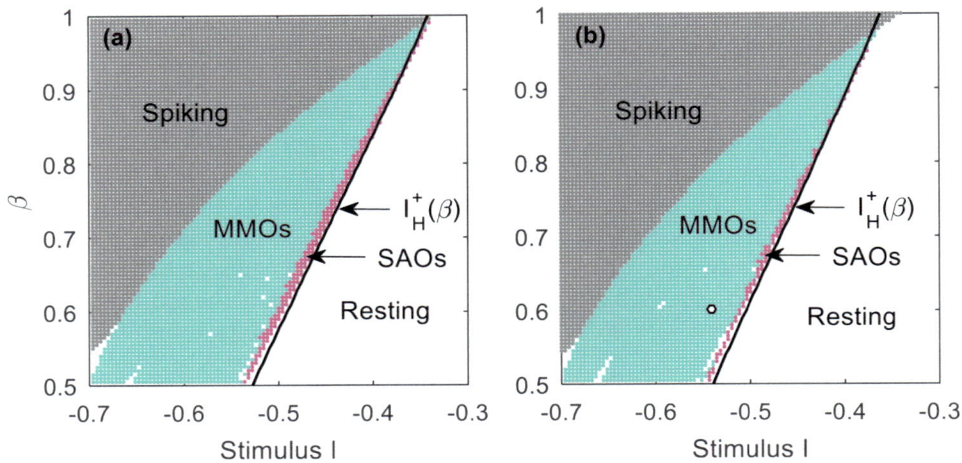

Figure 10.
Dynamical regimes maps in (I, β) space superimposed on the stability boundaries $I_H^+(\beta)$ obtained analytically: (a) $\alpha \to 1$; (b) $\alpha = 0.9$. Gray colored region: Spiking; cyan colored region: MMOs; magenta colored region: Small amplitude oscillations (SAOs); white region: Resting. The circle marker in (b) show the parameters used for ***Figure 11****.*

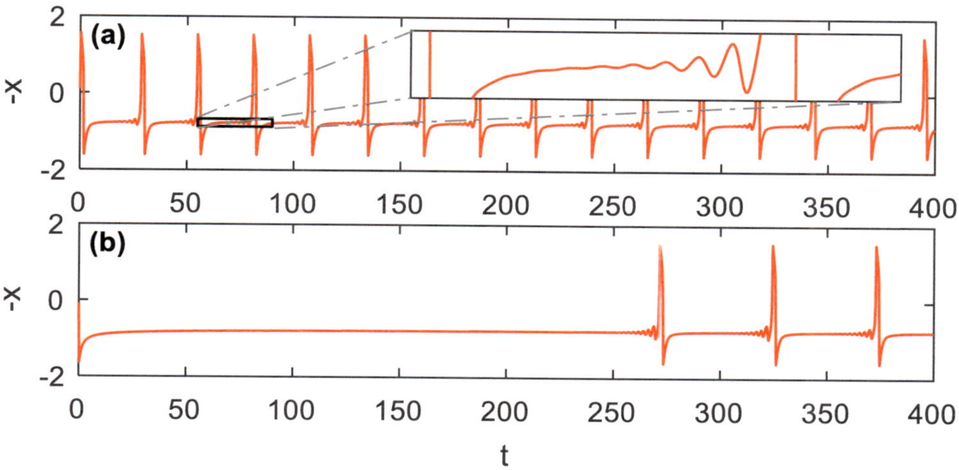

Figure 11.
Time series $-x(t)$ showing two different behaviors of the fractional-order FHN model for $\alpha = 0.9$, $\beta = 0.6$, and $I = -0.540$ and different initial conditions: (a) $(x(0), y(0)) = (0, 0)$; (b) $(x(0), y(0)) = (0, 1)$.

to singular Hopf bifurcation-induced MMOs observed in integer-order systems [16]. Note that unlike the integer-order FHN model for which in the unstable equilibrium region, the dynamics of the system converges to a unique limit cycle no matter the initial conditions, the dynamics of its fractional-order counterparts is sensitive to initial conditions [11]. The same phenomenon was observed in the fractional-order van der Pol oscillator with constant forcing [10] which is closely related to the FHN model. So, changing the initial conditions, it is possible to uncover new dynamical states. For example, we have change the set of initial conditions $(x(0), y(0))$ in **Figure 11** from $(0, 0)$ to $(0, 1)$, and we have found that the active state is preceded by a static-like regime. This static-like transient state was found for the first time in the context of dynamical systems by us, while studying a fractional-order van der Pol oscillator with constant forcing [10]. In the context of neuroscience, this quiescent transient regime is known as first spike latency. Teka *et al.* [22] found that the fractional derivation also induces the occurrence of first spike latency in a leaky integrate-and-fire model, and that this first spike latency is reinforced by decreasing value of the fractional derivative. On the other hand, still regarding the type MMOs shown in **Figure 11**, the instantaneous spike frequency (the inverse of the duration of the corresponding interspike interval), also known as the firing rate, increases as a function of the interspike interval number and approaches an asymptotic value after a certain number of intervals. Also, the increasing rate of the firing rate decreases with decreasing value of the fractional derivatives orders. This property of neurons known as upward spike timing adaptation, observed experimentally [21], has proved to appear only in fractional-order models [21, 22].

It was found in Ref. [10] that the lifetime of this static-like transient and the pseudo-period (corresponding to the interspike interval) of the asymptotic MMOs increase exponentially with the closeness to the Hopf bifurcation point. The same result is found in the present work. **Figure 12** shows the behavior of the first spike latency (FSL) and of the fiftieth interspike interval (ISI) – the period of the asymptotic MMOs after spike timing adaptation - when the value of the stimulus current I varies, coming close to the bifurcation point I_H^+. The results from the direct numerical simulation of Eq. (9) (depicted in **Figure 12**) have been fitted using the following exponential functions in EzyFit Matlab toolbox:

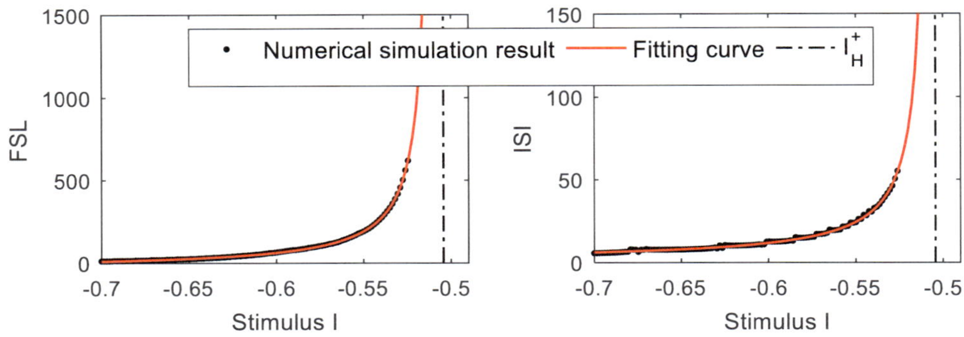

Figure 12.
First spike latency (FSL) and interspike interval (ISI) versus stimulus current I, for $\alpha = 0.9$ and $\beta = 0.6$. With $I_H^+ \approx 0.504497$.

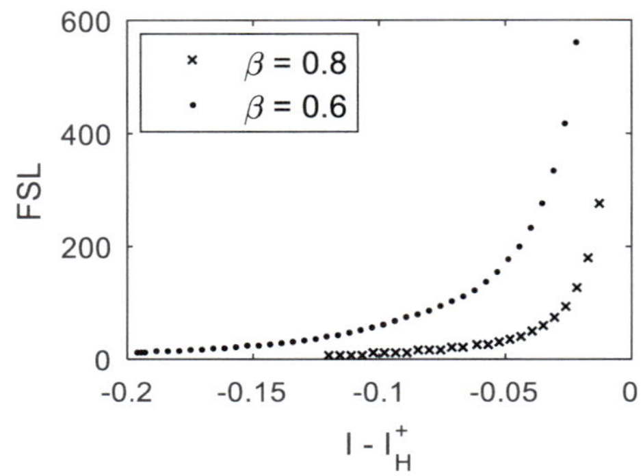

Figure 13.
First spike latency (FSL) versus displacement of the stimulus current I from the Hopf-like bifurcation point I_H^+, for $\alpha = 0.9$ and different values of β. Note that $I_H^+(\beta = 0.6) \approx -0.504497$ and $I_H^+(\beta = 0.8) \approx -0.431551$.

$$FSL(I) = a \exp\left(\frac{b}{I - I_H^+}\right) + c, \quad ISI(I) = a' \exp\left(\frac{b'}{I - I_H^+}\right) + c'. \tag{22}$$

For the two cases of fitting, the value of the correlation coefficient is over 0.999, indicating good fittings. Thus, this confirms the exponential growth of the first spike latency and of the interspike interval. Furthermore, as shown in **Figure 13**, the features of the first spike latency and of the interspike interval are reinforced by decreasing values of the orders of fractional derivatives.

Although the specific role of MMOs among the plethora of slow-fast dynamical behaviors occurring in neural systems has not yet been determined, it has been nevertheless suggested that the dynamical richness of a neuron expands its computational capacity by increasing its coding capacity [12, 15, 36]. On the other hand, it has been suggested that first spike latency could code for stimulus recognition in several sensory systems [37–40]. Indeed, the variation of first spike latency with stimulus parameters contains considerable information about those parameters [40]. The first spike latency has also been suggested as a source of information for accurate decisions [22]. In addition, the behavior of the firing rate (or interspike interval) of neurons also provides information on the stimulus statistics [21]. The fractional-order FHN model is a mathematically simple model with complex dynamics features, thus increasing the amount of information that can describe the input. So, the fractional FHN model is more computationally efficient than its integer-order counterpart.

6. Conclusion

In this work, the dynamical behavior of an incommensurate fractional-order FitzHugh-Nagumo model of neuron has been investigated in details. First of all, the considered fractional-order model has been derived from the fractional-order Hodgkin-Huxley model obtained taking advantage of the powerfulness of fractional derivatives in modeling the dielectric losses in cell membranes, and the anomalous diffusion of particles in ion channels. Then, it has been shown that the fractional-order FitzHugh-Nagumo model can be simulated by a simple electrical circuit. Then, the local stability of the incommensurate fractional-order model has been studied with a particular attention granted to the effect of the fractional derivatives. It has been found that the fractional derivatives enhance the local stability of the model. At last, the dynamical behavior of the fractional-order models has been explored numerically, which has confirmed the results of local stability. It has been found that the fractional-order FitzHugh-Nagumo exhibits a lot of complex dynamical features that are not observed in the behavior of its integer-order counterpart, and that cannot be observed in integer-order two-variable systems. Among others, the fractional-order model exhibits mixed mode oscillations, phasic spiking, first spike latency, and spike timing adaptation. These complex features of the dynamical behavior of the fractional-order model increase the computational capacity of the FitzHugh-Nagumo model. An outlook of this work would be the study of the mechanism(s) underlying the formation of mixed mode oscillations and first spike latency as effects of fractional derivation in two-variable systems.

Conflict of interest

The author declares no conflict of interest.

Author details

Serge Gervais Ngueuteu Mbouna
Laboratory of Modeling and Simulation in Engineering, Biomimetics and Prototypes, Faculty of Science, University of Yaoundé I, Yaoundé, Cameroon

*Address all correspondence to: ngueut@yahoo.fr

References

[1] Hodgkin AL, Huxley AF. A quantitative description of membrane current and its application to conduction and excitation in nerve. Journal of Physiology. 1952;**117**:500-544. DOI: 10.1113%2Fjphysiol.1952.sp004764

[2] Rieke F, Warland D, de Ruyter van Steveninck R, Bialek W. Spikes: Exploring the Neural Code. Cambridge, MA: MIT Press; 1999

[3] Izhikevich EM. Neural excitability, spiking, and bursting. International Journal of Bifurcation and Chaos. 2000; **10**:1171-1266. DOI: 10.1142/S0218 127400000840

[4] Izhikevich EM, Desai NS, Walcott EC, Hoppensteadt FC. Bursts as a unit of neural information: Selective communication via resonance. Trends in Neurosciences. 2003;**26**:161-167. DOI: 10.1016/s0166-2236(03)00034-1

[5] Lisman J. Bursts as a unit of neural information: Making unreliable synapses reliable. Trends in Neurosciences. 1997; **20**:38-43. DOI: 10.1016/s0166-2236(96) 10070-9

[6] Desroches M, Guckenheimer J, Krauskopf B, Kuehn C, Osinga HM, Wechselberger M. Mixed-mode oscillations with multiple time scales. SIAM Review. 2012;**54**:211-288. DOI: 10.1137/100791233

[7] Muratov V-EE. Noise-induced mixed-mode oscillations in a relaxation oscillator near the onset of a limit cycle. Chaos. 2008;**18**:015111. DOI: 10.1063/1.2779852

[8] Borowski P, Kuske R, Li Y-X, Cabrera JL. Characterizing mixed mode oscillations shaped by noise and bifurcation structure. Chaos. 2010;**20**: 043117. DOI: 10.1063/1.3489100

[9] Makarov VA, Nekorkin VI, Velarde MG. Spiking behavior in a noise-driven system combining oscillatory and excitatory properties. Physical Review Letters. 2001;**86**:3431-3434. DOI: 10.1103/PhysRevLett.86.3431

[10] Ngueuteu GSM, Yamapi R, Woafo P. Quasi-static transient and mixed mode oscillations induced by fractional derivatives effect on the slow flow near folded singularity. Nonlinear Dynamics. 2014;**78**:2717-2729. DOI: 10.1007/ s11071-014-1620-x

[11] Abdelouahab M-S, Lozi R, Chen G. Complex canard explosion in a fractional-order FitzHugh-Nagumo model. International Journal of Bifurcation and Chaos. 2019;**29**:1950111. DOI: 10.1142/S0218127419501116

[12] Teka W, Stockton D, Santamaria F. Power-law dynamics of membrane conductances increase spiking diversity in a Hodgkin-Huxley model. PLoS Computational Biology. 2016;**12**:e1004776. DOI: 10.1371/journal.pcbi.1004776

[13] Shi M, Wang Z. Abundant bursting patterns of a fractional-order Morris-Lecar neuron model. Communications in Nonlinear Science and Numerical Simulation. 2014;**19**:1956-1969. DOI: 10.1016/j.cnsns.2013.10.032

[14] Mondal A, Sharma SK, Upadhyay RK, Mondal A. Firing activities of a fractional-order FitzHugh-Rinzel bursting neuron model and its coupled dynamics. Scientific Reports. 2019;**9**:15721. DOI: 10.1038/s41598-019-52061-4

[15] Teka WW, Upadhyay RK, Mondal A. Spiking and bursting patterns of fractional-order Izhikevich model.

Communications in Nonlinear Science and Numerical Simulation. 2018;**56**: 161-176. DOI: 10.1016/j.cnsns.2017.07.026

[16] Curtu R. Singular Hopf bifurcations and mixed-mode oscillations in a two-cell inhibitory neural network. Physica D: Nonlinear Phenomena. 2010;**239**: 504-514. DOI: 10.1016/j.physd.2009.12.010

[17] Nagumo J, Arimoto S, Yoshizawa S. An active pulse transmission line simulating nerve axon. Proceedings of the IRE. 1962;**50**:2061-2070. DOI: 10.1109/JRPROC.1962.288235

[18] Westerlund S. Dead matter has memory! Physica Scripta. 1991;**43**: 174-179. http://iopscience.iop.org/1402-4896/43/2/011

[19] Westerlund S, Ekstam L. Capacitor theory. IEEE Transactions on Dielectrics and Electrical Insulation. 1994;**1**:826-839. DOI: 10.1109/94.326654

[20] Cole KS. Alternating current conductance and direct current excitation of nerve. Science. 1934;**79**: 164-165. DOI: 10.1126/science.79.2042.164

[21] Lundstrom BN, Higgs MH, Spain WJ, Fairhall AL. Fractional differentiation by neocortical pyramidal neurons. Nature Neuroscience. 2008;**11**: 1335-1342. DOI: 10.1038%2Fnn.2212

[22] Teka W, Marinov TM, Santamaria F. Neuronal spike timing adaptation described with a fractional leaky integrate-and-fire model. PLoS Computational Biology. 2014;**10**: e1003526. DOI: 10.1371/journal.pcbi.1003526

[23] Goychuk I, Hänggi P. Fractional diffusion modeling of ion channel gating. Physical Review E. 2004;**70**:051915. DOI: 10.1103/PhysRevE.70.051915

[24] Liebovitch LS, Sullivan JM. Biophysical Journal. Fractal analysis of a voltage-dependent potassium channel from cultured mouse hippocampal neurons. 1987;**52**:979-988. DOI: 10.1016/S0006-3495(87)83290-3

[25] Millhauser GL, Salpeter EE, Oswald RE. Diffusion models of ion-channel gating and the origin of power-law distributions from single-channel recording. Proceedings of the National Academy of Sciences of the United States of America. 1988;**85**:1503-1507. DOI: 10.1073/pnas.85.5.1503

[26] Caponetto R, Dongola R, Fortuna L, Petráš I. Fractional Order Systems: Modeling and Control Applications. Singapore: World Scientific Publishing Co. Pte. Ltd.; 2010

[27] Li C, Deng W. Remarks on fractional derivatives. Applied Mathematics and Computation. 2007;**187**:777-784. DOI: 10.1016/j.amc.2006.08.163

[28] Faraji S, Tavazoei MS. The effect of fractionality nature in differences between computer simulation and experimental results of a chaotic circuit. Central European Journal of Physics. 2013;**11**:836-844. DOI: 10.2478/s11534-013-0255-8

[29] Schäfer I, Krüger K. Modelling of lossy coils using fractional derivatives. Journal of Physics D: Applied Physics. 2008;**41**:045001. DOI: 10.1088/0022-3727/41/4/045001

[30] FitzHugh R. Impulses and physiological states in theoretical models of nerve membrane. Biophysical Journal. 1961;**1**:445-466. DOI: 10.1016/S0006-3495(61)86902-6

[31] Ahlfors LV. Complex Analysis. 2nd ed. New York: McGraw-Hill; 1966

[32] Tavazoei MS, Haeri M, Attari M, Bolouki S, Siami M. More details on analysis of fractional-order Van der pol oscillator. Journal of Vibration and Control. 2009;**15**:803-819. DOI: 10.1177%2F1077546308096101

[33] Tavazoei MS, Haeri M. A proof for non existence of periodic solutions in time invariant fractional order systems. Automatica. 2009;**45**:1886-1890. DOI: 10.1016/j.automatica.2009.04.001

[34] Henriquez HR, Pierri M, Taboas P. On *S*-asymptotically ω-periodic functions on Banach spaces and applications. Journal of Mathematical Analysis and Applications. 2008;**343**: 1119-1130. DOI: 10.1016/j.jmaa.2008.02.023

[35] Diethelm K, Ford NJ, Freed D. A predictor-corrector approach for the numerical solution of fractional differential equations. Nonlinear Dynamics. 2002;**29**:3-22. DOI: 10.1023/A:1016592219341

[36] Ghosh S, Mondal A, Ji P, Mishra A, Dana SK, Antonopoulos CG, et al. Emergence of mixed mode oscillations in random networks of diverse excitable neurons: The role of neighbors and electrical coupling. Frontiers in Computational Neuroscience. 2020;**14**: 49. DOI: 10.3389/fncom.2020.00049

[37] Gollisch T, Meister M. Rapid neural coding in the retina with relative spike latencies. Science. 2008;**319**:1108-1111. DOI: 10.1126/science.1149639

[38] Johansson RS, Birznieks I. First spikes in ensembles of human tactile afferents code complex spatial fingertip events. Nature Neuroscience. 2004;**7**: 170-177. DOI: 10.1038/nn1177

[39] Chase SM, Young ED. First-spike latency information in single neurons increases when referenced to population onset. Proceedings of the National Academy of Sciences of the United States of America. 2007;**104**:5175-5180. DOI: 10.1073/pnas.0610368104

[40] Heil P. First-spike latency of auditory neurons revisited. Current Opinion in Neurobiology. 2004;**14**: 461-467. DOI: https://psycnet.apa.org/doi/10.1016/j.conb.2004.07.002

Chapter 4

Some Solvability Problems of Differential Equations in Non-standard Sobolev Spaces

Bilal Bilalov, Sabina Sadigova and Zaur Kasumov

Abstract

In this chapter an m-th order elliptic equation is considered in Sobolev spaces generated by the norm of a grand Lebesgue space. Subspaces are determined in which the shift operator is continuous, and local solvability (in the strong sense) is established in these subspaces. It is established an interior and up-to boundary Schauder-type estimates with respect to these Sobolev spaces for m-th order elliptic operators, the trace of functions and trace operator are determined, the boundedness of trace operator and the extension theorem are proved, the properties of the Riesz potential are studied regarding these Sobolev spaces, etc. It is considered a second-order elliptic equation, and we study the Fredholmness of the Dirichlet problem in the Sobolev space generated by a separable subspace of the grand Lebesgue space. It is also considered one spectral problem for a discontinuous second-order differential operator and proved the theorem on the basicity of eigenfunctions of this operator in subspace of Morrey space, in which the infinitely differentiable functions with compact support are dense.

Keywords: non-standard function spaces, grand-Sobolev spaces, space of traces, Schauder type estimates, Riesz potentials, elliptic equations, Fredholmness, spectral problem, basicity, Morrey space

1. Introduction

Differential (also elliptic) equations play an especial role in the study of various processes and phenomena in natural science. Solvability problems of elliptic equations have a very rich history and remarkable monographs by various famous mathematicians are devoted to them. The theory of elliptic equations was developed in an comprehensive way in Hölder classes (solution in the classical sense) and in Hilbertian Sobolev spaces $W_2^k(\Omega)$. In the above mentioned case, depending on the nature of the problem, there are various methods of solution (for instance, the method of potentials, the periodic case, the method of the theory of functions, spectral method, etc.), which cannot be said for the non-Hilbert case $W_p^k(\Omega), p \neq 2$, in which each method faces certain difficulties. All considered spaces are separable Banach spaces and infinitely differentiable and finite functions are dense in them.

In the study of solvability of differential equations these facts are significant. Note that one of the methods to solve differential equations is a spectral method. To justify the solution by this method, one should study the basis properties of the root vectors of the considered spectral problem in the appropriate Banach function space.

In connection with applications in problems of mechanics, mathematical physics and pure mathematics, the so-called non-standard spaces of functions have greatly increased and the list of such spaces includes Lebesgue spaces with a variable summability index, Morrey spaces, grand Lebesgue spaces, Orlicz spaces, etc. For more details one can see the monographs [1–6]. Compared with other areas of mathematics, the apparatus of harmonic analysis has been fairly well studied in relation to these spaces. The problems of analysis and approximation theory have been relatively well studied in Lebesgue spaces with variable summability index and Morrey spaces (see [7–14]). The above mentioned problems have begun to be studied in Grand Lebesgue spaces, and valuable results have been obtained in this direction (see [15, 16]). The solvability problems of partial differential equations have also begun to be studied in the Sobolev spaces generated by these spaces (see [17–27]). Morrey-Sobolev and grand-Sobolev spaces are not separable and therefore infinitely differentiable and finite functions are not dense in them, in this reason the study the problems of solvability of differential equations in these spaces are of special scientific interest. Therefore, it is necessary to extract reasonable subspaces dictated by differential equations and develop an instruments for studying the solvability of differential equations in these subspaces.

An m-th order elliptic equation is considered in Sobolev spaces generated by the norm of a grand Lebesgue space. Subspaces are determined in which the shift operator is continuous, and local solvability (in the strong sense) is established in these subspaces. It is established an interior and up-to boundary Schauder-type estimates with respect to these Sobolev spaces for m-th order elliptic operators, the trace of functions and trace operator are determined, the boundedness of trace operator and the extension theorem are proved, the properties of the Riesz potential are studied regarding these Sobolev spaces, etc. It is considered a second-order elliptic equation and we study the fredholmness of the Dirichlet problem in the Sobolev space generated by a separable subspace of the grand Lebesgue space. It is considered one spectral problem for a discontinuous second order differential operator and proved the theorem on the basicity of eigenfunctions of this operator in subspace of Morrey space, in which the infinitely differentiable functions with compact support are dense.

2. Needful information

2.1 Standard notation

Z_+ will be the set of non-negative integers. $\overline{1,n} = 1, 2, \dots, n$. $\mathbb{C}$ is the set of complex numbers. $B_r(x_0) = \{x \in R^n : |x - x_0| < r\}$ will denote the open ball in R^n centered at x_0, where $|x| = \sqrt{x_1^2 + \dots + x_n^2}, x = (x_1, \dots, x_n)$. $\langle \cdot ; \cdot \rangle$ is a scalar product in R^n. $mes(M)$ will stand for the Lebesgue measure of the set M; $\partial\Omega$ will be the boundary of the domain Ω; $\overline{\Omega} = \Omega \bigcup \partial\Omega$; *diam* Ω will stand for the diameter of the set Ω; $f/_M$ denotes the restriction of f to M. $C_0^\infty(\Omega)-$ will denote the space of infinitely differentiable and finite functions in Ω and $C^{(m)}(\Omega)-$will stand for the space of m-th order continuously differentiable functions in the domain Ω. $C_0^m(\Omega)$—will stand for the space of m-th order continuously differentiable and finite functions in the domain Ω. D_L—will stand for the domain of the operator L; R_T—will stand for the range of the operator T; $KerT$— is the kernel of the operator T; T^* is the adjoint of T; $[X;Y]$ is a Banach space of bounded operators acting from X to Y; $[X] = [X;X]$. Throughout this paper, q' will denote the conjugate of a number, i.e. $\frac{1}{q} + \frac{1}{q'} = 1$.

2.2 Elliptic operator of m-th order and some necessary facts

Let $\Omega \subset R^n$ be some bounded domain with the rectifiable boundary $\partial\Omega$. We will use the notations of [19]. $\alpha = (\alpha_1, \dots, \alpha_n)$ will be the multiindex with the coordinates $\alpha_k \in Z_+, \forall k = \overline{1,n}$; $\partial_i = \frac{\partial}{x_i}$ will denote the differentiation operator, $\partial^\alpha = \partial_1^{\alpha_1}\partial_2^{\alpha_2} \dots \partial_n^{\alpha_n}$. For every $\xi = (\xi_1, \dots, \xi_n)$ we assume $\xi^\alpha = \xi_1^{\alpha_1}\xi_2^{\alpha_2} \dots \xi_n^{\alpha_n}$. Let L be an elliptic differential operator of m-th order

$$L = \sum_{|p| \leq m} a_p(x)\partial^p, \tag{1}$$

where $p = (p_1, \dots, p_n), p_k \in Z_+, \forall k = \overline{1,n}, a_p(\cdot) \in L_\infty(\Omega)$ are real functions. Consider the elliptic operator L_0:

$$L_0 = \sum_{|p|=m} a_p^0\, \partial^p, \tag{2}$$

with the constant coefficients a_p^0 and denote by $J(\cdot)$ a fundamental solution of Eq. (2) [28].

Let L be an elliptic operator and consider a “tangential operator”

$$L_{x_0} = \sum_{|p|=m} a_p\,(x_0)\partial^p, \tag{3}$$

at every point $x_0 \in \Omega$. Denote by $J_{x_0}(\cdot)$ the fundamental solution of the equation $L_{x_0}\varphi = 0$. The function $J_{x_0}(\cdot)$ is called a parametrics for the equation $L\varphi = 0$ with a singularity at the point x_0. Let

$$S_{x_0}\varphi = \psi(x) = \int J_{x_0}(x-y)\,\varphi(y)dy, \tag{4}$$

and

$$T_{x_0} = S_{x_0}(L_{x_0} - L). \tag{5}$$

Denote the operators S_{x_0}, L_{x_0} and T_{x_0}, corresponding to the point $x_0 = 0$, by S_0, L_0 and T_0, respectively.

Let us give the definition of smooth boundary.

Definition 1.1 We will say that the boundary $\partial\Omega$ of a domain $\Omega \subset R^n$ belongs to class $C^{(k)}$ if each sufficiently small piece of it can be mapped onto a segment of the hyperplane $x_n = 0$ using a coordinate transformation $y(x) = (y_1(x); \dots; y_n(x))$ with a positive Jacobian so that $y_i \in C^{(k)}, \forall i = \overline{1,n}$.

2.3 Grand-Sobolev spaces $W^m_{q)}(\Omega)$ and $WN^m_{q)}(\Omega)$

Define the grand-Lebesgue space $L_{q)}(\Omega)$, $1<q<+\infty$ (throughout this paper we will assume that this condition holds on q). Grand-Lebesgue space $L_{q)}(\Omega)$ is a Banach space of (Lebesgue) measurable functions f on Ω with the norm

$$\|f\|_{q)} = \sup_{0<\varepsilon<q-1} \left(\varepsilon \int_{\Omega} |f|^{q-\varepsilon} dx \right)^{\frac{1}{q-\varepsilon}}. \tag{6}$$

The following continuous embeddings hold

$$L_q(\Omega) \subset L_{q)}(\Omega) \subset L_{q-\varepsilon}(\Omega), \tag{7}$$

where $\varepsilon \in (0, q-1)$ is an arbitrary number. The space $L_{q)}(\Omega)$ is not separable.

Below in this section we will assume that every function defined on Ω is extended by zero to $R^n \backslash \overline{\Omega}$. Let T_δ be a shift operator, i.e. $(T_\delta f)(x) = f(\delta + x), \forall x \in \Omega$ and $\forall \delta \in R^n$. Let

$$N_{q)}(\Omega) = \left\{ f \in L_{q)}(\Omega) : \|T_\delta f - f\|_{q)} \to 0, \delta \to 0 \right\}. \tag{8}$$

The space $N_{q)}(\Omega)$ is a Banach space with the norm $\|\cdot\|_{q)}$, (i.e. is the subspace of $L_{q)}(\Omega)$.)

The following lemma is true (see [17]).

Lemma 1.2 $\overline{C_0^\infty(\Omega)} = N_{q)}(\Omega)$ (the closure is taken in $L_{q)}(\Omega)$).

Let us include the following lemma without proof.

Lemma 1.3 *The embeddings*

$$L_q(\Omega) \subset N_{q)}(\Omega) \subset L_{q)}(\Omega) \subset L_1(\Omega), \tag{9}$$

hold and every inclusion is strict.

Denote by $W^m_{q)}(\Omega)$ the grand-Sobolev space generated by the norm

$$\|f\|_{W^m_{q)}} = \sum_{k=0}^{m} \left\| f^{(k)} \right\|_{q)}. \tag{10}$$

Let

$$WN^m_{q)}(\Omega) = \left\{ f \in W^m_{q)}(\Omega) : \|T_\delta f - f\|_{W^m_{q)}} \to 0, \delta \to 0 \right\}. \tag{11}$$

Consider the following singular kernel

$$k(x) = \frac{\omega(x)}{|x|^n}, \tag{12}$$

where $\omega(x)$ is a positive homogeneous function of degree zero, which is infinitely differentiable and satisfies

$$\int_{|x|=1} \omega(x) d\sigma = 0, \tag{13}$$

$d\sigma$ being a surface element on the unit sphere. By K we will denote the corresponding singular integral

$$(Kf)(x) = k * f(x) = \int_{\Omega} f(y) k(x-y) dy. \tag{14}$$

The following theorem is valid for the operator K (see [4]).

Theorem 1.4 [4] *The inclusion* $K \in [L_{q)}(\Omega)], 1<q<+\infty$ *is valid, i.e.* $\exists c>0$*:*

$$\|Kf\|_{q)} \le c\|f\|_{q)}, \forall f \in L_{q)}(\Omega) \tag{15}$$

The validity of the following lemma is given in [17].

Lemma 1.5 [17] $N_{q)}(\Omega), 1<q<+\infty$, is an invariant subspace of the singular operator K in $L_{q)}(\Omega)$.

Considering the expression for the norm $N_{q)}$, it is not hard to prove the following.

Proposition 1.6 *Let* $\Omega \subset R^n$ *–be a bounded domain and L be an elliptic operator with coefficients* $a_p \in L_\infty(\Omega), \forall |p| \le m$*. Then it is valid* $L \in \left[WN^m_{q)}(\Omega); N_{q)}(\Omega)\right]$*, i.e. the following inequality*

$$\|Lu\|_{N_{q)}(\Omega)} \le C\|u\|_{WN^m_{q)}(\Omega)}, \forall u \in WN^m_{q)}(\Omega), \tag{16}$$

holds, where $C>0$ is a constant independent of u.

In the sequel, when $\Omega = B_r$ the spaces $L_{q)}(\Omega), N_{q)}(\Omega), W_{q)}(\Omega)$ and $WN_{q)}(\Omega)$ will be redenoted by $L_{q)}(r), N_{q)}(r), W^m_{q)}(r)$ and $WN^m_{q)}(r)$, respectively. Along with $WN^m_{q)}(\Omega)$, consider the following space of functions $N^m_{q)}(\Omega)$ equipped with the norm

$$\|f\|_{N^m_{q)}(\Omega)} = \sum_{|p|\le m} d_\Omega^{|p|-\frac{n}{q}} \|\partial^p f\|_{L_{q)(\Omega)}}, \tag{17}$$

where $d_\Omega = diam\ \Omega$, and we will assume $N^0_{q)}(\Omega) = N_{q)}(\Omega)$. The closure of $C_0^\infty(\Omega)$ in $N^m_{q)}(\Omega)$ $(N^m_q(\Omega))$ we will denote by $\circ N^m_{q)}(\Omega)$ $(\circ N^m_q(\Omega))$.

3. Main lemma

3.1 Solvability in the small

Introduce the following

Definition 1.7 We will say that the operator L has the property $P_{x_0})$ if its coefficients satisfy the conditions: i) $a_p \in L_\infty(B_r(x_0))$, $\forall |p| \le m$, for some $r>0$; ii) $\exists r>0$: for $|p|=m$ the coefficient $a_p(\cdot)$ *coincides a.e. in* $B_r(x_0)$ *with some function bounded and continuous at the point* x_0.

It is absolutely clear that if $a_p \in C(\Omega)$, $\forall |p| \leq m$, then L has the property $P_{x_0})$ for $\forall x_0 \in \Omega$.

In establishing the interior Schauder-type estimate for grand-Sobolev spaces $N^m_{q)}(\Omega)$ the following Main Lemma, proved in [18] plays a key role.

Main Lemma. *Let the m-th order elliptic operator L have the property $P_{x_0})$ at the point x_0. Let $\varphi \in N^m_{q)}(B_r(x_0))$ and φ vanish in a neighborhood of $|x - x_0| = r$. Then for $q > 1$ it holds*

$$\|T_{x_0}\varphi\|_{N^m_{q)}(B_r(x_0))} \leq \sigma(r)\|\varphi\|_{N^m_{q)}(B_r(x_0))}, \tag{18}$$

where the function $\sigma(r) \to 0, r \to 0$, depends only on the ellipticity constant L_{x_0}, on the coefficients of L and their moduli of continuity.

Let us consider the m-th order elliptic operator L with the coefficients $a_p(x)$ defined by (1), and the corresponding operator T_{x_0} defined by (5). Using Main Lemma, it is proved the following local existence theorem.

Theorem 1.8 Let L be an m-th order elliptic operator which has the property $P_{x_0})$ at some point $x_0 \in \Omega$ and $f \in G_{q)}(\Omega), 1 < q < +\infty$. Then, for sufficiently small r, there exists a solution of the equation $Lu = f$ belonging to the class $N_{q)}(B_r(x_0))$.

4. Interior Schauder type estimates

Let $\omega(\cdot)$ be an infinitely differentiable function on $[0, 1]$ such that for $0 \leq t < \frac{1}{3}$, $\omega(t) \equiv 1$ and for $\frac{2}{3} < t \leq 1$, $\omega(t) \equiv 0$. For $0 < R_1 < R_2$ we put

$$\xi(x) = \begin{cases} 1, |x| \leq R_1, \\ \omega\left(\dfrac{|x| - R_1}{R_2 - R_1}\right), R_1 < |x| \leq R_2. \end{cases} \tag{19}$$

Regarding this function it holds the following.

Lemma 1.9 *There is a constant $C > 0$ depending only on R_2 and $\omega(\cdot)$, such that for $\forall R_1 : 0 < R_1 < R_2$, there is*

$$\|\xi\|_{C^m(R_2)} \leq C\left(1 - \frac{R_1}{R_2}\right)^{-m}. \tag{20}$$

Accept the following property with respect to the domain Ω.

Property α). *We say that the domain Ω admits the continuation of functions of the space $N^k_{q)}(\Omega)$ if there exists a domain $\Omega' \supset \overline{\Omega}$ and a linear mapping θ of the space $N^k_{q)}(\Omega)$ into $N^k_{q)}(\Omega')$ such that*

$$\theta u = u \text{ in } \Omega,$$
$$\|\theta u\|_{N^k_{q)}(\Omega')} \leq const\ \|u\|_{N^k_{q)}(\Omega)}, \tag{21}$$

holds.

So, the following lemma is true.

Lemma 1.10 *Let the domain* Ω *have the Property* $\alpha)$ *with respect to space* $N^k_{q)}(\Omega)$. *Then* $\exists C > 0$ *depending only on* n, q *and on a constant from* (21), *which holds*

$$\|\varphi\|_{N^k_{q)}(\Omega)} \leq \varepsilon \|\varphi\|_{N^{k+1}_{q)}(\Omega)} + C\varepsilon^{-k}\|\varphi\|_{L_{q)}(\Omega)}, \tag{22}$$

for $\forall k = \overline{1, m-1}$ and $\forall \varepsilon > 0$.

The main result of this section is the following Schauder type estimate.

Theorem 1.11 *Let the coefficients of m-th order elliptic operator L satisfy the following conditions: i)* $a_p(\cdot) \in C(\overline{\Omega})$, $\forall p : |p| = m$; *ii)* $a_p(\cdot) \in L_\infty(\Omega)$, $\forall p : |p| < m$; *where* $\Omega \subset R^n$ *– bounded domain with boundary* $\partial\Omega$. *Let* $\Omega_0 \subset \Omega$ *be an arbitrary compact. Then for* $\forall u \in WG^m_{q)}(\Omega)$, $1 < q < +\infty$, *the following a priori estimate holds*

$$\|u\|_{N^m_{q)}(\Omega_0)} \leq C\Big(\|Lu\|_{L_{q)}(\Omega)} + \|u\|_{L_{q)}(\Omega)}\Big), \tag{23}$$

where the constant C depends only on the ellipticity constant m, Ω, Ω_0 of L, on the coefficients of the operator L.

5. Extension of functions from $N^m_{q)}(\Omega)$Compactness

Consider the question of the possibility of extension of a function f from class $N^m_{q)}(\Omega)$ to a wider class $N^m_{q)}(\Omega')$ with $\Omega' \supset \overline{\Omega}$. Following the classical case (see, monograph [29]), first consider the case when Ω' is a cube with an edge $2a > 0 : K_a = \{|y_i| < a, i = \overline{1, n}\}$, and Ω is a parallelepiped $K_a^+ = K_a \bigcap \{y_n > 0\}$.

The following lemma is true.

Lemma 1.12 *For* $\forall f \in WN^k_{q)}(K_a^+)$ *there exists an extension* $F \in WN^k_{q)}(K_a)$ *and, in addition, inequality*

$$\|F\|_{N^m_{q)}(K_a)} \leq C\|f\|_{N^m_{q)}(K_a^+)}, \tag{24}$$

holds.

It is completely analogous to the monograph [29, p. 129], it is proved the following

Lemma 1.13 Let $f \in WN^k_{q)}(\Omega)$ and for $\forall \xi \in \partial\Omega$ there exists a function $F_\xi(x)$, *defined in a ball* $B_r(\xi)$ of some radius $r = r(\xi) > 0$, such that $F_\xi(x) = f(x), \forall x \in \Omega \cap B_r(\xi)$ *and* $F_\xi \in WN^k_{q)}(B_r(\xi))$. Besides

$$\|F_\xi\|_{WN^k_{q)}(B_r(\xi))} \leq C\|f\|_{WN^k_{q)}(\Omega)}, \tag{25}$$

is true, where $C > 0$ is a constant independent of f. Then, for any $\rho > 0$, the function f has an extension F to the domain $\Omega_\rho = \bigcup_{x \in \Omega} B_\rho(x)$ with the properties $F \in WN^k_{q)}(\Omega_\rho), F(x) = 0, \forall x \in \Omega_\rho \backslash \Omega_{\rho/2}$: and the inequality

$$\|F\|_{WN^k_{q)}(\Omega_\rho)} \leq C\|f\|_{WN^k_{q)}(\Omega)}, \tag{26}$$

holds, where the constant $C > 0$, is dependent only on domain Ω and ρ.

Using Main Lemma and Lemmas 1.12, 1.13, similarly to [29, p. 130] it is proved the following extension.

Theorem 1.14 *Let Ω, Ω' be bounded domains in R^n, $\bar{\Omega} \subset \Omega'$ and $\partial\Omega \in C^{(m)}$. Then for $\forall f \in N^m_{q)}(\Omega)$ there exists a finite extension $F \in N^m_{q)}(\Omega')$ in Ω' and the following estimate*

$$\|F\|_{N^m_{q)}(\Omega')} \leq C\|f\|_{N^m_{q)}(\Omega)}, \tag{27}$$

is valid, where $C > 0$ is a constant independent of f.

Consider the compactness question of the family in $N_{q)}(\Omega)$. The following theorem is true.

Theorem 1.15 Let $\Omega \subset R^n$ be a bounded domain with a boundary $\partial\Omega \in C^{(1)}$. *Then a set, bounded in $N^1_{q)}(\Omega)$, is compact in $N_{q)}(\Omega)$.*

Analogously to Theorem 1.15, the following theorem is also proved.

Theorem 1.16 Let $\Omega \subset R^n$ be a bounded domain with a boundary $\partial\Omega \in C^{(k)}$.
If the set of functions is bounded in $N^k_{q)}(\Omega)$, then the set of their traces on $(n-1)$-dimensional surface $\Gamma \subset \bar{\Omega}$ from the class $C^{(k)}$ is compact in $W^r_{q)}(\Omega)$, $\forall r = \overline{0, k-1}$.

6. Trace of functions from the grand-Sobolev space $N^m_{q)}(\Omega)$

In this section, we will define a concept of the trace for functions from the grand-Sobolev space $N^m_{q)}(\Omega)$ on an $(n-1)$-dimensional differentiable surface.

Based on the embedding $N^m_{q)}(\Omega) \subset N^1_{q)}(\Omega)$, $\forall m \geq 2$, it is sufficient to define this concept regarding the functions from $N^1_{q)}(\Omega)$. So, assume $S \subset \overline{\Omega}$: $S \in C^{(1)}$ is some $(n-1)$-dimensional surface. Let $x_0 \in S$ be an arbitrary point. Then it is obvious that there exists a sufficiently small neighborhood of this point $S_{x_0} \subset S$, such that uniquely projected onto some domain D of the plane $x_n = 0$ in R^n and it has the equation

$$x_n = \varphi(x') \in C^{(1)}(\bar{D}), x' = (x_1, \dots, x_{n-1}) \in D. \tag{28}$$

Ω is bounded domain and we will consider that it is placed inside a cube $\{0 < x_i < a, i = \overline{1, n}\}$, with an edge $a > 0$. Let $f \in \circ C^\infty(\Omega)$ be some finite function in Ω. For $\forall (x'; \varphi(x')) \in S_{x_0}$ we have

$$f(x)/_{S_{x_0}} = f(x'; \varphi(x')) = \int_0^{\varphi(x')} \frac{\partial f(x'; \xi_n)}{\partial \xi_n} d\xi_n. \tag{29}$$

Let $\varepsilon \in (0, q-1)$—be an arbitrary number, $q_\varepsilon = q - \varepsilon$ and $\frac{1}{q_\varepsilon} + \frac{1}{q'_\varepsilon} = 1$. Applying Hölder's inequality from (29), we obtain

$$\left| f/_{S_{x_0}} \right|^{q_\varepsilon} \leq |\varphi(x')|^{\frac{q_\varepsilon}{q'_\varepsilon}} \int_0^{\varphi(x')} \left| \frac{\partial f(x'; \xi_n)}{\partial \xi_n} \right|^{q_\varepsilon} d\xi_n \leq a^{\frac{q_\varepsilon}{q'_\varepsilon}} \int_0^{\varphi(x')} \left| \frac{\partial f(x'; \xi_n)}{\partial \xi_n} \right|^{q_\varepsilon} d\xi_n. \tag{30}$$

Let $C = \max\{a; 1\}$. Consequently

$$\varepsilon \left| f/_{S_{x_0}} \right|^{q_\varepsilon} \leq C^{q_\varepsilon} \varepsilon \int_0^{\varphi(x')} \left| \frac{\partial f(x'; \xi_n)}{\partial \xi_n} \right|^{q_\varepsilon} d\xi_n, \tag{31}$$

where C is a constant independent of f and ε. Multiplying by $\sqrt{1+\varphi_{x_1}^2 + \dots + \varphi_{x_{n-1}}^2}$ and integrating over D we obtain

$$\left\| \varepsilon^{\frac{1}{q_\varepsilon}} f \right\|_{L_{q_\varepsilon}(S_{x_0})}^{q_\varepsilon} \leq C \left\| \varepsilon^{\frac{1}{q_\varepsilon}} \frac{\partial f}{\partial x_n} \right\|_{L_{q_\varepsilon}(\Omega)}^{q_\varepsilon}. \tag{32}$$

Since the surface S can be covered by a finite number of surfaces of type S_{x_0}, then summing the corresponding inequalities (from (32)) we establish

$$\left\| \varepsilon^{\frac{1}{q_\varepsilon}} f \right\|_{L_{q_\varepsilon}(S)}^{q_\varepsilon} \leq C \left\| \varepsilon^{\frac{1}{q_\varepsilon}} \frac{\partial f}{\partial x_n} \right\|_{L_{q_\varepsilon}(\Omega)}^{q_\varepsilon}, \tag{33}$$

where $C > 0$ is a constant independent of f and ε. This immediately implies

$$\left\| \varepsilon^{\frac{1}{q_\varepsilon}} f \right\|_{L_{q_\varepsilon}(S)} \leq C \sum_{|p|=1} d_\Omega \left\| \varepsilon^{\frac{1}{q_\varepsilon}} \partial^p f \right\|_{L_{q_\varepsilon}(\Omega)} \leq C \left\| \varepsilon^{\frac{1}{q_\varepsilon}} \partial^p f \right\|_{N^1_{q_\varepsilon}(\Omega)}. \tag{34}$$

Taking first $\sup\limits_{0<\varepsilon<q-1}$ on the right and then the same sup on the left, from this estimate for $\forall f \in \overset{0}{C}\infty(\Omega)$, we have

$$\|f\|_{L_{q)}(S)} \leq C \|f\|_{N^1_{q)}(\Omega)}. \tag{35}$$

If $\partial\Omega \in C^{(1)}$, then Theorem 1.14 implies that the inequality (35) holds for $\forall f \in C^{(1)}(\overline{\Omega})$.

Let $f \in N^1_{q)}(\Omega)$ be an arbitrary function. Then $\exists \{f_n\} \subset C^\infty(\overline{\Omega})$:

$$\| f_n - f \|_{N^1_{q)}(\Omega)} \to 0, n \to \infty. \tag{36}$$

It follows directly from (35) that the sequence $\{f_n\}$ is fundamental in $L_{q)}(S)$:

$$\| f_n - f_m \|_{L_{q)}(S)} \to 0, n, m \to \infty. \tag{37}$$

From the completeness of $L_{q)}(S)$ it follows that $\exists f_S \in L_{q)}(S)$:

$$\| f_n - f_S \|_{L_{q)}(S)} \to 0, n \to \infty. \tag{38}$$

Similarly the classical case, it is proved that f_S does not depend on the choice of the sequence $\{f_n\}$.

f_S is called the trace of the function $f \in N^1_{q)}(\Omega)$ on S and we will denote it by the operator $\Gamma_S : \Gamma f = f/_S$.

Based on the concept of Γ_S, we define the following linear space

$$N_{q)}^1(S) = N_{q)}^1(\Omega)/_S = \left\{ f \in L_{q)}(S) : \exists u \in N_{q)}^1(\Omega) \Rightarrow f = \Gamma_S u = u/_S \right\}. \tag{39}$$

For the case of $S = \partial\Omega$, the operator Γ_S will be simply denoted by $\Gamma : \Gamma_{\partial\Omega} = \Gamma$. The following lemma is true.

Lemma 1.17 *Let $\Omega \subset R^n$ be an bounded domain and $\partial\Omega \in C^{(1)}$. Then the linear spaces $\mathcal{F}_{q)}^1$ and $N_{q)}^1(\partial\Omega)$ are isomorphic, where*

$$N_{q)}^1(\partial\Omega) \equiv N_{q)}^1(\Omega)/_{\partial\Omega} = \left\{ f \in L_{q)}(\partial\Omega) : \exists u \in N_{q)}^1(\Omega) \Rightarrow f = \Gamma u = u/_{\partial\Omega} \right\}. \tag{40}$$

Based on this lemma, we define the norm in $N_{q)}^1(\partial\Omega)$

$$\|f\|_{N_{q)}^1(\partial\Omega)} = \left\|\Gamma^{-1}f\right\|_{\mathcal{F}_{q)}^1}, \forall f \in N_{q)}^1(\partial\Omega). \tag{41}$$

Since $\mathcal{F}_{q)}^1$ is a Banach space with respect to the norm $\|\cdot\|_{\mathcal{F}_{q)}^1}$, then this lemma immediately implies that $N_{q)}^1(\partial\Omega)$ is also Banach with respect to the norm (41).

The space $N_{q)}^m(\partial\Omega)$ is defined similarly and the corresponding lemma is true for the spaces $\mathcal{F}_{q)}^m$, where

$$N_{q)}^m(\partial\Omega) \equiv N_{q)}^m(\Omega)/_{\partial\Omega} = \left\{ f \in L_{q)}(\partial\Omega) : \exists u \in N_{q)}^m(\Omega) \Rightarrow f = \Gamma u = u/_{\partial\Omega} \right\}. \tag{42}$$

The following theorem is true (regarding the proof see [30]).

Theorem 1.18 Let $\Omega \subset R^n$ be a bounded domain with a boundary $\partial\Omega \in C^{(m)}$. *If the set of functions is bounded in $N_{q)}^m(\Omega), m \geq 1$,* then the set of their traces on the $(n-1)$-dimensional surface $S \subset \overline{\Omega}$ *from the class $C^{(m)}$ is compact in $L_{q)}(S)$.*

7. Schauder-type estimate up to the boundary

Using the results obtained in the previous sections, it is established a Schauder-type estimate up to the boundary for a second-order elliptic operator with nonsmooth coefficients. The following theorem is true.

Theorem 1.19 *Let $\Omega \subset R^n$ be a bounded domain with a boundary $\partial\Omega \in C^{(2)}$ and L be a second-order elliptic operator (i.e. $m = 2$) with coefficients $a_p \in C(\bar{\Omega}), \forall p : |p| = m$ and $a_p \in L_\infty(\Omega), \forall p : |p| < m$. Then for $\forall u \in N_{q)}^2(\Omega)$ the following estimate*

$$\|u\|_{N_{q)}^2(\Omega)} \leq C\left(\|Lu\|_{N_{q)}(\Omega)} + \|u\|_{N_{q)}^2(\partial\Omega)} + \|u\|_{N_{q)}(\Omega)}\right), \tag{43}$$

holds true, where $C > 0$ is a constant independent of u, but depends on the norms of the coefficients of L in $L_\infty(\Omega)$.

8. Solvability of the Dirichlet problem for a second-order elliptic operator

Let us apply the estimates established in the previous sections to the solvability question (in the strong sense) of the Dirichlet problem for a second-order elliptic type equation in classes $N^2_{q)}(\Omega)$. So, let $\Omega \subset R^n$ be a domain with a boundary $\Omega \in C^{(2)}$. Assume that $f \in N_{q)}(\Omega)$ is a given function and $a_{ij} \in C(\bar{\Omega})$; $a_i; a \in L_\infty(\Omega)$ $i;j = \overline{1,n}$. Consider the equation

$$Lu = \sum_{i;j=1}^{n} a_{ij}(x)\frac{\partial^2 u}{\partial x_i \partial x_j} + \sum_{i=1}^{n} a_i(x)\frac{\partial u}{\partial x_i} + a(x)u = f(x), x \in \Omega. \tag{44}$$

In the sequel we will assume that the following uniformly ellipticity condition holds a.e. in Ω

$$\nu|\xi|^2 \leq \sum_{i,j=1}^{n} a_{ij}(x)\xi_i\xi_j \leq \nu^{-1}|\xi|^2, \forall \xi \in R^n, \tag{45}$$

where $\nu \in (0,1]$ is some constant.

Under the solution of the Eq. (44) we mean a function $u \in N^2_{q)}(\Omega)$ for which equality (44) holds a.e. $x \in \Omega$. Let $\varphi \in N^2_{q)}(\partial\Omega)$ be a given function. Let us define the boundary condition

$$\Gamma u = u/_{\partial\Omega} = \varphi \tag{46}$$

where $\Gamma : N^2_{q)}(\Omega) \to N^2_{q)}(\partial\Omega)-$ is a trace operator.

We will say that the domain Ω has a property (δ_q) is class $\circ N^2_{q)}(\Omega)$, if the Dirichlet problem (46) is correctly solvable for the Poisson equation, i.e. the problem

$$\begin{cases} \Delta u = f, in\ \Omega, \\ \Gamma u = 0, on\ \partial\Omega, \end{cases} \tag{47}$$

has a unique solution for $\forall f \in N_{q)}(\Omega)$ in class $\circ N^2_{q)}(\Omega)$.

In order to solve the problem (44), (46) we apply the parameter continuation method (see e.g. [28, p. 247]).

Furthermore, assume that the operator L satisfies the following inequality

$$\|u\|_{L_{q)}(\Omega)} \leq C\|Lu\|_{L_{q)}(\Omega)}, \forall u \in \circ N^2_{q)}(\Omega), \tag{48}$$

where the constant C depends only on the ellipticity constants of the operator L, on the sup norms of the coefficients L, on domain Ω and is independent of the function $u \in \circ N^2_{q)}(\Omega)$.

We will say that the operator L has property (A) if an inequality (48) holds for an operator L.

The question of whether inequality (48) holds (i.e. property (A)) we will consider later.

Thus, the following main theorem is true.

Theorem 1.20 *Let $\Omega \subset R^n$ be a bounded domain with a boundary $\partial\Omega \in C^{(2)}$ and L is a second-order elliptic differential operator defined by* expression (44) *with* coefficients *$a_{ij} \in C(\bar{\Omega})$; a_i; $a \in L_\infty(\Omega)$, $i;j = \overline{1,n}$. Assume that the domain Ω has property $(\Delta_{q)})$ in class $\circ N^2_{q)}(\Omega)$ and the operator L has property (A). Then the equation $Lu = f$ is uniquely solvable for $\forall f \in N_{q)}(\Omega)$ in class $\circ N^2_{q)}(\Omega)$, i.e. $\circ L : N^2_{q)}(\Omega) \leftrightarrow N_{q)}(\Omega)$ is an isomorphism and it is obvious that the estimate*

$$\|u\|_{N^2_{q)}(\Omega)} \le C\|f\|_{N_{q)}(\Omega)}, \forall f \in N_{q)}(\Omega), \tag{49}$$

holds true, where $C > 0$ is a constant independent of f.

Now consider a homogeneous equation $Lu = 0$ in Ω with a nonhomogeneous boundary condition $\Gamma u = u/_{\partial\Omega} = \varphi$, where $\varphi \in N^2_{q)}(\Omega)$–is given function. From the results of Section 6 it follows that $\exists \Phi \in N^2_{q)}(\Omega) : \Gamma\Phi = \varphi$. Suppose $v = u - \Phi$ and let $f = -L\Phi$. It is clear that $\Gamma v = v/_{\partial\Omega} = 0$ and $Lv = f$ in Ω. If $a_{ij}; a_i; a \in L_\infty(\Omega)$, $i;j = \overline{1,n}$, then, as follows from Proposition 1.6 that $f \in N_{q)}(\Omega)$. Therefore, we can apply the Theorem 1.20 to the problem

$$\begin{cases} Lv = f, a.e. in\ \Omega, \\ v/_{\partial\Omega} = 0. \end{cases} \tag{50}$$

If all the conditions of Theorem 1.20 are satisfied, then this problem is uniquely solvable in class $N^2_{q)}(\Omega)$ and for the solution it is valid the following estimate

$$\|v\|_{N^2_{q)}(\Omega)} \le C\|f\|_{N_{q)}(\Omega)}, \tag{51}$$

where $C > 0$ is a constant independent of f. It is quite obvious that then the problem

$$\begin{cases} Lu = 0, a.e. in\ \Omega, \\ u/_\Gamma = \varphi, \end{cases} \tag{52}$$

is also uniquely solvable in $N^2_{q)}(\Omega)$.

Taking into account expression (41) for the norm in $N^2_{q)}(\partial\Omega)$, we obtain

$$\|u\|_{N^2_{q)}(\Omega)} \le C\|\varphi\|_{N^2_{q)}(\partial\Omega)}. \tag{53}$$

Consider a nonhomogeneous equation with a nonhomogeneous boundary condition

$$\begin{cases} Lu = f\ a.e. in\ \Omega, \\ u/_{\partial\Omega} = \varphi, \end{cases} \tag{54}$$

where $f \in N_{q)}(\Omega)$ and $\varphi \in N^2_{q)}(\partial\Omega)$–are given functions. Representing the function u in the form $u = v + w$, where

$$\begin{cases} Lv = f, \\ v/_{\partial\Omega} = 0, \end{cases} \begin{cases} Lw = 0, \\ w/_{\partial\Omega} = \varphi, \end{cases} \tag{55}$$

from Theorem 1.20 and taking into account estimate (53), we arrive at the following conclusion.

Theorem 1.21 *Let the domain Ω and the operator L satisfy all the conditions of Theorem 1.20. Then for $\forall f \in N_{q)}(\Omega)$ and $\forall \varphi \in N^2_{q)}(\partial\Omega)$ the problem* (54) *is uniquely solvable in the space $N^2_{q)}(\partial\Omega)$ and regarding the solution the estimate*

$$\|u\|_{N^2_{q)}(\Omega)} \leq C\left(\|f\|_{N_{q)}(\Omega)} + \|\varphi\|_{L_{q)}(\partial\Omega)}\right), \tag{56}$$

is fulfilled, where $C > 0$ is a constant independent of f and φ.

9. Some properties of a Riesz potential

For obtaining main results we need some properties of a Riesz potential and embedding theorems regarding the spaces $N^m_{q)}$. In this section, we will give some properties of an integral operator with a weak singularity. These properties are used to study the properties of functions from class $W^k_{q)}(\Omega)$. Let us remember the Sobolev integral identity

$$u(x) = \sum_{|\alpha|=0}^{k-1} x^\alpha \int_\Omega b_\alpha(y)u(y)dy + \sum_{|\alpha|=k} \int_\Omega \frac{A_\alpha(x;y)}{r^{n-k}} \partial^\alpha u(y)dy, \forall u \in C^k(\bar{\Omega}), \tag{57}$$

where $b_\alpha \in C(\bar{\Omega})$, $A_\propto \in L_\infty(\Omega \times \Omega)$ (generally speaking, for $x \neq y$: $A_\propto(x;y)$ is infinitely differentiable). In establishing many properties of a function from Sobolev classes the representation (57) plays a key role. In accordance with (57) consider the integral operator (Riesz potential).

$$(K\rho)(x) = V(x) = \int_\Omega \frac{A(x,y)}{r^\lambda} \rho(y)dy, \tag{58}$$

where $r = |y - x|$; $x \in \Omega \subset R^n$ is a bounded domain, $0 \leq \lambda < n$; $A \in L_\infty(\Omega \times \Omega)$. The following theorem is true.

Theorem 1.22 *Let* $\Omega \subset R^n$ be a bounded domain, $\rho \in L_{q)}(\Omega)$, $A \in C(\bar{\Omega} \times \bar{\Omega})$ and $\lambda q' < n$. Then operator (58) acts compactly from $L_{q)}(\Omega)$ to $C(\bar{\Omega})$.
It is true the following classical analogue

Theorem 1.23 Let $\lambda q' \geq n$ and an integer s satisfy $n - (n - \lambda) < s \leq n$. Then the integral (58) defines a function that, on any intersection Ω_s of the set Ω by a plane of dimension s, is defined almost everywhere in the sense of the Lebesgue measure in R^s. The operator K defined by formula (58) is bounded as an operator from $L_{q)}(\Omega)$ *to* $L_r(\Omega_s)$ *(also from* $L_{q)}(\Omega)$ *to* $L_{r)}(\Omega_s)$), for $\forall r$: $1 < r < r_0 = \frac{sq}{n-(n-\lambda)q}$.
It is valid the following

Theorem 1.24 If $\lambda q' \geq n$, then the operator K, defined by expression (58), acts compactly from $L_{q)}(\Omega)$ *to* $L_r(\Omega)$ *(also from* $L_{q)}(\Omega)$ *to* $L_{r)}(\Omega)$), for $\forall r : 1 < r < r_0 = \frac{nq}{n-(n-\lambda)q}$.

10. Embedding theorems

To obtain Sobolev-type embedding theorems in spaces $W^k_{q)}(\Omega)$ we will use the results obtained in the previous section. Throughout this section, we assume that the domain $\Omega \subset R^n-$ is bounded and stellar relative to some sphere. Remember that a domain is called stellar relative to some point if any ray outgoing from this point has one and only one common point with the boundary of this domain. A domain is stellar with respect to some set if it is stellar at every point of this set. The following theorem is true.

Theorem 1.25 If $qk > n$, then $W^k_{q)}(\Omega)$ compactly embedded in $C(\bar{\Omega})$.

The following theorem is true.

Theorem 1.26 Let $qk \leq n$ and $\Omega_s \subset \Omega-$be a piecewise smooth manifold of s dimensions, where $n-kq < s \leq n$. Then $W^k_{q)}(\Omega)$ *continuously embedded in* $L_r(\Omega_s)$ *(also in* $L_{r)}(\Omega_s)$), where $1 < r < r_0 = \frac{sq}{n-kq}$.

The following theorems are proved in a completely similar way.

Theorem 1.27 If $qk \leq n$, then $W^k_{q)}(\Omega)$ compactly embedded in $L_r(\Omega)$ (also in $L_{r)}(\Omega)$), where $1 \leq r < \frac{nq}{n-kq}$.

The following theorem is also true.

Theorem 1.28 Let $u \in W^k_{q)}(\Omega)$. Then it has all possible generalized derivatives of any order $l < k$ in Ω. At the same time $W^k_{q)}(\Omega)$ *compactly embedded in* $C^l(\bar{\Omega})$, *if* $(k-l)q > n$ and in $W^l_r(\Omega)$*(also in* $W^l_{r)}(\Omega)$*), if* $(k-l)q \leq n$ and $1 \leq r < \frac{nq}{n-kq}$.

Let us give some equivalent norms in the grand-Sobolev spaces $W^k_{q)}(\Omega)$ (then in $N^k_{q)}(\Omega)$). Let a function $f(\cdot)$ continuous in R^r have the following properties

$$\begin{aligned} &\alpha) f(t) \geq 0 \wedge f(t) = 0 \Leftrightarrow t = 0; \\ &\beta) f(\lambda t) = |\lambda| f(t), \forall \lambda \in R \wedge \forall t \in R^r; \\ &\gamma) f(t+\tau) \leq f(t) + f(\tau), \forall t; \tau \in R^r. \end{aligned} \tag{59}$$

In a completely analogous way to the classical case, the following theorem is proved.

Theorem 1.29 *Let r denote the number of distinct monomials of degree* $\leq k-1$ *and let* $l_1, \ldots, l_r$ *be linear functionals bounded on* $W^k_{q)}(\Omega)$ *that do not simultaneously vanish on any polynomial of degree* $\leq k-1$*, except for the identically zero. Let* $f(\cdot)$ *be a continuous function in* R^r*, having the properties of a norm* $\alpha)-\gamma)$*. Then the norm*

$$\|u\|^*_{q),k} = f(l_1 u; l_2 u; \ldots; l_r u) + \sum_{|\alpha|=k} \|\partial^\alpha u\|_{L_{q)}(\Omega)}, \tag{60}$$

is equivalent to the norm $\|\cdot\|_{W^k_{q)}(\Omega)}$.

The following lemma is true.

Lemma 1.30 *Let $\Omega \subset R^n$ be a bounded domain with sufficiently smooth boundary $\partial\Omega$. Then the norm $\|\cdot\|^*_{q),k}$, defined by expression*

$$\|u\|^*_{q),k} = \sum_{|\alpha|=0}^{k-1} \left| \int_{\partial\Omega} \partial^\alpha u d\sigma \right| + \sum_{|\alpha|=k} \|\partial^\alpha u\|_{L_{q)}(\Omega)}, \tag{61}$$

is equivalent to $\|\cdot\|_{W^k_{q)}(\Omega)}$ in $W^k_{q)}(\Omega)$. In this case for $u \in \circ W^k_{q)}(\Omega)$ the following inequality

$$\|u\|_{L_{q)}(\Omega)} \leq C \sum_{|\alpha|=k} \|\partial^\alpha u\|_{L_{q)}(\Omega)}, \tag{62}$$

holds, C is a constant independent of u.

11. About the property (a). Fredholmness

Let us get back to the question of whether property (A) is satisfied. Let $\Omega \subset R^n$ be a bounded domain. Let $S \subset \bar{\Omega}$ be some $(n-1)$-dimensional surface. Define the following class of functions. Let $S \subset \bar{\Omega}$ belong to class $C^{(k)}$ and $\varepsilon > 0$ be some number. Put $\Omega_\varepsilon(S) = \{x \in \Omega : \rho(x; S) > \varepsilon\}$.

We say that the function f belongs to class $C^k_0(S)$ (i.e. f vanishes in some neighborhood S), if $f \in C^k(\bar{\Omega})$ and $\exists \varepsilon > 0 : f(x) = 0$, $\forall x \in \Omega \backslash \Omega_\varepsilon(S)$. *Denote by $\circ N^k_{q)}(\Omega; S)$ the closure $C^k_0(S)$ in $N^k_{q)}(\Omega)$.*

Thus, it is clear that $\circ N^k_{q)}(\Omega; \partial\Omega) = \circ N^k_{q)}(\Omega)$. Denote by $\mathcal{F}^m_{q)}(\Omega; S)$ the factor space $N^m_{q)}(\Omega)/\circ N^m_{q)}(\Omega; S)$. Thus, $\mathcal{F}^m_{q)}(\Omega; \partial\Omega) = \mathcal{F}^m_{q)}$. Each function $f \in N^1_{q)}(\Omega)$ (also $f \in N^k_{q)}(\Omega)$) has a (unique) trace $f/_S$ on S, and $f/_S \in L_{q)}(S)$. Consider the following class of functions

$$N^k_{q)}(S) = N^k_{q)}(\Omega)/_S = \left\{ f \in L_{q)}(S) : \exists u \in N^k_{q)}(\Omega) \Rightarrow f = u/_S \right\}. \tag{63}$$

The following theorem is true.

Lemma 1.31 *Let $S \subset \bar{\Omega} \wedge S \in C^{(k)}$ be $(n-1)$-dimensional surface. Then the linear spaces $\mathcal{F}^k_{q)}(\Omega; S)$ and $N^k_{q)}(S)$, $k \geq 1$, are isomorphic.*

It is not hard to see that if $f \in \circ N^k_{q)}(\Omega; S)$, then $f/_S = 0$, $\forall k \geq 1$. Applying this lemma, completely similar to Theorem 1.19, we can prove the following.

Theorem 1.32 *Let $\Omega \subset R^n$ be a bounded domain with a boundary $\partial\Omega \in C^{(2)}$. Let L be a second-order elliptic operator with coefficients $a_{ij} \in C(\bar{\Omega})$, $a_i, a \in L_\infty(\Omega)$, $\forall i,j = \overline{1,n}$ defined by* expression (44). *Let $S \subset \bar{\Omega} \wedge S \in C^{(2)}$ be some $(n-1)$-dimensional surface* *and $\Omega' \subset \subset \Omega \bigcup S$ (i.e. $\bar{\Omega'} \subset \Omega \bigcup S$) be an arbitrary domain. Then the following estimate holds true for $\forall u \in \circ N^2_{q)}(\Omega; S)$:*

$$\|u\|_{N^2_{q)}(\Omega')} \leq C\Big(\|Lu\|_{N_{q)}(\Omega)} + \|u\|_{N_{q)}(\Omega)}\Big), \tag{64}$$

where the constant C depends only on the ellipticity constants of operator L, on the norms of the coefficients of L in $L_\infty(\Omega)$, on S and Ω' (is independent of u).

It is not hard to see that Theorem 1.19 is a particular case of this theorem, for this it is sufficient to take $S = \partial\Omega$. By Theorem 1.32 completely analogous to Theorem 9.14 of the monograph [31, p. 240] the following theorem is proved.

Theorem 1.33 *Let $\Omega \subset R^n$ be a bounded domain with a boundary $\partial\Omega \in C^{(2)}$ and L be an elliptic operator (44) with coefficients $a_{ij} \in C(\bar{\Omega})$, $a_i, a \in L_\infty(\Omega)$, $i,j = \overline{1,n}$. Then the following estimate holds for $\forall u \in \circ N^2_{q)}(\Omega)$:*

$$\|u\|_{N_{q)}(\Omega)} \leq C\|Lu - \sigma u\|_{N_{q)}(\Omega)}, \tag{65}$$

for $\forall \sigma \geq \sigma_0$, where $C; \sigma_0 > 0$ are some constants that independent of u.
The following theorem is the result of Theorems 1.20 and 1.33.

Theorem 1.34 Let L be an elliptic operator (44) with coefficients $a_{ij} \in C(\bar{\Omega})$, a_i, $a \in L_\infty(\Omega)$, $i,j = \overline{1,n}$. Let $\Omega \subset R^n$ be a bounded domain with a boundary $\partial\Omega \in C^{(2)}$, *which has a property* $(\Delta_{q)})$. Then $\exists \sigma_0 > 0$: the equation $Lu - \sigma u = f$ is uniquely solvable for $\forall f \in N_{q)}(\Omega)$ *in class* $\circ N^2_{q)}(\Omega)$, $\forall \sigma \geq \sigma_0$.

The Fredholm alternatives hold for the equation Lu = f, i.e. the following main theorem is true.

Theorem 1.35 Let L be an elliptic operator (44) with coefficients $a_{ij} \in C(\bar{\Omega})$, a_i, $a \in L_\infty(\Omega)$, $i,j = \overline{1,n}$ and $\Omega \subset R^n$ be a bounded domain with a boundary $\partial\Omega \in C^{(2)}$, *which has a property* $(\Delta_{q)})$. Then: i) if $KerL = 0$ in $\circ N^2_{q)}(\Omega)$, then the boundary value problem $Lu = f$, $u/_\Gamma = \varphi$, has a unique solution for $\forall \varphi \in \mathcal{F}^2_{q)}(\Omega)$ and $\forall f \in N_{q)}(\Omega)$; ii) $KerL$ is a finite-dimensional subspace in $\circ N^2_{q)}(\Omega)$.
Regarding the proof of all these facts one can see the works [4, 6].

12. On one spectral problem in Morrey-Smirnov space

In this section we consider one spectral problem in Morrey-Smirnov space. Such spectral problems arise in the problem of vibrations of a loaded string with fixed ends is solved by applying the Fourier method (see [32–34]). Morrey space is also non separable space and we define its subspace in which the infinitely differentiable functions are dense. We prove that eigenfunctions of the considered spectral problem form a basis in this subspace after eliminating an arbitrary term from them.

We need some facts from the theory of Morrey-type spaces. Let Γ be some rectifiable Jordan curve on the complex plane C. By $|M|_\Gamma$ we denote the linear

Lebesgue measure of the set $M \subset \Gamma$. By the Morrey- Lebesgue space $L^{p,\alpha}(\Gamma), 0 \le \alpha \le 1, p \ge 1$, we mean a normed space of all functions $f(\cdot)$ measurable on Γ equipped with a finite norm $\|f\|_{L^{p,\alpha}(\Gamma)}$:

$$\|f\|_{L^{p,\alpha}(\Gamma)} = \sup_B \left(|B \cap \Gamma|_\Gamma^{\alpha-1} \int_{B\cap\Gamma} |f(\xi)|^p |d\xi| \right)^{\frac{1}{p}} < +\infty. \tag{66}$$

$L^{p,\alpha}(\Gamma)$ is a Banach space and $L^{p,1}(\Gamma) = L_p(\Gamma), L^{p,0}(\Gamma) = L_\infty(\Gamma)$. The embedding $L^{p,\alpha_1}(\Gamma) \subset L^{p,\alpha_2}(\Gamma)$ is valid for $0 \le \alpha_1 \le \alpha_2 \le 1$. Thus $L^{p,\alpha}(\Gamma) \subset L_p(\Gamma), \forall \alpha \in [0,1], \forall p \ge 1$. The case of $\Gamma \equiv [a,b]$ will be denoted by $L^{p,\alpha}(a,b)$.

Denote by $\tilde{L}^{p,\alpha}(a,b)$ the linear subspace of $L^{p,\alpha}(a,b)$ consisting of functions whose shifts are continuous in $L^{p,\alpha}(a,b)$, i.e. $\|f(\cdot+\delta) - f(\cdot)\|_{L^{p,\alpha}(a,b)} \to 0$ as $\delta \to 0$. The closure of $\tilde{L}^{p,\alpha}(a,b)$ in $L^{p,\alpha}(a,b)$ will be denoted by $M^{p,\alpha}(a,b)$. In [35] the following theorem is proved.

Theorem 1.36 The exponential system $\{e^{i\,nt}\}_{n\in Z}$ is the basis in $M^{p,\alpha}(-\pi,\pi)$, $1 < p < +\infty, 0 < \alpha \le 1$.

Using this theorem, it is easy to obtain the following

Theorem 1.37 Each of the trigonometric systems $\{sinnx\}_{n=1}^{\infty}$ and $\{cosnx\}_{n=0}^{\infty}$ forms the basis for $M^{p,\alpha}(0,\pi), 1 < p < +\infty, 0 < \alpha \le 1$.

Consider a sample eigenvalue problem for the discontinuous second-order differential operator

$$-y''(x) = \lambda y(x), \quad x \in \left(0, \frac{1}{3}\right) \cup \left(\frac{1}{3}, 1\right), \tag{67}$$

with the boundary conditions

$$\left.\begin{array}{l} y(0) = y(1) = 0, \\ y\left(\frac{1}{3} - 0\right) = y\left(\frac{1}{3} + 0\right), \\ y'\left(\frac{1}{3} - 0\right) - y'\left(\frac{1}{3} + 0\right) = \lambda m y\left(\frac{1}{3}\right), \end{array}\right\} \tag{68}$$

where λ is the spectral parameter, m is a non-zero complex number.

Let us give some results from [36], which we will need throughout the paper.

Lemma 1.38 [36] *The spectral problem* (67), (68) *has two series of eigenfunctions which are given by the following expressions*

$$y_{1,n}(x) = \sin 3\pi n x, x \in [0,1], \quad n = 1, 2, \ldots, \tag{69}$$

$$y_{2,n}(x) = \begin{cases} \sin \rho_{2,n}\left(x - \frac{1}{3}\right) + \sin \rho_{2,n}\left(x + \frac{1}{3}\right), x \in \left[0, \frac{1}{3}\right], \\ \sin \rho_{2,n}(1-x), x \in \left[\frac{1}{3}, 1\right], n = 0, 1, 2, \ldots. \end{cases} \tag{70}$$

Let us construct the operator L, which linearizes the problem (67), (68) in the direct sum $L_p(0,1)\oplus C$. Denote by $W_p^2(0,\frac{1}{3})\oplus W_p^2(\frac{1}{3},1)$ the space of functions whose restrictions to intervals $(0,\frac{1}{3})$ and $(\frac{1}{3},1)$ belong to Sobolev spaces $W_p^2(0,\frac{1}{3})$ and $W_p^2(\frac{1}{3},1)$, respectively, where $1<p<\infty$. Let us define the operator L in the following way. As its domain D_L we take the manifold

$$D_L=\left\{\hat{y}=\left(y(x),my\left(\frac{1}{3}\right)\right):y(x)\in W_p^2\left(0,\frac{1}{3}\right)\oplus W_p^2\left(\frac{1}{3},1\right),\right.$$
$$\left.y(0)=y(1)=0,y\left(\frac{1}{3}-0\right)=y\left(\frac{1}{3}+0\right)\right\}, \tag{71}$$

and for $\hat{y}\in D_L$ the operator L is defined by the relation

$$L\hat{y}=\left(-y'';y'\left(\frac{1}{3}-0\right)-y'\left(\frac{1}{3}+0\right)\right). \tag{72}$$

The following lemma holds true.

Lemma 1.39 The operator L defined by expressions (71), (72) is a densely defined closed operator with a completely continuous resolvent. The eigenvalues of the operator L and the problem (67), (68) coincide. If $y(x)$ *is the eigenfunction (associated function) of problem* (67), (68), then $\hat{y}=\left(y(x);my\left(\frac{1}{3}\right)\right)$ is the eigenvector (associated vector) of the operator L.

In order to obtain the main results, we need some concepts and facts from the theory of bases in a Banach space.

Recall the following definition.

Definition 1.40 *The basis $\{u_n\}_{n\in N}$ of Banach space X is called a p-basis, if for any $x\in X$ one has the inequality*

$$\left(\sum_{n=1}^{\infty}|\langle x,\vartheta_n\rangle|^p\right)^{\frac{1}{p}}\leq M\|x\|, \tag{73}$$

where $\{\vartheta_n\}_{n\in N}$ is the biorthogonal system for $\{u_n\}_{n\in N}$.

Definition 1.41 *The sequences $\{u_n\}_{n\in N}$ and $\{\varphi_n\}_{n\in N}$ of Banach space X are called p-close, if*

$$\sum_{n=1}^{\infty}\|u_n-\varphi_n\|^p<\infty. \tag{74}$$

We will also use the following results from [37, 38] (see also [39, 40]).

Theorem 1.42 [37] *Let $\{x_n\}_{n\in N}$ form a q-basis for the space X, and the system $\{y_n\}_{n\in N}$ is p- close to $\{x_n\}_{n\in N}$, where $\frac{1}{p}+\frac{1}{q}=1$. Then the following properties are equivalent:*

1. $\{y_n\}_{n\in N}$ is complete in X;

2. $\{y_n\}_{n\in N}$ is minimal in X;

3. $\{y_n\}_{n\in N}$ forms an isomorphic basis to $\{x_n\}_{n\in N}$ for X.

Let $X_1 = X\oplus C^m$ and $\{\hat{u}_n\}_{n\in N}\subset X_1$ be some minimal system, and $\{\hat{\vartheta}_n\}_{n\in N}\subset X_1^* = X^*\oplus C^m$ be its biorthogonal system:

$$\hat{u}_n = (u_n;\alpha_{n1}, \dots, \alpha_{nm});\hat{\vartheta}_n = (\vartheta_n;\beta_{n1}, \dots, \beta_{nm}). \tag{75}$$

Let $J = \{n_1, \dots, n_m\}$ be some set of m natural numbers. Suppose

$$\delta = \det\left\|\beta_{n_i j}\right\|_{i,j=\overline{1,m}}. \tag{76}$$

The following theorem holds true.

Theorem 1.43 [38] *Let the system $\{\hat{u}_n\}_{n\in N}$ form a basis for X_1. In order to the system $\{u_n\}_{n\in N_J}$, where $N_J = N\backslash J$, form a basis for X it is necessary and sufficient that the condition $\delta\neq 0$ is satisfied. In this case the biorthogonal system to $\{u_n\}_{n\in N_J}$ is defined by*

$$\vartheta_n^* = \frac{1}{\delta}\begin{vmatrix} \vartheta_n & \vartheta_{n1} & \dots & \vartheta_{nm} \\ \beta_{n1} & \beta_{n_1 1} & \dots & \beta_{n_m 1} \\ \dots & \dots & \dots & \dots \\ \beta_{nm} & \beta_{n_1 m} & \dots & \beta_{n_m m} \end{vmatrix} \tag{77}$$

For $\delta = 0$ the system $\{u_n\}_{n\in N_J}$ is not complete and is not minimal in X.
Let X be a Banach space and $\{u_{kn}\}_{k=\overline{1,m};\, n\in N}$ be some system in X.
Let $a_{ik}^{(n)}, i, k = \overline{1,m}, n\in N$, be some complex numbers. Put

$$A_n = \left(a_{ik}^{(n)}\right)_{i,k=\overline{1,m}} \textit{ and } \Delta_n = \det A_n, n\in N. \tag{78}$$

Let us consider the following system in space X

$$\hat{u}_{kn} = \sum_{i=1}^{m} a_{ik}^{(n)} u_{in}, k = \overline{1,m}; n\in N. \tag{79}$$

Theorem 1.44 *If the system* $\{u_{kn}\}_{k=\overline{1,m};\, n\in N}$ forms a basis for X and

$$\Delta_n\neq 0, \forall n\in N, \tag{80}$$

then the system $\{\hat{u}_{kn}\}_{k=\overline{1,m};\, n\in N}$ forms a basis with parentheses for X. If in addition the conditions

$$\sup_n\{\|A_n\|, \|A_n^{-1}\|\} < \infty, \sup_n\{\|u_{kn}\|, \|\vartheta_{kn}\|\} < \infty, \tag{81}$$

hold, where $\{\vartheta_{kn}\}_{k=\overline{1,m};\, n\in N}\subset X^*$ is biorthogonal system to $\{u_{kn}\}_{k=\overline{1,m};\, n\in N}$, then the system $\{\hat{u}_{kn}\}_{k=\overline{1,m};\, n\in N}$ forms the usual basis for X.

The following theorem holds true.

Theorem 1.45 The system of eigen and associated vectors of the operator L forms the basis for space $M^{p,\alpha}(0,1)\oplus C, 1<p<\infty, 0<\alpha\leq 1$.

Now, let us consider the basicity of the system $\{y_0\}\cup\{y_{i,n}\}_{i=1,2;\,n\in N}^{\infty}$ with a removed function in space $M^{p,\alpha}(0,1)$.

Theorem 1.46 *If from the system of eigen and associated functions of problem* (67), (68) $\{y_0\}\cup\{y_{i,n}\}_{i=1,2;\,n\in N}^{\infty}$ *we eliminate any function* $y_{2,n_0}(x)$, *corresponding to a simple eigenvalue, then the new system forms a basis for* $M^{p,\alpha}(0,1), 1<p<\infty,\ 0<\alpha\leq 1$. *And if we eliminate any function* $y_{1,n_0}(x)$ *from this system, then the obtained system does not form a basis in* $M^{p,\alpha}(0,1)$; *moreover, in this case this system is not complete and is not minimal in this space.*

Proof. For the eigenfunctions $\{z_0\}\cup\{z_{i,n}\}_{i=1,2;\,n\in N}^{\infty}$ of the adjoint problem we have $z_{1,n}\left(\frac{1}{3}\right)=0$ for any $n\in N$ and $z_{2,n}\left(\frac{1}{3}\right)\neq 0$. On the other hand, the eigenvectors of the adjoint operator L^* are defined by $\hat{z}_n=\left(z_n,\overline{m}z_n\left(\frac{1}{3}\right)\right), n=0,1,\ldots,$. Applying Theorem 1.43 to the system $\{\hat{y}_0\}\cup\{\hat{y}_{i,n}\}_{i=1,2;\,n\in N}^{\infty}$, we notice that $\delta=\overline{m}z_{1,n}\left(\frac{1}{3}\right)=0$ for any $n\in N$ and $\delta=\overline{m}z_{2,n}\left(\frac{1}{3}\right)\neq 0$ for any eigenfunction corresponding to a simple eigenvalue, and the statements of the theorem follow from the corresponding statements of Theorem 1.43.
Theorem is proved.

Conflict of interest

The authors declare no conflict of interest.

Author details

Bilal Bilalov[1*†], Sabina Sadigova[1,2†] and Zaur Kasumov[1†]

1 Institute of Mathematics and Mechanics of NAS of Azerbaijan, Baku, Azerbaijan

2 Khazar University, Baku, Azerbaijan

*Address all correspondence to: b_bilalov@mail.ru

† These authors contributed equally.

References

[1] Cruz-Uribe DV, Fiorenza A. Variable Lebesgue spaces. Birkhauser, Springer; 2013

[2] Adams DR. Morrey Spaces. Switzherland: Springer; 2016

[3] Kokilashvili V, Meskhi A, Rafeiro H, Samko S. Integral operators in non-standard function spaces. In: Variable Exponent Lebesgue and Amalgam Spaces. Vol. Volume 1. Springer; 2016

[4] Kokilashvili V, Meskhi A, Rafeiro H, Samko S. Integral operators in non-standard function spaces. In: Variable Exponent Holder, Morrey–Campanato and Grand Spaces. Vol. Volume 2. Springer; 2016

[5] Harjulehto P, Hasto P. Orlicz Spaces Generalized Orlicz Spaces, 169 p. Springer; 2019

[6] Castillo RE, Rafeiro H. An Introductory Course in Lebesgue Spaces. Springer; 2016

[7] Bilalov BT, Gasymov TB, Guliyeva AA. On solvability of Riemann boundary value problem in Morrey-hardy classes. Turk. J. of Math. 2016;**40**(50):1085-1101. DOI: 10.3906/mat-1507-10

[8] Bilalov BT, Guseynov ZG. Basicity of a system of exponents with a piece-wise linear phase in variable spaces. Mediterranean Journal of Mathematics. 2012;**9**(3):487-498. DOI: 10.1007/s00009-011-0135-7

[9] Bilalov BT, Guliyeva AA. On basicity of the perturbed systems of exponents in Morrey-Lebesgue space. International Journal of Mathematics. 2014;**25**(1450054):1-10

[10] Bilalov BT. The basis property of a perturbed system of exponentials in Morrey-type spaces. Sib. Math. Journ. 2019;**60**(2):323-350

[11] Israfilov DM, Tozman NP. Approximation in Morrey-Smirnov classes. Azerb. J. Math. 2011;**1**(1): 99-113

[12] Sharapudinov II. On direct and inverse theorems of approximation theory In variable Lebesgue and Sobolev spaces. Azerb. J. Math. 2014;**4**(1):55-72

[13] Bilalov BT, Huseynli AA, El-Shabrawy SR. Basis properties of trigonometric Systems in Weighted Morrey Spaces. Azerb. J. Math. 2019; **9**(2):200-226

[14] Bilalov BT, Seyidova FS. Basicity of a system of exponents with a piecewise linear phase in Morrey-type spaces. Turkish Journal of Mathematics. 2019; **43**:1850-1866. DOI: 10.3906/mat-1901-113

[15] Zeren Y, Ismailov MI, Karacam C. Korovkin-type theorems and their statistical versions in grand Lebesgue spaces. Turkish Journal of Mathematics. 2020;**44**:1027-1041. DOI: 10.3906/mat-2003-21

[16] Zeren Y, Ismailov M, Sirin F. On basicity of the system of eigenfunctions of one discontinuous spectral problem for second order differential equation for grand-Lebesgue space. Turkish Journal of Mathematics. 2020;**44**(5):1995-1612. DOI: 10.3906/mat-2003-20

[17] Bilalov BT, Sadigova SR. On solvability in the small of higher order elliptic equations in grand-Sobolev spaces. Complex Variables and Elliptic Equations. 2020;**66**:2117-2130. DOI: 10.1080/17476933.2020.1807965

[18] Bilalov BT, Sadigova SR. Interior Schauder-type estimates for higher-order elliptic operators in grand-Sobolev spaces. Sahand Communications in Mathematical Analysis. 2021;**18**:129-148

[19] Palagachev DK, Softova LG. Singular integral operators, Morrey spaces and fine regularity of solutions to PDE's. Potential Analysis. 2004;**20**:237-263

[20] Chen Y. Regularity of the solution to the Dirichlet problem in Morrey space. J. Partial Differ. Eqs. 2002;**15**:37-46

[21] Fario GD, Ragusa MA. Interior estimates in Morrey spaces for strong solutions to nondivergence form equations with discontinuous coefficients. Journ. of Func. Anal. 1993; **112**:241-256. DOI: 10.1006/jfan.1993.1032

[22] Softova LG. The Dirichlet problem for elliptic equations with VMO coefficients in generalized Morrey spaces. Operator Theory: Advances and Applications. 2013;**229**:371-386. DOI: 10.1007/978-3-0348-0516-2_21

[23] Palagachev DK, Ragusa MA, Softova LG. Regular obligue derivative problem in Morrey spaces, Elec. Jour. of Diff. Eq. 2020;**2000**(39):1-17 https://digital.library.txstate.edu/handle/10877/9133

[24] Byun SS, Palagachev DK, Softova LG. Survey on gradient estimates for nonlinear elliptic equations in various function spaces. St. Petersbg. Math. J. 2020;**31**(3):401-419 and Algebra Anal. 31, No. 3, 10-35 (2019)

[25] Caso L, D'Ambrosio R, Softova L. Generalized Morrey spaces over unbounded domains. Azerb. J. Math. 2020;**10**(1):193-208

[26] Di Fazio G, Palagachev DK, Ragusa MA. Global Morrey regularity of strong solutions to the Dirichlet problem for elliptic equations with discontinuous coefficients. Journal of Functional Analysis. 1999;**166**(2):179-196. DOI: 10.1006/jfan.1999.3425

[27] Di Fazio G. On Dirichlet problem in Morrey spaces. Differential and Integral Equations. 1993;**6**(2):383-391

[28] Bers L, John F, Schechter M. Partial differential equations. Moscow: Mir; 1966 (in Russian)

[29] Mikhaylov VP. Partial differential equations. Moscow: Nauka; 1976 (in Russian)

[30] Bilalov BT, Sadigova SR. On the Fredholmness of the Dirichlet problem for a second-order elliptic equation in grand-Sobolev spaces. Ricerche di Matematica. 2021. DOI: 10.1007/s11587-021-00599-9

[31] Gilbarg D, Trudinger NS. Elliptic Partial Differential Equations of Second Order. Berlin-Heidelberg-New York-Tokyo: Springer-Verlag; 1983

[32] Atkinson FV. Discrete and Continuous Boundary Problems. Moscow: Mir; 1968

[33] Collatz L. Eigenvalue Problems. Moscow: Fizmatgiz; 1968 504 p (in Russian)

[34] Tikhonov AN, Samarskii AA. Equations of Mathematical Physics. Dover, New York: Mosk. Gos. Univ., Moscow (1999); 2011

[35] Bilalov BT, Quliyeva AA. On basicity of exponential systems in Morrey type spaces. International Journal of Mathematics. 2014;**25**(6):1-10

[36] Gasymov TB, Maharramova GV. On completeness of eigenfunctions of the

spectral problem. Caspian Journal of Applied Mathematics, Ecology and Economics. 2015;**3**(2):66-76

[37] Bilalov BT. Bases of exponentials, sines, and cosines. Differentsial'nye Uravneniya. 2003;**39**(5):619-623

[38] Gasymov TB. On necessary and sufficient conditions of basicity of some defective systems in Banach spaces. Transactions of National Academy of Sciences Azerbaijan Series of Physical Technical and Mathematical Sciences. 2006;**26**(1):65-70

[39] Bilalov BT. Some Problems of Approximation. Baku, Elm; 2016 380 p. (in Russian)

[40] Bilalov BT, Guseynov ZG. K-Bessel and K-Hilbert systems and K-bases. Doklady Mathematics. 2009;**80**(3): 826-828

Section 2

Spatial-Temporal Analysis

Chapter 5

Chaos Analysis Framework: How to Safely Identify and Quantify Time-Series Dynamics

Markus Vogl

Abstract

Within this chapter, a practical introduction to a nonlinear analysis framework tailored for time-series data is provided, enabling the safe quantification of underlying evolutionary dynamics, which describe the referring empirical data generating process. Furthermore, its application provides the possibility to distinct between underlying chaotic versus stochastic dynamics. In addition, an optional combination with (strange) attractor reconstruction algorithms to visualize the denoted system's dynamics is possible. Since the framework builds upon a large variety of algorithms and methods, its application is by far trivial, especially, in hindsight of reconstruction algorithms for (strange) attractors. Therefore, a general implementation and application guideline for the correct algorithm specifications and avoidance of pitfalls or other unfavorable settings is proposed and respective (graphical) empirical examples are shown. It is intended to provide the readers the possibility to incorporate the proposed analysis framework themselves and to conduct the analyses and reconstructions properly with correct specifications and to be knowledgeable about misleading propositions or parameter choices. Finally, concluding remarks, future avenues of research and future refinements of the framework are proposed.

Keywords: nonlinear dynamics, attractor reconstruction, time-series quantification, chaos analysis framework, financial markets

1. Introduction

Following current estimates, the predictive analytics market is expected to grow from around 10.5 Bn. USD at the end of 2021 to around 28–30 Bn. USD until 2030, thus, stating the immense relevance of successful forecasting capabilities for the technological advancement in our digitalized, fully connected and global economy [1–3]. Therefore, respective fields of applications for predictive analytics (or related methodologies) can be represented by any real-world system interacting with practitioners or researchers alike [4]. For example, climate, hydrological cycles, ecosystems, the human brain, neuroscientific observations, the universe, engineering applications, economic systems or financial markets can be seen as such real-world systems [5]. Nonetheless, previously denoted examples are all classified as complex systems [5].

The meaning of referring to complex systems renders itself obvious once one regards contemplation of real-life confrontations, in which similar scenarios tend to evolve similarly and occur repeatedly [6]. The latter similar repetition of scenarios leads to the association of a predefined level of determinism in said real-life systems due to the timely evolution of memory and experience effects [6, 7]. Henceforth, quantitative modeling via deterministic differential equations proposes itself as a suitable methodology to cope with these kind of systems, since the respective entireties can be characterized by equivalent mathematical differential equations [6, 8]. Presupposing that the initial conditions of the referring systems are exactly disclosed, the respective differential equations enable the predictions of the systems' final states to an indefinite level of precision and time span due to the deterministic characteristics of the systems [6]. In terms of predictive analytics and forecasting attempts of such systems, presupposing said deterministic scenario, would illustrate the prerequisite of the future evolutions of the systems to be completely explicable via the current states, principally indicating a 'plainness' in terms of the predictability of such systems [9, 10].

However, scrutinizing one of the previously denoted examples as a representation of such systems with deterministic real-world characteristics leads to the emergence of unexpectedly drastic insights with vast practical implications [11]. The before mentioned real-world systems, such as financial markets, viewed as complex systems and driven by inherent or underlying empirical properties (i.e. stylized facts[1]), result in a contrastingly challenging effort in terms of predictability and mathematical modeling in comparison to the before assumed 'plainness' of deterministic forecasts [6, 11]. Consequently, the determination of the true data generating process (DGP) of time-series, which are empirical observations of the underlying complex systems such as a stock price series for financial markets, with respect to stylized facts and other innovations is advantageous for the systems' observers, researchers or other involved entities [20, 21]. Under the presupposition of complex dynamical systems, seemingly conceptual differences are the basis for the discussion on the underlying nature and essential functioning of the emerging dynamics of, for example, financial markets or other defined real-world systems [6]. A deeper understanding of such assumed underlying laws of dynamical motions would facilitate the thorough application of chaos analysis in such real-world systems [22].

Substantial literature about testing underlying systems' dynamics and chaos in such real-world systems (e.g. financial markets) provides strong evidence of nonlinearity and as a consequence, a special class of models, namely chaos models, arose [11, 23]. Chaos institutes a deeper rationale for the above-mentioned essential characteristics and the underlying nature of the evolutionary processes driving a complex (real-world) system, which is affected by nonlinearities [10]. The first property, or distinctive feature of chaotic dynamical systems, is that even though deterministic, these systems characterize themselves via sensitivity to initial conditions[2], implying

[1] Stylized facts, in particular, on financial markets can be volatility dynamics (e.g. [12]), nonlinearity (e.g. [13]), asymmetry (e.g. [3]), long memory (e.g. [14]), multifractality (e.g. [15]) and momentum driven trend characteristics, which clearly contradict the efficient market hypothesis [16]. Furthermore, studying stylized facts requires considering the heterogeneity of actors (e.g. [17]), resulting in multifractal timescales and behavioral patterns (e.g. [18]). All these properties occur at different timescales simultaneously, indicating the existence of stated nonlinearities (e.g. [19]) within the complex system of financial markets. Note that other complex systems may yield a similar variety of empirical characteristics to be regarded in respective predictive endeavors.

[2] Deviations from a trajectory of the system's phase (or state) space.

slight fluctuations or even marginal perturbations of the initial conditions to be capable to render precise predictions on a long time scale meaningless and futile in their totality due to exponentially increasing error terms [9, 22]. In addition, data measurement limitations[3] in regard to current initial conditions, state an upper bound of the predictability, even if the model is completely disclosed [22]. Second to elaborate on, is the recurrence property, reflecting upon the dynamical behavior of such systems, which can potentially be exploited for the characterization of underlying dynamical evolutionary rules (or empirical DGPs) as presented later [6, 8].

Recent trends within chaotic dynamical analysis have led to a proliferation of publications, stating structural nonlinear models to be capable of displaying instabilities and chaos to be able to mimic empirical time-series properties[4] [22]. Henceforth, a crucial pillar in nonlinear forecasting for over 40 years is the revelation of whether the considered time-series data sets are generated via deterministic or stochastic[5] dynamical systems since their respective mathematical operations differ noticeable (see the bibliometric review of Vogl [25]) [23, 24, 26]. Speaking in a mathematical sense, a chaotic dynamical system has a dense collection of points with periodic orbits, sensitivity to initial conditions and topological transitivity, which is discussed in Eckmann and Ruelle [27], Devaney [28], BenSaïda and Litimi [29]. Chaos further refers to bounded steady-state behavior, which neither represents an equilibrium point, nor a quasi-periodic nor a periodic point nor indicates that nearby points separate exponentially in finite time, resulting in those chaotic systems revealing very complex and seemingly random evolutions out of the view of standard statistical tests [22].

Hence, chaos reveals apparent randomness of (chaotic) complex system realizations, yielding underlying patterns, interconnectedness, feedback loops, recurrence, self-similarity (fractality) and self-organization capabilities [30–32]. For example, in financial systems, hyperchaotic[6] phenomena potentially evolve into crises, denying any form of system control [37]. Referring to scientific literature, the first tests of chaotic behavior for complex time-series systems were executed following the Brock-Dechert-Scheinkman (BDS) test of Brock et al. [38], yet, revealed its omnipotence, since it is unable to differentiate, whether the revealed nonlinearities originate from stochastic or chaotic dynamics [39]. Unfortunately, even comparisons between the most powerful tests (e.g. close-return test, BDS test and Lyapunov exponent[7]) do not result in conclusive findings [39]. In point of fact, several propositions toward a more conclusive solution in the scientific literature were brought to light, with no further positive indications [39].

The former statement is an allegory for the vast dilemma concerning the determination of the true, mostly unknown nature of complex dynamical (real-world) systems – whether it be stochasticity or chaoticity [40]. These systems are almost graphically similar and cannot be differentiated by respective statistical standard tests [29, 41]. Following Aguirre and Billings [41], a verification of strong noise influence on the identifiability of chaotic dynamics is provided, leading to misspecifications

[3] In terms of measurement errors, sampling frequency and data accuracy, among others.

[4] Thus, vast disseminations of literature about deterministic chaotic behavior and the design of (economic) models in the regime of chaotic behavior from a theoretical view arose [24].

[5] Originating from pure randomness.

[6] Hyperchaos is considered, if more than two positive Lyapunov exponents exist (e.g. [33–35]). If a discrete nonlinear system is dissipative (spontaneously symmetry breaking), a positive maximum Lyapunov exponent is an indication of chaotic dynamics within the system under regard [36].

[7] A positive maximum Lyapunov exponent can occur even in non-chaotic series, due to inadequate application on noisy data sets [39].

of chaotic dynamics as stochastic dynamics due to noise-disturbance, rendering the discovery of evolutionary chaotic processes very difficult [42]. The majority of empirical time-series data is often small and noisy in comparison to its laboratory-based 'physics' data counterparts', suggesting a preclusion of dynamical identification if the noise levels are greater than a predefined critical threshold value [10, 41, 43]. Therefore, the great controversy of the nonlinear empirical literature as stated above is whether a complex system is characterizable via (low-dimensional) deterministic chaos or generated via stochastic dynamics and if those chaotic complex systems are controllable [10, 43].

To solve this ongoing debate, Vogl and Rötzel [40] and Vogl [44] successfully proposed distinct analysis frameworks, enabling the clear and safe quantification and determination of the underlying (empirical) DGPs of time-series data sets. Nonetheless, due to publication technical reasons, Vogl and Rötzel [40] present a framework specification tailored toward solely stationary time-series, while Vogl [44] supposes additional customization for non-stationary data. However, these specifications originate from one original, singular and holistic chaos analysis framework for nonlinear time-series, which will be presented in its totality hereinafters. Furthermore, since the determination of the (empirical) DGP by the proposed framework is by far trivial in its application, owing to a large variety of advanced algorithms to be implemented, this chapter is purposed to provide a clear and distinct practical guideline on how to successfully implement the denoted analysis framework. Particularly, the determination of 'scaling regions' via correlation sum and correlation dimensional schemes (refer to [45–47]) and the reconstruction algorithms of (strange, fractal) attractors of complex dynamical systems with chaotic traits (see [10, 48]) represent one of the main emphases, among others, since many erroneous conductions are possible and are widely dispersed throughout the scientific literature [25, 40, 45]. The distinct goal and aim of this chapter are to provide the researcher and practitioner with an empirical-practical guide on how to implement the presented chaos analysis framework successfully, thus, determining the (empirical) DGP and reconstructing potentially existing system attractors out of a scalar time-series given [40, 44]. Moreover, the insights provided are mostly independent of the framework, generalizable and abstractable to any other kind of subsequent or related empirical analyses conducted.

Therefore, given in Section 2, the framework will be introduced completely and its inherent parts briefly reconciled, while the example data and subsequent correct specifications and selections to conduct a correct analysis are proposed in Section 3. Furthermore, Section 4 levels around the avoidance of pitfalls and the empirical results of misspecifications via practical examples, before concluding remarks and future avenues are discussed in Section 5. Please note that mathematical definitions or formulas are neglected and the reader is referred to the stated literature instead. If no further explication is granted, the literature is seen as a prerequisite for arguments and propositions, since the focus is purely on practical applicability in a theoretical-scientific context.

2. Framework overview

Before elaborating on the analysis framework itself, it is relevant to reconcile the contribution and relevance of the propagated inherent paradigm shift in quantitative modeling, namely, the previous determination of the empirical DGP and its characteristics before selecting referring mathematical procedures and models [40, 44]. Referring back to the introductory mentioned expected market increase by around 20 Bn. USD in predictive analytics, it is crucial for successful forecasting to be informed about the underlying evolutionary building mechanics of the to be analyzed time-series data, before deploying cost-intensive predictive applications. One may imagine deploying planned-out long-term predictive solutions on time-series systems, which however are chaotic and, thus, only predictable over a short time scale. This would either lead to disastrous outcomes and very poor predictive accuracy or render model performance and the totality of forecasts futile due to exponentially increasing error terms owing to said chaotic mechanics [49, 50]. Instead, the proposed paradigm shift underlying the framework states the initial determination of the empirical DGP with inherent empirical characteristics, leading to exploitable knowledge about the predictive time horizons, hidden system properties and, therefore, the minimum model capability requirements, before practical implementation and roll-outs are conducted [40]. The model selections, thus, follow the insights of the determined empirical DGP [44]. Regarding the scientific side of the framework, existing literature and research do not execute sufficient theoretical precaution within respective applications and interpretations, leading to fragmentation and dispersion of methodology and modeling, thereby representing the rationale for the framework's creation [40, 45].

Hereinafter, the chaos analysis framework is presented in detail. In Section 2.1, the framework in general and its components are elucidated, while sections 2.2 through 2.7 level around the contents of each analysis step, while also introducing the inherently applied algorithms and methods in a sparse and nonmathematical as well as practical-error avoidant oriented manner.

2.1 Chaos analysis framework

The holistic chaos analysis framework presented in **Figure 1**, consists of six steps and will be elucidated hereinafter. Before elaborating on the steps in detail, the brief course of analysis will be outlined. First, it is mandatory to analyze given noise contamination and its respective levels and the nature of the potential noise [43]. Noise is capable of disturbing the identification of the underlying dynamics and, thus, is regarded as analysis destructive [45]. Furthermore, it is deemed favorable to gather basic statistical insights from the (denoised) datasets under analysis via determination of standard statistical tests, which incorporate tests for stationarity, nonlinearity and correlations, among others [40, 45]. It is possible to determine the applicability of reconstructions solely based on these insights. Second, several chaos measures and nonlinear metrics are calculated such as the sample entropy (see [51]), Lyapunov exponents (refer to [52, 53]) or the Hurst exponent (see [54]). These insights are relevant to determining the nature of the underlying dynamical system based upon mathematical procedures. Third, if applicable, (strange) attractor reconstruction algorithms can be implemented to reconstruct the system's attractor visually. Fourth, an independent recurrence quantification analysis (RQA) paired with discrete wavelet transformations (DWT, refer to [55]) can be applied to (a) determine the existence of various sub-dynamics and (b) exploit denoted recurrence properties mathematically as well as visually [6, 8, 56]. This reveals hidden characteristics of the analyzed datasets. Fifth, spectral characteristics, especially exploitable in forecasting by applying fractional calculus, are analyzed via wavelet-based multiresolution analysis (MRA) [57–59].

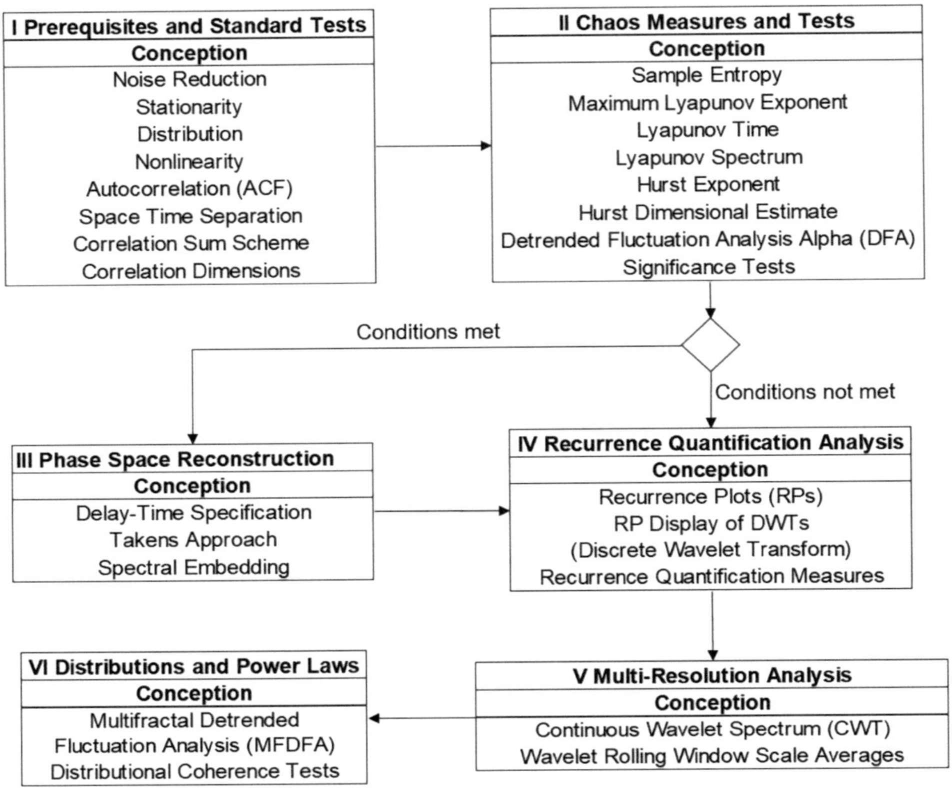

Figure 1.
Generalized chaos analysis framework for the determination of the empirical DGP and underlying empirical system characteristics based on scalar time-series data taken with permission of [40, 44]. Step I (prerequisites and standard tests) consists of testing prerequisites, which are required to conduct a nonlinear dynamical analysis. Therefore, noise reduction is mandatory, followed by tests for stationarity, Gaussianity (distribution in general), nonlinearity, and space-time separations, which can prevent an analysis. Step II (chaos measures and tests) encompasses a collection of effective nonlinear dynamical or chaotic measures. First, a correlation sum scheme is applied to determine and test significant 'scaling regions'. Moreover, the dimensionalities and properties of the system are tested (e.g. correlation dimension, Lyapunov exponents). Furthermore, information content via sample entropy is analyzed, among others. Step III (phase space reconstruction) involves the proper reconstruction of the system and a graphical representation using embedding theorems such as the traditional Takens' embedding. Step IV (recurrence quantification analysis) is an independent confirmation of the previous steps, namely, the ability to describe and quantitatively measure the characteristics of the underlying dynamics, optionally, with the application of rolling window scale averages and a subsequent discrete wavelet transformation (DWT) application to determine the potential existence of sub-dynamics within the data. Furthermore, the quantification is not dependent on the prerequisites of steps I–III. Step V (multiresolution analysis) elaborates on the spectral properties of the data and is elucidated via continuous wavelet power spectra (CWT). Step VI (distributions and power-Laws) is to determine power-law characteristics via multifractal detrended fluctuation analysis (MFDFA) and distributional coherence tests.

Finally, (multi)scaling and (multi) fractal characteristics are elaborated on via the conduction of a multifractal analysis. The multifractal analysis includes a multifractal detrended fluctuation analysis (MFDFA, see [60]) as basis, which renders (locally) minimum and maximum Hurst exponents graphically visible, while subsequently providing inherent scaling coefficients. In particular, generalized Hurst exponents, multifractal scaling exponents and the multifractal scaling spectrum can be derived, thereinafter. Moreover, distributional coherence tests can be applied to validate, the 'less worst' distributional fit and, whether a power law is present in the data (refer to [61]).

In total, holistic insight into the underlying empirical DGP and inherent characteristics is obtained, with which one may select appropriate models thereinafter.

2.2 Prerequisites and standard tests

Initially, the prerequisites and standard tests are presented briefly. The time-series has mandatorily to be denoised properly to ensure a descent beyond a predefined threshold level, best via nonlinear filter schemes [43, 45, 58]. Most time-series are contaminated by noise due to measurement errors and microstructure noise occurrences [39, 62]. Following Aguirre and Billings [41], noise exerts a mentionable (negative) influence on the identifiability of processes inherited via chaotic dynamics. Henceforth, if a certain level of noise is exceeded, accurate estimations of dynamic models and subsequent analyses are voided in their entirety[8] [41]. The only feasible approach, therefore, is the drastic reduction of noise levels to 'workable levels', since contaminating noise in evolutionary dynamics may be dynamical noise in either additive or multiplicative specification, thus, disrupting the dynamical identification on several even small scales [26, 63]. Regarding the nonlinear filter structures, two criteria have to be met, namely, (1) the applied filters are required to be unbiased and (2) the residual variance of the filters levels the noise variance [41]. Please note that some nonlinear filters such as the median filter Introduce (artificial) autocorrelations in the data, which should be avoided, thus, wavelet filters are deemed to be favorable for nonlinear denoising (refer to [55, 65, 66]).

Once the time-seriess is successfully denoised, standard statistics can be applied to elaborate on primal insights into the underlying mechanics [40]. Within the framework, the first and – destructive of reconstruction algorithms if missing – property is stationarity[9] [40]. One has to exert special strictness in terms of stationarity, thus, proposing a 1% significance level for two successive tests is deemed favorable [44, 45]. For the framework, the augmented Dickey-Fuller (ADF) test and the more powerful Kwiatkowski-Phillips-Schmidt-Shin (KPSS) test are executed (refer to [67, 68]), which both have to concur to be regarded as valid results in terms of stationarity. To elaborate on the initial test of distributional characteristics, a Kolmogorov-Smirnov (KS) test for a Gaussian specification (refer to [69]) is conducted, yet, other specifications are possible. Nonetheless, a 1% level of significance is recommended to adhere to the strictness of the presuppositions of the analysis. Moreover, to test for the existence of nonlinearity, which is a prerequisite for the existence of chaotic dynamics, the BDS test (refer to [38]) is executed. Please note that due to its stated omnipotence, it is only applied to identify nonlinearity in general and specifically not to distinguish stochasticity versus chaoticity [39]. Further note that sufficiently enough embedding dimensions have to be selected for the BDS test as well as subsequent methodologies to meet practical insights.

Lastly, correlation structures have to be elucidated, beginning with the calculation of autocorrelation functions (ACFs, see [70]) with sufficiently large lags (e.g. 100–300). The ACFs serve as the basis for the validation of potential reconstructions (see Section 2.4) and indicate, whether analysis disturbance is given. Moreover, following Kantz

8 For example, too much noise, leads to test rejections, disrupts the Grassberger-Procaccia algorithm (see [46, 47]), thus, the correlation dimensional estimates and alters the Lyapunov exponent calculations [63, 64].

9 Even if scientifically debatable, 'brute-force' methods such as logarithmic distances will provide sufficient results, since the sole purpose is analysis not forecasting, thus, no drawbacks are to be expected.

and Schreiber [45], nonzero autocorrelations are deemed problematic owing to trajectory vectors being closely located in phase space due to continuously evolving time, which is also known as temporal correlation. To determine a relevant 'scaling region' by the application of correlation sum schemes (see [46, 47]), the absence of temporal correlations is mandatory due to fitting issues in regional curve shapes and the lack of invariance of said correlation sums as depicted in [45]. Hence, dynamically correlated time-series violate the estimation requirements and if sufficiently large or worst oscillating, the analysis is futile [45].

To analyze temporal correlations, Provenzale et al. [71] propose estimates of the correlation time by applying time separation plots, presupposing pairs of points in phase space to be dependent on threshold distance and, additionally, on elapsed time in between respective measurements. Henceforth, the contour curves of said plots have to saturate and remain at an acceptable and non-oscillating boundary level [71]. The existence of a sufficient 'scaling region' is the premise for successful reconstructions [45]. Building upon the correlation sum scheme, determining the slopes of each correlation sum curve per selected embedding dimension results in an estimate of the (fractal) correlation dimension[10], which is plotted by itself and has to saturate as well (refer to [45]). Novelty within the framework to determine the validity of the underlying 'scaling region' is a step difference test proposed by Vogl and Rötzel [40], which tests step differences of the correlation sum curves in a Student's t-test against zero and graphically examines the resulting p-value heatmap. One can select the minimum embedding dimension by selecting the one, which has no p-value above 1% significance. Note that the existence of an ongoing 'scaling region' is also directly visible in the heatmap.

2.3 Chaos measures and tests

In addition to the prerequisites, several singular chaos metrics are worth determining to gather more initial insights into the potential underlying nature of the time-series dynamics under analysis [40]. First, the sample entropy as proposed in Richman and Moorman [51] is calculated, reflecting information content and self-similarity characteristics, thus, delivering insights into the presence of fractality within the data.

Furthermore, various Lyapunov exponents are determined, namely, (1) the maximum Lyapunov exponent, (2) the Lyapunov spectrum and (3) the Lyapunov time. Lyapunov exponents measure chaotic strength in a dynamical system by measuring the exponential convergence or divergence of nearby trajectories in phase space [45, 73]. It is possible to calculate Lyapunov exponents equaling the number of phase space dimensions, i.e. the number of the estimated embedding dimension, leading to the Lyapunov spectrum, which indicates the nature of the underlying dynamical systems, whether it be conservative or dissipative [74]. The largest exponent is labeled as the maximum Lyapunov exponent, depicting the exponential divergence or convergence of close trajectories and can be determined via the algorithm of Rosenstein et al. [75]. Note that a positive maximum Lyapunov exponent in combination with a negative Lyapunov spectrum sum is mostly seen as a sign of chaos, yet, is critiqued by the lack of distributional tests [11]. Therefore, a distributional rationale in form of the Bask-Gençay bootstrapping test is favorable, since it provides

[10] In finite scalars like time-series, according to Ramsey et al. [72], correlation dimensional estimates tend to return artificially smaller values than the theoretically assumable fractal dimension.

a significance indication, particularly, in cases of small positive, beyond zero maximum Lyapunov exponents at a sufficient level of significance [76]. Please note that several ten to a hundred thousand of bootstrapping steps are advisable to obtain reliable results. Moreover, the Lyapunov time represents the inverse of the maximum Lyapunov exponent and, thus, implies the time-span, the system requires to render itself chaotic and non-predictable, i.e. the time in which the exponential growing errors remain in a 'forecastable' range, before diverging too far [53]. The Lyapunov time is interpretable in either time-series units or in SI units [seconds] for real-world applications [53].

Finally, the Hurst exponent or in the case of non-stationary data, the detrended fluctuation analysis (DFA) alpha value is calculated to obtain in-depth information about the evolutionary nature of the dynamical system [54, 77, 78]. In an ongoing debate, the interpretation of the Hurst exponent and its initial interpretation by Benoit Mandelbrot (see [79, 80]) is challenged [16, 25, 44].

The Hurst exponent is interpreted as follows, namely, (1) the system is representing a Wiener process[11], should the Hurst exponent equal exactly 0.5, (2) the system is revealing long memory effects if it is exceeding 0.5 and (3) is being a mean-reverting system, should the exponent value be below 0.5 [16]. Nonetheless, recent empirical studies (refer to [16, 44]) state that the exceedance of 0.5 by the Hurst exponent reveals measurable fractal trends (or trending characteristics), which are an explicative rationale for momentum effects on financial markets. Within the setting of this analysis, the latter, novel indication is more suitable. The exceedance of 0.5 indicates persistency and the existence of a power-law, resulting in the denoted fractal characteristics [16, 44, 81]. Additionally, the Hurst exponent can be applied to determine the fractal dimensionality estimation (2-H) [82]. Finally, as an additional novelty, is adapting the Bask-Gençay test to the Hurst exponent as depicted in Vogl [44], ensuring the said exponent to be tested on significance [83]. In total, the second step enables the elucidation of the dynamical systems' properties directly, thus, providing a solid indication of its underlying evolutionary nature.

2.4 Phase space reconstruction

An important step toward the conduction of successful predictions of nonlinear time-series systems is the method of attractor reconstruction, leading back to the 1920s (refer to [84]) and the ideas of Packard et al. [85], Ruelle [86] and Takens [48], which represent the calculation of various invariant quantities required to characterize the underlying system [87]. This is mostly the presupposition for the nonlinear dynamical analysis of a time-series and state space model implementations [88]. The main contribution of reconstructions is given by the reconstruction of phase space, which is capable of preserving geometrical invariants (e.g. eigenvalues, fixed points or fractal dimension) of referring system attractors, including the Lyapunov exponent of according trajectories [88]. To phrase it differently, attractor reconstruction can be seen as a method to recreate the full deterministic state space based upon a lower dimensional time-series (i.e. a scalar) [87]. Thence, state space reconstruction is the generation of a multidimensional, deterministic state space out of the underlying, sampled time-series data [88]. Furthermore, embedding is, thus, the mathematical

[11] Only in this scenario the efficient market hypothesis taken out of quantitative finance holds and is violated else.

process by which an attractor is reconstructable presupposing a given set of scalar measurements, i.e. time-series datasets owing to dimensional preservation characteristics [43].

The resulting accuracy of the attractor reconstruction is directly dependent on the methodology applied to the reconstruction process and also influences the Lyapunov spectrum [87]. Therefore, several problems may occur, since the Lyapunov exponent cannot be labeled as invariant toward initial conditions, thus, stating a dependence on sample size within the reconstruction of time-series trajectories in phase space [11]. Following Nichols and Nichols [87], several methods for delay-time and embedding dimension selection exist for the standardized delay coordinate reconstruction, namely, the comparison between ACFs and the probabilistic concept of mutual information, while false nearest neighbor approaches are feasible to minimize said delay vectors. Nonetheless, the most common procedure is the delay-time reconstruction in combination with various embedding dimensions [89]. A non-exhaustive overview is proposed in **Table 1**. To be more detailed, the delay-time is defined as the time-span between two neighboring points applied to reconstruct the attractor, while the referring embedding dimension represents an estimate of the true dimension of the assumed phase space, which is intended to be reconstructed [97]. To refer back to the denoising presupposition given in Section 2.2, the underlying scientific theory requires noise-free data, on which natural processes timely evolve, which else leads to difficulties in state variable estimations [87].

According to Takens [48], in absence of noise contaminants, it is always feasible to embed a scalar time-series into a state space. Assuming the existence of noise, two trajectories of the same initial condition, potentially evolve differently and converge to different asymptotic behavior, thus, even the exact knowledge of said initial conditions does not guarantee the predictability of the system's final state [88, 98]. Therefore, noise has to be treated as an influential source of unpredictability, which cannot be fully disclosed via the deployment of conventional methodologies of nonlinear dynamical analysis such as exit bases or uncertainty exponents [98].

Algorithm	Short Description	Reference
TE	Takens' delay-time embedding, implies shifting a timely delayed comb through the data to generate 3D coordinates	[48]
SE	Spectral embedding in combination with a k-nearest neighbor algorithm, principal component analysis and Laplacian Eigenmaps	[10]
LRNN & CRBP	Locally recurrent neural networks with casual recursive backpropagation learning by applying algebraic observability through Takens' theorem	[90]
HH	The reconstruction is conducted by a combination of hyperhelices	[91]
ESN	The reconstruction is conducted by echo state networks	[92]
SVM	The reconstruction is conducted by support vector machines	[93]
HT	State space reconstruction by Hilbert transformations	[94]
MI	Mutual information in the probabilistic method	[95]
TM	Trajectory matrix for singular system analysis	[96]

Table 1.
Overview of existing reconstruction algorithms within the scientific literature.

To revert to practical implementations, it is relevant to determine the choice of delay- times (τ), which directly influences the success and accuracy of reconstruction algorithms [87]. Hence, a too small selection results in vectors to be very near and almost identical, thus, carrying redundant information and leading the attractor to collapse onto the 45° line in state space [87]. In contrast, a too large selection will produce uncorrelated (unrelated) coordinates owing to exponentially growing errors in chaotic regimes, resulting in decorrelated vectors in hindsight of the underlying time-series [87]. Henceforth, the two boundary scenarios have to be well-balanced to receive a proper reconstruction, yielding maximal independence, while preserving dynamically related coordinate properties [87]. The most commonly applied variant is the ACF delay, with several possibilities, namely, (1) the first zero crossing, (2) crossing of 0.1, or 0.5 and (3) not exceeding 1/e [87]. Please note that ACFs propose linear time evolutionary calculations and may, thus, be misleading [87]. Due to experiments, the most accurate representations by the author were achieved by selecting variant (1), i.e. the first zero crossing or in modification, the first zero crossing, while subsequent coefficients additionally stay insignificant. Moreover, the embedding dimension as shown by Sauer et al. [99] has to be topologically equivalent to the true attractor, if the embedding dimension is chosen to be larger than two times the fractal dimension of said attractor. Note that once the embedding dimension is selected sufficiently high, a reconstruction resembles almost always an embedding, independent of parameter selections [88]. Mostly, delay-coordinates are selected, yet, there exist the families of derivatives and principal component reconstructions, as depicted combinatorial in the spectral embedding (see [10]) [88]. Within practical applications, the author deems a combination of (1) Takens delay-time embedding, which, unfortunately, resembles a 'spaghetti monster' in most cases and (2) the more sophisticated variant by Song et al. [10], applying a spectral embedding in combination with a k-nearest neighbors algorithm (k-NN), principal component analysis (PCA) and Laplacian Eigenmaps as very suitable. In the author's empirical experiments, the PCA components are best selected to equal the embedding dimension, while the number of neighbors for the k-NN can be best determined by the following heuristic, namely, 0.01len(data)*1.5τ [40]. Moreover, to receive a correct reconstruction the properties stated in **Table 2** are strict to be adhered to.

Parameter	Favorable	Negative	Impact
Stationarity	stationary	non-stationary	very high
Nonlinearity	nonlinear	linear	medium
ACFs	not significant	significant	high
Space-Time Separation	very low, non-oscillating	high, oscillating	high
Correlation Sum	significant 'scaling region'	no 'scaling region'	very high
Correlation Dimension	saturating	not saturating	high
Maximum Lyapunov Exponent	significant and positive	negative	high
Lyapunov Spectrum Sum	negative	positive	high

Table 2.
Overview of parameter selections for attractor reconstruction specifications.

2.5 Recurrence quantification analysis

Independently from previous attractor reconstruction and other prerequisites, the RQA bases itself upon the introductory denoted exploitation of the recurrence property[12] of dynamical systems, thus, is applicable to any time-series data (e.g. [6]) [100]. The RQA is conducted by quantifying the recurrence plot (RP) as introduced in Eckmann et al. [101]. The RP and RQA analysis benefit from the preservation[13] of time-ordering information contents given in the analyzed data as well as contained spatial structure [22]. With the RQA one may detect fundamentally given characteristics underlying a dynamical system, namely, the recurrence states, resulting in a 'robust to noise and data limitations' method of quantifying and identifying (chaotic) dynamics [22]. Thus, respective trajectories and transitions are rendered visible, in combination with the degree of complexity, i.e. the fractal structures, which may be inherent in the analyzed data [8]. To determine the RQA a threshold level has to be decided on, which determines whether nearby points are counted as recurrent or not [7]. Following van den Hoorn et al. [102], propose several threshold determination methods, yet, traditionally, according to Koebbe and Mayer-Kress [103] as well as Zbilut and Webber Jr. [104], the threshold value should not exceed 10%. Additionally, the threshold value should not be lower than five times the sample noise [6]. Furthermore, it is common to exclude the identity line (proportional to the maximum Lyapunov exponent) of the RP from the analysis [105]. First, the RPs can be interpreted visually, which is presented in Marwan et al. [6], p. 251. Second, there exist two different types of measures taken out of a RQA, namely, minima-dependent versus single-value measures [6]. One may plot the minima dependency for several selections and choose an appropriate value to quantify the RPs. The length of diagonal lines represents the duration of trajectory local evolutions, while vertical (horizontal) lines mark time durations, in which the underlying dynamics are trapped (labeled as intermittency or laminar state) [7]. The commonly applied measures are depicted in Vogl and Rötzel [40], **Table 3**. Finally, to distinguish the results from stochastic, chaotic or other systems, one may either apply a Wiener process realization, a mathematical chaotic system realization or respective surrogate [14] datasets (see [45]). Paired with the conceptions taken out of Section 2.6, signal theoretical decompositions can be applied to identify potential hidden sub-dynamics [56]. To gather and obtain the most information out of the analyzed data signal, wavelets with better localization properties are commonly proposed in form of a DWT filter bank [44]. The resulting low pass and high pass decompositions can then be applied as novel datasets to the RQA analysis and sub-RPs can be created and quantified to demonstrate potential sub-dynamics [56]. Note that the process can be repeated as often as required, should more than residual noise remain after one respective decomposition or iteration.

2.6 Multi-resolution analysis

To be very brief, time-series data are localized in the time domain, yet, may also yield exploitable frequency components, which in case of non-stationarity or

[12] The recurrence property originates from a topological approach and is given by the Poincaré recurrence theorem.

[13] Presupposing the existence of a low-dimensional attractor, presence of dependence on initial conditions and the manifestation of said recurrence property.

[14] I.e. destroying given determinism by shuffling via FTs. Then, comparison with original data.

Parameters	Generic	Mars	S&P500	Implication
Denoising	no	yes	yes	Denoising with level 6 cascade wavelet filter and 'Bior 2.8' wavelet, since 'biorthogonal' states the best localization characteristics for non-tailored wavelet functions.
Stationarity	yes	yes	yes	ADF and KPSS (c, ct) tests with 1% significance each
Gaussianity	no	no	no	KS test with 1% significance
Nonlinearity	yes	yes	no	BDS test for embedding dimension of five and 1% significance
Correlation Sum	scaling	no scaling	no scaling	Build upon correlation sum graphs and step-test with 1% significance
Correlation Dimension	saturating	dropping	not saturating	Based upon different correlation dimensions for embedding dimension
ACFs	very low	depending on lag	very low	ACFs with 300 lags and 1% significance
Temporal Correlations	none	very strong	oscillating	Graphical via space-time separation plots for min. 100 steps
Sample Entropy	very low	low	very low	Very low entropy is seen as a sign of self-similarity in terms of information contents
Maximum Lyapunov	low positive	low positive	low positive	Tested for significance via Bask-Gençay test to propose the distributional theory
Lyapunov sum	negative	negative	negative	Negative sum indicates dissipative system in combination with maximum Lyapunov exponent positivity
Hurst Exponent	trending	trending	mean reverting	Trending: Hurst >0.5; mean reverting: Hurst <0.5
Result	**positive**	**negative**	**negative**	**Reconstruction invalid for real-world datasets, even if (non)linear dissipative systems**

Table 3.
Overview of steps one and two for all datasets with implication.

non-periodicity or other unfavorable characteristics will not be extractable via classical Fourier transformations (FT) [13, 106, 107]. Therefore, wavelets (i.e. tailored or bi-orthogonal) applied in a multi-resolution analysis (MRA) are well suited to extract underlying frequency information by retaining as much time localization information as possible [13, 16]. Thus, for filtering or denoising activities of time-series, discrete cascade filter banks with wavelet shrinkage (see [55, 108]) are applicable at length for various scales [40]. Moreover, to obtain insights into time-frequency localizations of to-be-analyzed datasets, one may apply a continuous wavelet transformation (CWT), resulting in a spectrum [57].

2.7 Distributions and power-laws

To elaborate on power-law characteristics, it is important to denote the interconnections between chaotic dynamics, strange attractors, fractals and power-laws. In short, a dissipative (chaotic) dynamical system will reveal its phase space over timely evolution to deflate onto its own strange attractor, which is characterized via a fractal set [109, 110]. Generally, a fractal set yields a non-integer (non-Euclidean, thus, generalized) dimension, namely, the Hausdorff-Besicovitch dimension and is further characterized via self-similarity, i.e. multi-scaling, in addition to irregularities, non-differentiability and recursiveness [111]. Henceforth, a (multi) fractal system requires a local power-law contributing to the mentioned scaling properties [111]. Therefore, a power-law is defined as the scalar relationship between two quantities and, thus, is characterized via scale invariance [112]. A fractal system with one scaling exponent is labeled monofractal, yet, multifractal systems require a singularity spectrum of exponents [111]. Referring back to a dissipative dynamical system, which deflates onto its strange attractor, thus, is represented by a fractal set. The fractal set of a strange attractor is rendered visible via its Poincaré sections, which show intersections of said strange attractor [110, 111]. To be more detailed, the intersections of strange attractors are fractal sets, which are described via multi-scaling and, thus, via powerlaws [111].

Analyzing time-series enables not only the reconstruction of potential (strange) attractors, yet, opens the way to mathematically determine given power-laws (i.e. the multi-scaling characteristics) of its underlying (multi) fractal properties [113]. Following Yuan et al. [114], state two rationales for time-series multifractality, namely, (1) the existence of fat-tailed probability distributions and (2) nonlinear temporal correlations. To draw out the multifractal spectrum, one may apply a multifractal analysis, which is built upon the MFDFA. The MFDFA visually depicts the scaling properties, as well as the (local) maximum and minimum Hurst exponents, also supporting the fractal trending interpretation discussed earlier [60]. Moreover, the generalized Hurst exponents, multifractal scaling exponents and the previously denoted multifractal scaling spectrum can be derived from the MFDFA output quantities [60]. In addition, calculating complementary cumulative distribution functions (CCDFs) and comparing them with power-law or other potential theoretical distribution types, enables the more or less save determination of power-law or other distribution fits [61]. However, as a word of absolute caution, the determination of coherence tests for various distributions has to be interpreted very carefully. The coherence tests are calculated via paired distributional fitting comparisons based upon log-likelihood measures, alongside other parameters [61]. These serve the purpose of achieving insights into suitable distributions, which may describe the datasets best, or to phrase it realistically, which at least represent the 'less worst' fit [61]. The coherence tests, thus, represent a comparison and no goodness of fit, which as indicated requires the reader to exert special care with the interpretation. It is advisable to fall back on graphical displays on log-log plots, which revealed as a useful guide for the practical implementations of the author. Concluding the powerlaws, the analysis is complete and the interpretation can carefully be exerted.

3. Correct empirical specifications

For each step of the analysis several algorithms are to be determined and a larger variety yield graphical insights, which can be either quantified or applied as a visual aid to deduce further insights and implications. Since a complete analysis as shown in Vogl and Rötzel [40] or Vogl [44] would vastly exceed the page limitations of this guide, the didactics of the practical display are as follows. First, this section will provide an idealistic outcome of a generic and mathematically tailored time-series, based upon the Lorenz system (refer to [115]) to demonstrate the 'best-case' scenario as the generalized point of reference, while two additional real-world datasets are presented as a comparison, namely, (1) the change rate of wind speed of Mars and (2) S&P500 1-minute tick return data. Second, the real-world datasets will be elaborated on in Section 4, since some hindrances are given and, thus, require analysis of potential misspecifications. The preliminary elaborations via steps one and two are depicted in **Table 3**, while **Figure 2** presents the correlation sum, the correlation dimensional scheme as well as the correlation structures for the generic dataset. Note that for attractor reconstructions, **Table 2** already proposes the favorable characteristics to enable the correct implementation of reconstruction algorithms. Steps three

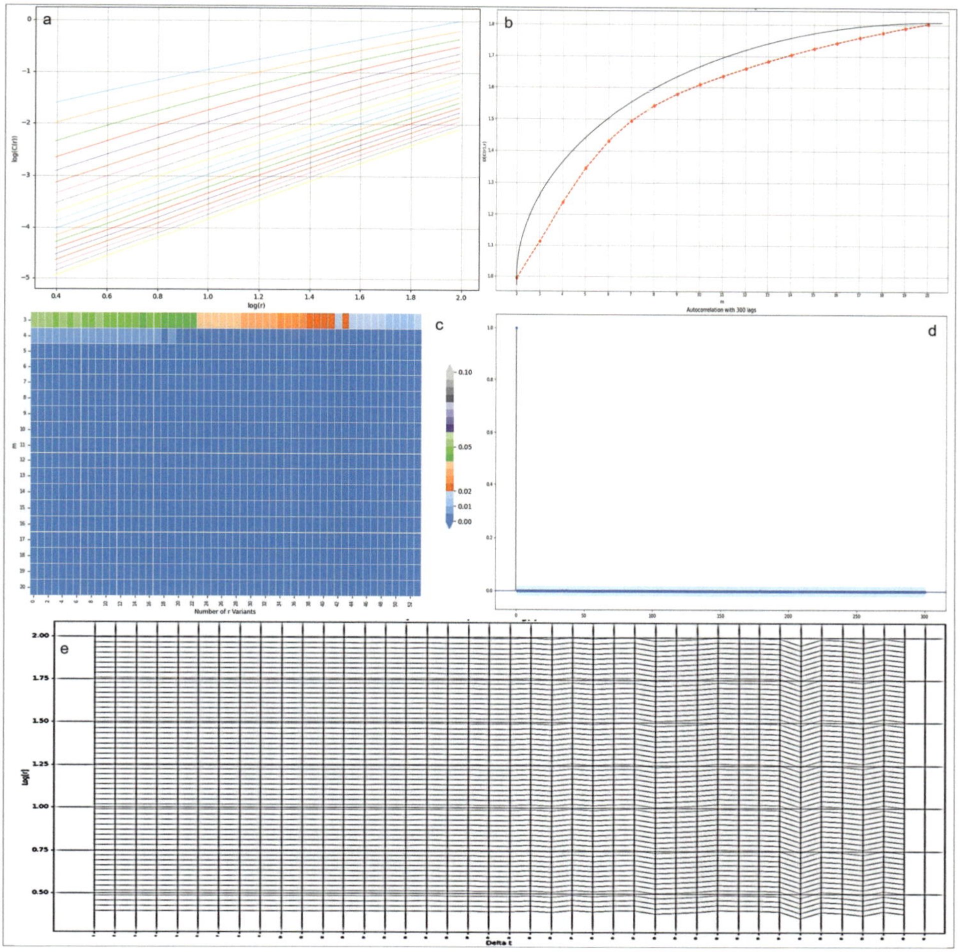

Figure 2.
Generalized point of orientation via the generic dataset for determination of reconstruction possibility. (a) Shows the correlation sum plots, which visually depict 'scaling regions', (b) shows the correlation dimensional plot, which ideally saturates as shown, (c) states the heatmap of p-values for the step-test of the lines of (a), (d) states the ACFs for 300 lags, which are insignificant and (ideally) stay that way and (e) shows the space-time separation plot for 100 steps, which is very low and non-oscillating. Note that for step two of the analysis, the minimum embedding dimension can be taken out of the heatmap, namely, by picking the first row with only 1% significance or lower p-values. During the existing experiments of the author, the heuristic of two times the fractal dimension as stated in the main text is also given by applying this selection method.

to six are only stated in Section 4 for the real-world examples. In general, the stricter the interpretation and analysis, the better the results of the reconstruction, correct specification of underlying empirical DGPs and subsequent modeling.

Moreover, while steps one and two as presented above (in addition to potential reconstructions of step three), suffice to quantify the nature of the underlying system (i.e. whether it is dissipative, reconstructable or potentially chaotic), regarding the analysis of steps four through six, provide the exact quantified details of the system's characteristics and serve as double confirmation procedure. Mainly, the RQA measure quantification provides exact details about the underlying empirical DGP, namely, a percentage comparison with surrogate data or pure stochastic (e.g. a Wiener process) or pure deterministic chaotic systems (e.g. a Lorenz system) (refer to [6, 40]). Thence, one is capable of pinpointing whether the underlying system is pure chaotic, pure stochastic or a mixture of both and in which margins [40]. Subsequently, deriving frequency information, information about sub-dynamics, the existence of power-laws and multifractal spectra enables the correct model selection as final outcome of the quantification. Nonetheless, a proper differentiation can be achieved after step four already. Since the RQA is too spacious it is neglected for this display, thus, the reader is referred to [6, 8, 40, 56].

4. Empirical negative examples

Continuing the previous section, hereinafter, the results for the two real-world datasets are presented in **Figures 3** and **4**, while the remaining indications are provided in **Table 4**. The invalid reconstruction algorithm results via Takens delay-time embedding (refer to [48]) and via spectral embedding (refer to [10]) are proposed in **Figure 5** to clarify the relevance of proper prerequisites analysis. Regarding the Mars wind speed change rates, clear deterministic traits and sub dynamics are observable by the RQA, yet, a clear identity as a chaotic system as well as a distinct reconstruction is not possible. This is illustrated by the lack of a clear 'scaling region', dropping correlation dimensions, high ACFs and temporal correlations, which render this analysis step invalid. Furthermore, no scale independent multifractal scaling spectrum is visible and a nested power-law-exponential distribution is proposed as 'less worst' distribution via the coherence tests (see [61]). In addition, no frequency information is determinable via CWT. Thus, the only insight generated is that it is a potentially chaotic, deterministic and dissipative system, while the exact modeling metrics are extractable out of the RQA quantification tables (refer to [6]). Regarding the (invalid) Takens reconstruction may suggest a non-chaotic attractor, since the results resemble a valid one in parts, yet, this is an invalid approach nonetheless. Spurious chaotic measure results are obtained by the S&P500 1-minute return series, since according to the BDS test nonlinearity is excluded, while the Hurst exponent indicates a clear mean-reversion. Furthermore, no 'scaling region' by correlation sums and a non-saturating correlation dimension in combination with oscillating temporal correlation voids any other step of the analysis or reconstruction. Regarding the reconstruction by Takens, the linear nature is determinable. Moreover, the system has frequency information, yet, no power-law nor multifractal scaling characteristics (in agreement with the Hurst exponent indication of mean-reversion). Following the RQA, sub dynamics and low levels of determinism are given, while vast stochastic characteristics are dominant. A final concluding remark at this point, considers

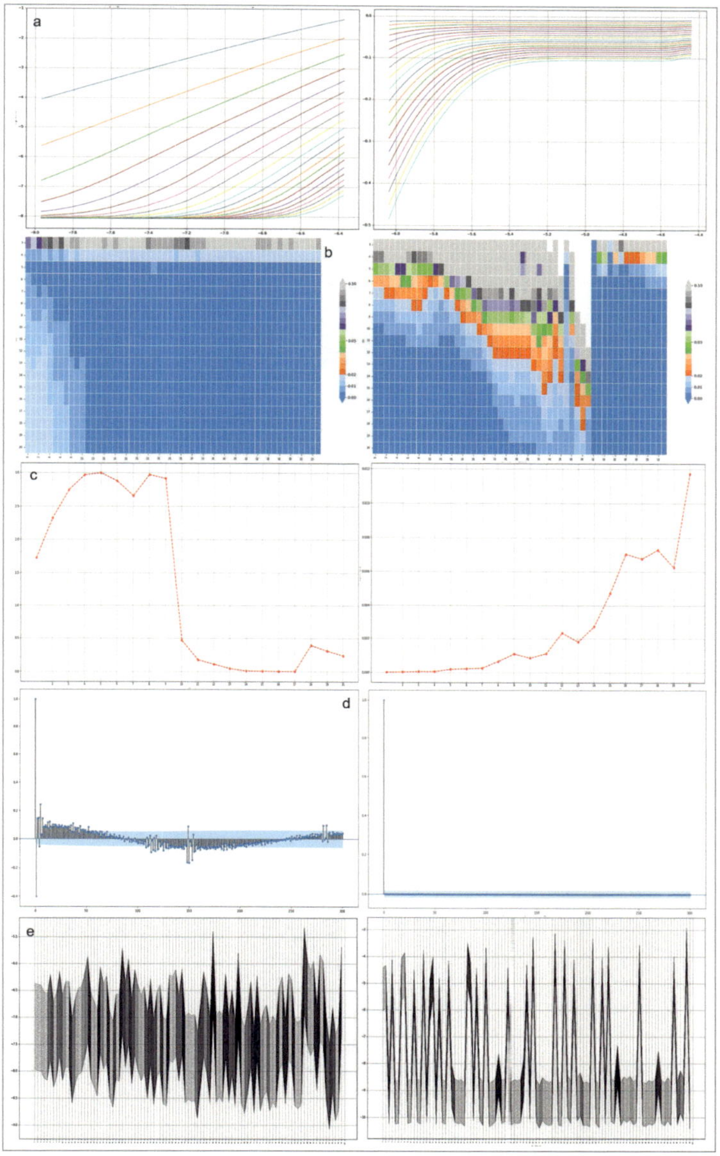

Figure 3.
Results for the dataset wind speed change rate Mars (left) and S&P500 1-minute ticks (right). (a) States the correlation sum scheme, (b) the p-value heatmap, (c) the correlation dimensions, (d) the ACFs for 300 lags and (e) the space-time separation plots with 100 steps. Comparing with the generic datasets visually already reveals the conceptual differences and problems inherent in the analyzed data.

the frequency of the data samples, namely, Vogl and Rötzel [40] observed chaos in daily S&P500 returns, while in S&P500 1-minute tick return data, mean-reversion is present, leading to insights proposed by BenSaïda [39], namely, that the same system at different frequency levels, may propose different dynamics, revealing a scale dependence of the underlying empirical DGP.

Referring back to step three, namely, the attractor reconstruction, one may see various outcomes based upon false pretenses in the reconstruction results. In terms

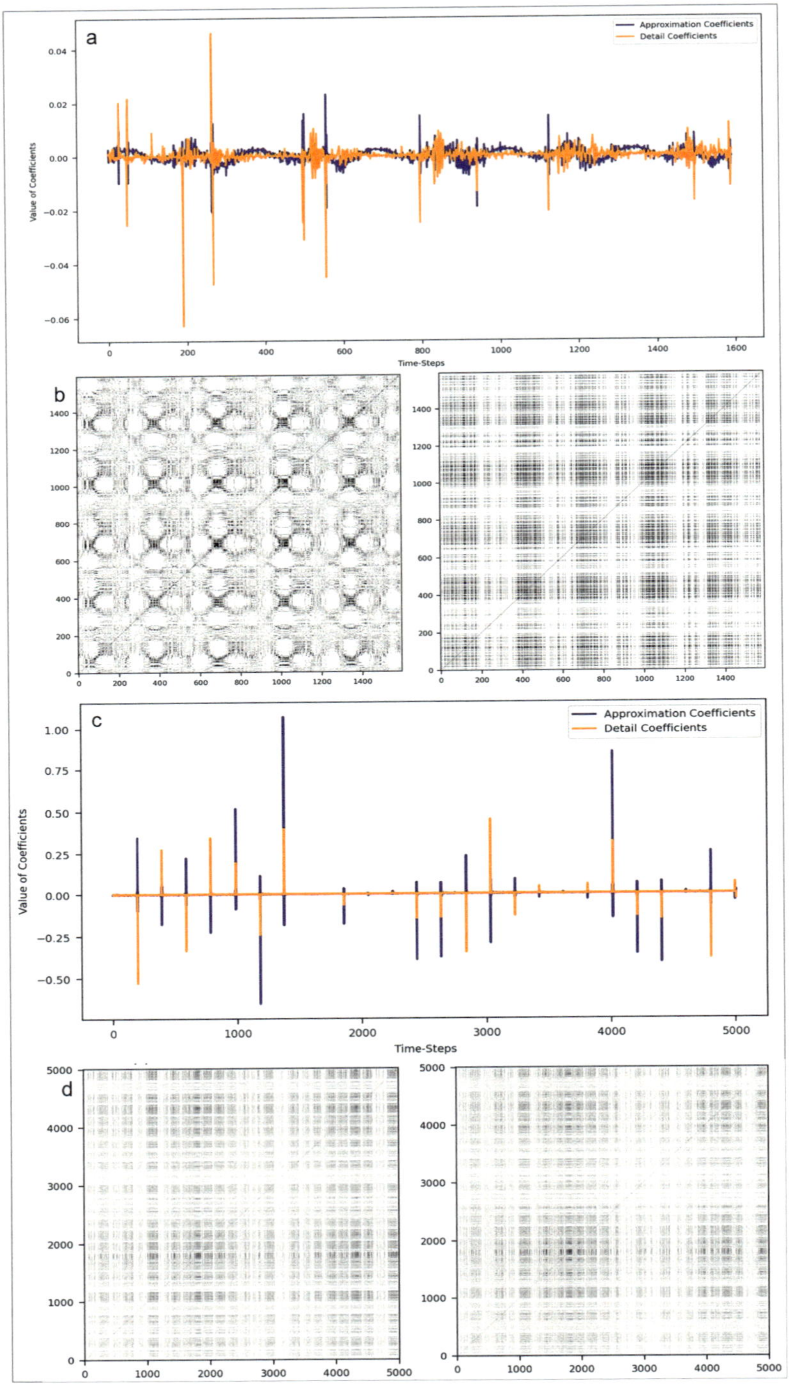

Figure 4.
Overview of RQA-DWT results for the display of RPs and sub-dynamics. (a) Is the DWT for the wind speed change rates of Mars and (c) for the S&P500 1-minute return ticks. In addition, (b) represents the approximation (left) and detail (right) coefficients for (a), while (d) represents the same for (c). It is denotable that both series consists of sub-dynamics. Note that the Mars detail coefficients may resemble a hidden chaotic subsystem, which can be separately analyzed.

Parameters	Generic	Mars	S&P500	Implication
RQA	strong determinism, chaotic	determinism	low determinism, stochastic	RQA measures versus comparative data provide characteristics for quantitative modeling
Sub-Dynamics	no	yes	yes	DWT with one iteration based upon a 'Bior 2.8' wavelet filter bank, the resulting high and low pass data are inserted into RQA
Frequency	yes, distinct	no	yes, recurrent	Based upon a CWT with a 'Shannon' wavelet, 1024 scales
Multifractal	yes, clear	scale-dependent	no	Based upon a MFDFA analysis
Power law	yes	nested exponential with fat tails	no	Based upon CCDFs versus theoretical power-law backed with coherence tests on 1% significance
Result	**Chaotic, Multifractal System**	**Stochastic, Deterministic System**	**Mean Reverting system**	**Very careful interpretation advisable**

Table 4.
Overview of steps four to six for all datasets with implications.

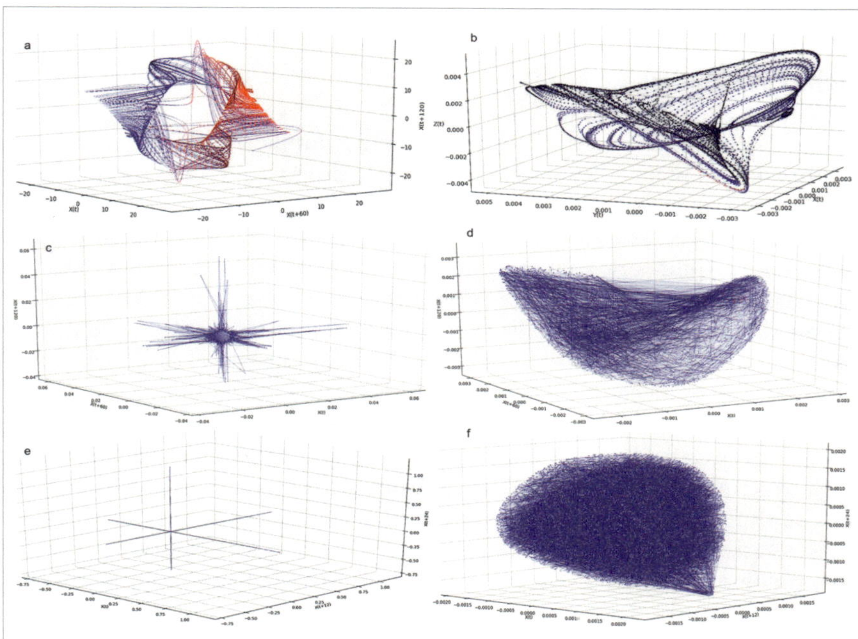

Figure 5.
Display of attractor reconstructions for the generic dataset based upon a Lorenz set (a, b), the wind speed change rates of Mars (c, d) and the S&P500 1-minute return ticks (e, f). (a), (c) and (e) represent the Takens delay-time embedding, yet, (c) and (e) are proven to be not reconstructable. (b), (d) and (f) display the spectral embedding in combination with a k-NN algorithm and a PCA with Laplacian Eigenmaps. As with (c) and (e), note that the analysis shows that (d) and (f) are not to be reconstructable. It is visible that a violation of reconstruction prerequisites results in very poor reconstructions since those are not to be conducted in the first place.

of extreme high ACFs or temporal correlations, the attractor is dispersed and flattened, while a lack of scaling characteristics results in singular 'spaghetti like' lines. Furthermore, as stated in Nichols and Nichols [87], a deflation or stretching on the 45° line of the 3D space is also possible. Note that in the case of a linear system such as the S&P500 1-minute tick returns, the Takens embedding only states straight lines, which clearly indicates the absence of nonlinearity. A proper reconstruction shows a closed and dense area and visible attractor-like structures. For reference, as stated in Vogl and Rötzel [40], a pure stochastic system such as a Wiener process will end up resembling a 'ball' with no trajectory structure. In regards to time-series data with higher dimensional estimates, exceeding 3D spaces, the reconstructed graphical display may appeal 'deformed', owing to the lack of degrees of freedom in the visual display. On a final note, the exertion of particular care regarding the prerequisites of the reconstruction is highly advised, since violations result in poor representations and false characteristics, which will build the groundwork for subsequent quantitative modeling attempts. Furthermore, it is advisable to alter the delay-times and dimension estimates in several iterations to be sure to hit the most 'representable' form of the time-series system under analysis, especially, in more complex applications such as spectral embedding. Finally, the framework only provides the most basic intuitions or the minimum set of knowledge for analysis to be possible at all, refinements are always encouraged. Taken together, the stated insights can be abstracted into a minimum set of requirements, which have to be fulfilled by potential model selections. Furthermore, one may reapply the whole analysis on the DWT sub-dynamics series to elaborate on potential hidden (strange) attractors.

5. Concluding remarks

Within this chapter, a practical guideline for the complete implementation of a combinatory, chaos analysis framework separately proposed in Vogl and Rötzel [40] for stationary and in Vogl [44] for non-stationary data is presented in its entirety. The framework is proposed as an integrated, holistic approach to analyzing the empirical DGP of nonlinear time-series data and provides the possibility to distinguish chaoticity from stochasticity while referring to underlying evolutionary dynamics. The analysis steps are elucidated, potential pitfalls and theoretical rationales stated and prerequisites discussed in detail. Moreover, an 'idealistic' versus 'negative' case is empirically and graphically introduced and debated based upon real-world time-series sets [15]. With this guide, the reader should be able to conduct the analysis themselves, without being prone to misspecifications and common errors present in the scientific literature.

Lastly, concluding remarks and current frontiers in the elaborated context are briefly to be stated. Current gaps in research and frontiers on the reconstruction of attractors is vastly seen in the application of neural network, evolutionary algorithms and other reconstruction methodologies to obtain sufficient and high-quality reconstructions and analysis insights (see [116, 117]). Nonetheless, the research field of time-series reconstruction and quantification of empirical DGPs is scarce and defined as a current gap in research, particularly, in hindsight of novel technological advancements such as artificial intelligence solutions. To conclude, Nieto et al. [98], states

[15] Even if not displayed in this chapter, during the preparation period, several different time-series have been analyzed, e.g. flood and river discharge series, wind power, energy prices, tweet-frequencies, nonlinear fluids and fundamental economic indicators, among others.

unpredictability as a 'fundamental topic' in the nonlinear scientific domain, owing to its consequences being rooted in the existence of sensitivity to initial conditions as the main trait of chaotic dynamics. Furthermore, no common understanding of unpredictability exists, since differing definitions may be applicable, e.g. problems in predicting trajectory evolutions may not be seen as a problem in hindsight of scattering problems, which only level around asymptotic behavior, thus, define problems only in the prediction of final system states [98]. Furthermore, predictability in subsequently implemented models is a vast topic, which is neglected for the discussion of this chapter, yet, deemed of utmost relevance to it.

The stated framework can be enhanced further and shows several limitations, namely, it is computationally expensive and consists of many algorithms and methods, which are time intensive. Moreover, the selected methods are chosen due to their vast application in the scientific literature and not on performance. Hence, no optimization has been conducted yet, owing to the goal dependence of the analysis framework, even if applicability to various time-series is given. Furthermore, there exists no way to resolve attractor reconstructions given the existence of high ACFs and high or oscillating temporal correlations. Moreover, the framework is graphically reliant, which is seen as a potential hindrance in terms of future automatization and application on larger data pools and automated decision rule generations. Nonetheless, to conclude, the presented framework is seen as the fundamental basis or minimal building block for future research, i.e. as the provision of a stepping stone toward more advanced, transparent and reliable insights originating from the scientific nonlinear dynamics community. The enablement to safely distinguish chaoticity from stochasticity paired with the detailed characterization of the empirical time-series DGP, in general, is expected to have a positive influence on the quantification, modeling and the future prospects of the field, solving a 40-year-old debate. Resolving stated debate, hopefully, opens the way to more coherent insights and persistent knowledge about time-series systems and the quantification of the real-world in various disciplines such as medicine, hydrology, economics and physics. The inherent paradigm shift is also expected to make model selection easier and more self-explanatory in the future of time-series predictions.

Author details

Markus Vogl
Executive Management, Markus Vogl {Business & Data Science}, Wiesbaden, Germany

*Address all correspondence to: markus.vogl@vogl-datascience.de

References

[1] Gaul V and Kumar V. Allied Market Research. 2020. [Online]. Available: https://www.alliedmarketresearch.com/predictive-analytics-market. [Accessed May 27, 2022].

[2] Markets and Markets. Markets and Markets. 2021. [Online]. Available: https://www.marketsandmarkets.com/Market-Reports/predictive-analytics-market-1181.html. [Accessed May 27, 2022].

[3] Aguilar-Rivera R, Valenzuela-Rendón M, Rodríguez-Ortiz J. Genetic algorithms and Darwinian approaches in financial applications: A survey. Expert Systems with Applications. 2015;**42**:7684-7697

[4] Poornima S, Pushpalatha M. A survey on various applications of prescriptive analytics. International Journal of Intelligent Networks. 2020;**1**:76-84

[5] Dudkowski D, Jafari S, Kapitaniak T, Kuznetsov N, Lenonov G, Prasad A. Hidden attractors in dynamical systems. Physics Reports. 2016;**637**:1-50

[6] Marwan N, Romano MC, Thiel M, Kurths J. Recurrence plots for the analysis of complex systems. Physics Reports. 2007;**438**:237-329

[7] Marwan N, Wessel N, Meyerfeldt A, Schirdewan A, Kurths J. Recurrence plot based measures of complexity and its application to heart rate variability data. Physical Reviews E. 2002;**66**(2):026702

[8] Marwan N, Kurths J. Line structures in recurrence plots. Physical Letters A. 2005;**336**(4-5):349-357

[9] Guégan D, Leroux J. Forecasting chaotic systems: The role of local Lyapunov exponents. Chaos, Solitons and Fractals. 2009;**41**:2401-2404

[10] Song X, Niu D, Zhang Y. The chaotic attractor analysis of DJIA based on manifold embedding and Laplacian Eigenmaps. Mathematical Problems in Engineering. 2016;**4**:1-10

[11] Fernández-Rodríguez F, Sosvilla-Rivero S, Andrada-Félix J. Testing chaotic dynamics via Lyapunov exponents. Journal of Applied Econometrics. 2005;**20**:911-930

[12] Adams Z, Füss R, Glück T. Are correlations constant? Empirical and theoretical results on popular correlation models in finance. Journal of Banking & Finance. 2017;**84**:9-24

[13] Alexandridis AK, Kampouridis M, Cramer S. A comparison of wavelet networks and genetic programming in the context of temperature derivatives. International Journal of Forecasting. 2017;**33**:21-47

[14] Shi Y, Ho K-Y. Long memory and regime switching: A simulation study on the Markov regime-switching ARFIMA model. Journal of Banking & Finance. 2015;**61**:189-204

[15] Kristoufek L. Multifractal height cross-correlation analysis: A new method for analyzing long-range cross correlation. Europhysics Letters. 2011;**95**(6):68001

[16] Berghorn W. Trend Momentum. Quantitative Finance. 2015;**15**:261-284

[17] Ramiah V, Xu X, Moosa IA. Neoclassical finance, behavioural finance and noise traders: A review and assessment of the literature. International Review of Financial Analysis. 2015;**41**:89-100

[18] Celeste V, Corbet S, Gurdgiev C. Fractal dynamics and wavelet analysis:

Deep volatility and return properties of bitcoin, Ethereum and ripple. The Quartely Review of Economics and Finance. 2019;**76**:310-324

[19] De Luca G, Dominique G, Giorgia R. Assessing tail risk for nonlinear dependence of MSCI sector indices: A copula three-stage approach. Finance Research Letters. 2019;**30**:327-333

[20] Beltratti A, Stulz RM. Why is contagion asymmetric during the European sovereign crisis? Journal of International Money and Finance. 2019;(99-C):102081

[21] Charfeddine L. True or spurious long memory in volatility: Further evidence on the energy futures markets. Energy Policy. 2014;(71-C):76-93

[22] Barkoulas JT, Chakraborty A, Ouandlous A. A metric and topological analysis of determinism in the crude oil spot market. Energy Economics. 2012;**34**:584-591

[23] Matilla-García M, Queralt R, Sanz P, Vázquez F. A generalized BDS statistic. Computational Economics. 2004;**24**:277-300

[24] Sandubete JE, Escot L. Chaotic signals inside some tick-by-tick financial time series. Chaos, Solitons and Fractals. 2020;**137**:109852

[25] Vogl M. Controversy in financial chaos research and nonlinear dynamics: A short literature review. Chaos, Solitons and Fractals. 2022;**162**:112444

[26] Çoban G, Büyüklü AH. Deterministic flow in phase space of exchange rates: Evidence of chaos in filtered series of Turkish lira-Dollar daily growth rates. Chaos, Solitons and Fractals. 2009;**42**(2):1062-1067

[27] Eckmann J, Ruelle D. Ergodic theory of chaos and strange attractors. Reviews of Modern Physics. 1985;**57**(3):617-656

[28] Devaney R. An Introduction to Chaotic Dynamical Systems. Cambridge: Addison Wesley; 1989

[29] BenSaïda A, Litimi H. High level chaos in the exchange and index markets. Chaos, Solitons and Fractals. 2013;**54**:90-95

[30] Abarbanel H, Brown R, Sidorowich J, Tsimring L. The analysis of observed chaotic data in physical systems. Reviews of Modern Physics. 1993;**65**:1331

[31] Fuh C-C, Tsai H-H, Yao W-H. Combining a feedback linearization controller with a disturbance observer to control a chaotic system under external excitation. Communications in Nonlinear Science and Numerical Simulation. 2012;**17**:1423-1429

[32] Sornette D. Critical Phenomena in Natural Sciences: Chaos, Fractals, Selforganization and Disorder: Concepts and Tools. Heidelberg: Springer Verlag; 2004

[33] Rössler O. An equation for hyperchaos. Physics Letters A. 1979;**71**:155-157

[34] Ma C, Wang X. Hopf bifurcation and topological horseshoe of a novel finance chaotic system. Communications in Nonlinear Science and Numerical Simulation. 2012;**17**:721-730

[35] Gao Q, Ma J. Chaos and Hopf bifurcation of a finance system. Nonlinear Dynamics. 2009;**58**:209

[36] Dechert WD, Gençay. The topological invariance of Lyapunov exponents in embedded dynamics. Physica D. 1996;**90**:40-55

[37] Jahanshahi H, Yousefpour A, Wei Z, Alcaraz R, Bekiros S. A financial

hyperchaotic system with coexisting attractors: Dynamic investigation, entropy analysis, control and synchronization. Chaos, Solitons and Fractals. 2019a;**126**:66-77

[38] Brock W, Dechert W, Scheinkman J, LeBaron B. A test for independence based on the correlation dimension. Econometric Reviews. 1996;**15**:197-235

[39] BenSaïda. Noisy chaos in intraday financial data: Evidence from the American index. Applied Mathematics and Computation. 2014;**226**:258-265

[40] Vogl M, Rötzel PG. Chaoticity versus stochasticity in financial markets: Are Daily S&P 500 return dynamics chaotic? Communications in Nonlinear Science and Numerical Simulation. 2022;**108**:106218

[41] Aguirre LA, Billings S. Identification of models for chaotic systems from noisy data: Implications for performance and nonlinear filtering. Physica D. 1995;**85**:239-258

[42] Kyrtsou C, Labys WC, Terraza M. Noisy chaotic dynamics in commodity markets. Empirical Economics. 2004;**29**:489-502

[43] Kostelich EJ. The analysis of chaotic time-series data. Systems & Control Letters. 1997;**31**:313-319

[44] Vogl M. Hurst Exponent Dynamics of S&P 500 Returns: Implications for Market Efficiency, Long Memory, Multifractality and Financial Crises Predictability by Application of a Nonlinear Dynamics Analysis Framework. Working Paper SSRN, Under Review. 2022. Available from: https://papers.ssrn.com/sol3/papers.cfm?abstract_id=3838850

[45] Kantz H, Schreiber T. Nonlinear Time Series Analysis. Cambridge: Cambridge University Press; 2003

[46] Grassberger P, Procaccia I. Characterizsation of strange attractors. Physica Review Letters. 1983a;**50**:346-394

[47] Grassberger P, Procaccia I. Measuring the strangeness of strange attractors. Physica 9D. 1983;**9**(1-2):189-208

[48] Takens F. Detecting strange attractors in fluid turbulence. In: Rand D, Young L-S, editors. Dynamical Systems and Turbulence. Berlin: Springer; 1981. pp. 366-381

[49] Cencini M, Cecconi E, Vulpiani A. Chaos from simple models to complex systems. World Scientific. 2010;**17**:1-480

[50] Tirandaz H, Aminabadi S, Tavakoli H. Chaos synchronization and parameter identification of a finance chaotic system with unknown parameters, a linear feedback controller. Alexandria Engineering Journal. 2018;**57**:1519-1524

[51] Richman J, Moorman J. Physiological time-series analysis using approximate entropy and sample entropy. American Journal of Physiology - Heart and Ciculatory Physiology. 2000;**278**(6):H2039-H2049

[52] Park JY, Whang Y-J. Random walk or chaos: A formal test on the Lyapunov exponent. Journal of Econometrics. 2012;**169**:61-74

[53] Shevchenko II. Lyapunov and diffusion timescales in the solar neighborhood. Working Paper with arXiv-ID: 1012. 2018;**3606v2**:1-22

[54] Hurst H. Long-term storage capacity of reservoirs. Transactions of the American Society of Civil Engineers. 1951;**116**:770

[55] Sundararajan D. Discrete Wavelet Transform - a Signal Processing Approach, India. Singapore: John Wiley & Sons; 2015

[56] Chen Y, Yang H. Multiscale recurrence analysis of long-term nonlinear and nonstationary time series. Chaos, Solitons and Fractals. 2012;**45**(7):978-987

[57] Mallat S. A theory for multiresolution signal decomposition: The wavelet representation. IEEE Transactions on Pattern Analysis and Machine Intelligence. 1989;**11**(7):674-693

[58] Mitra S. A wavelet filtering based analysis of macroeconomic indicators: The Indian evidence. Applied Mathematics and Computation. 2006;**175**:1055-1079

[59] López de Prado M. Advances in Financial Machine Learning. Hoboken: John Wiley & Sons Inc.; 2018

[60] Fan Q, Liu S, Wang K. Multiscale multifractal detrended fluctuation analysis of multivariate time series. Physica A. 2019;**532**:121864

[61] Alstott J, Bullmore E, Plenz D. Powerlaw: A python package for analysis of heavy-tailed distributions. PloSONE. 2014;**9**(4):e95816

[62] Bao D, Yang Z. Intelligent stock trading system by turning point confirming and probabilistic reasoning. Expert Systems with Applications. 2008;**34**:620-627

[63] Altan A, Karasu S, Bekiros S. Digital currency forecasting with chaotic meta-heuristic bio-inspired signal processing techniques. Chaos, Solitons and Fractals. 2019;**126**:325-336

[64] Argyris J, Andreadis I. The influence of noise on the correlation dimension of chaotic attractors. Chaos, Solitons and Fractals. 1998;**9**(3):343-361

[65] Donoho D. Denoising by soft threshold. IEEE Trans on Information Theory. 1995;**41**(3):613-627

[66] Donoho D, Johnstone I. Ideal spatial adaption by wavelet shrinkage. Biometrika. 1994;**81**(3):425-455

[67] MacKinnon J. Approximate asymptotic distribution functions for unit-root and cointegration tests. Journal of Business and Economic Statistics. 1994;**12**:167-176

[68] Kwiatkowski D, Phillips P, Schmidt P, Shin Y. Testing the null hypothesis of stationary against the alternative of a unit root. Journal of Econometrics. 1992;**54**:159-178

[69] Massey FJ Jr. The Kolmogorov-Smirnov test for goodness of fit. Journal of the Americal Statistical Association. 2012;**46**(253):68-78

[70] Andrews D. Heteroskedasticity and autocorrelation consistent covariance matrix estimation. Econometrica. 1991;**59**:817-858

[71] Provenzale A, Smith L, Vio R, Murante G. Distinguishing between low-dimensional dynamics and randomness in measured time series. Physica D. 1992;**58**:31-49

[72] Ramsey J, Sayers C, Rothman P. The statistical properties of dimension calculations using small data sets: Some economic applications. International Economic Review. 1990;**4**:991-1020

[73] Bajo-Rubio O, Fernandez-Rodriguez F, Sosvilla-Riverio S. Chaotic behaviour in exchange-rate series: First results for the peseta-U.S. Dollar case. Economics Letters. 1992;**39**:207-211

[74] Darbyshire A, Broomhead D. Robust estimation of tangent maps and Lyapunov spectra. Physica D. 1996;**89**:287

[75] Rosenstein M, Collins J, De Luca C. A practical method for calculating largest

Lyapunov exponents from small data sets. Physica D. 1993;**65**:117-134

[76] Bask M, Gençay R. Testing chaotic dynamics via Lyapunov exponents. Physica D. 1998;**114**:1-2

[77] Peng C-K, Buldyrev S, Havlin S, Simons M, Stanley H, Goldberger L. Mosaic organization of DNA nucleotides. Physical Review E. 1994;**49**(2):1685

[78] Hardstone R, Poil S-S, Schiavone G, Jansen R, Nikulin V, Mansvelder H, et al. Detrended fluctuation analysis: A scale-free view on neuronal oscillations. Frontiers in Physiology. 2012;**3**:450

[79] Mandelbrot BB, van Ness J. Fractional Brownian motions, fractional noises and applications. SIAM Review. 1968;**10**(4):422-437

[80] Mandelbrot BB, Wallis JR. Some long-run properties of geophysical records. Water Resources Research. 1969;**5**(2):321-340

[81] Opong K, Mulholland G, Fox A, Farahmand K. The behaviour of some UK equity indices: An application of Hurst and BDS tests. Journal of Empirical Finance. 1999;**6**(3):267-282

[82] Mandelbrot BB. Fractals and Chaos. New York: Springer; 2004

[83] Norouzzadeh P, Jafari G. Application of multifractal measures to Teheran price index. Physica A. 2005;**356**:609-627

[84] Yule G. On a method of investigating periodicities in disturbed series with special reference to wolfer's sunspot numbers. Philosophical Transactions of the Royal Society of London Series A. 1927;**226**:267-298

[85] Packard N, Crutchfield J, Farmer J, Shaw R. Geometry from a time series. Physical Review Letters. 1980;**45**:712-716

[86] Ruelle D. Chaotic Evolution and Strange Attractors. Cambridge: Cambridge University Press; 1989

[87] Nichols J, Nichols J. Attractor reconstruction for non-linear systems: A methodological note. Mathematical Biosciences. 2001;**171**:21-32

[88] Gibson J, Farmer J, Casdagli M, Eubank S. An analytic approach to practical state space reconstruction. Physica D. 1992;**57**:1-30

[89] Cao L. Practical method for determining the minimum embedding dimension of a scalar time series. Physica D. 1997;**110**(1-2):43-50

[90] Cannas B, Cincotti S. Neural reconstruction of Lorenz attractors by an observable. Chaos, Solitons and Fractals. 2002;**14**:81-86

[91] Toledo-Suárez C. Meta-chaos: Reconstructing chaotic attractors from the separation of nearby initial conditions on hyperhelices. Communications in Nonlinear Dynamics and Numerical Simulation. 2010;**15**:2249-2253

[92] Yeo K. Data-driven reconstruction of nonlinear dynamics from sparse observation. Journal of Computational Physics. 2019;**395**:671-689

[93] Asefa T, Kemblowski M, Lall U, Urroz G. Support vector machines for nonlinear state space reconstruction: Application to the great salt Lake time series. Water Resources Research. 2005;**41**:W12422

[94] Ma H-G, Zhang C-L, Li F. State space reconstruction of nonstationary time-series. Journal of Computational

and Nonlinear Dynamics. 2017;**12**:031009-031001

[95] Fraser A, Swinney H. Independent coordinates for strange attractors from mutual information. Physical Review A. 1986;**33**(2):1134-1140

[96] Broomhead D, King G. Extracting qualitative dynamics from experimental data. Physica D. 1986;**20**:217-236

[97] Rüdisüli M, Schildhauer T, Biollaz S, Van Ommen J. Measurement, monitoring and control of fluidized bed combustion and gasification. In: Fluidized Bed Technologies for Near-Zero Emission Combustion and Gasification. UK: Woodhead Publishing; 2013. pp. 813-864

[98] Nieto A, Seoane J, Sanjuán M. Final state sensitivity in noisy chaotic scattering. Chaos, Solitons and Fractals. 2021;**150**:111181

[99] Sauer T, Yorke J, Casdagli M. Embedology. Journal of Statistical Physics. 1991;**65**:579

[100] Pentari A, Tzagkarakis G, Tsakalides P, Simos P, Bertsias G, Kavroulakis E, et al. Changes in resting-state functional connectivity in neuropsychiatric lupus: A dynamic approach based on recurrence quantification analysis. Biomedical Signal Processing and Control. 2022;**72**:103285

[101] Eckmann J-P, Kamphorst S, Ruelle D. Recurrence plots of dynamical systems. Europhysics Letters. 1987;**5**:973-977

[102] Van den Hoorn W, Hodges P, van Dieen J, Kerr G. Reliability of recurrence quantification analysis of postural sway data. A comparison of two methods to determine recurrence threshold. Journal of Biomechanics. 2020;**107**:109793

[103] Koebbe M, Mayer-Kress G. Use of recurrence plots in the analysis of time-series data. In: Casdagli M, Eubank S, editors. Proceedings of SFI Studies in the Science of Complexity, Vol. XXI, Redwood City, 1992. Reading, MA: Addison-Wesley; 1992. pp. 361-378

[104] Zbilut J, Webber C Jr. Embeddings and delays as derived from quantification of recurrence plots. Physics Letters A. 1992;**171**(3-4):199-203

[105] Theiler J. Spurious dimension from correlation algorithms applied to limited time-series data. Physical Reviews A. 1986;**34**(3):2427-2432

[106] Cariolaro G. Unified Signal Theory. London: Springer; 2011

[107] Wojtaszczyk P. A Mathematical Introduction to Wavelets. Cambridge: Cambridge University Press; 1997

[108] Chang S, Grace S, Yu B, Vetterli M. Adaptive wavelet thresholding for image denoising and compression. IEEE Transactions. 2000;**9**(9):1532-1546

[109] Strogatz S. Nonlinear Dynamics and Chaos. Colorado: Westview Press; 2014

[110] Katok A, Hasselblatt B. Introduction to the Modern Theory of Dynamical Systems. Cambridge: Cambridge University Press; 1995

[111] Mandelbrot BB. The Fractal Geometry of Nature. USA: Freeman; 1977

[112] Cao G, He L-Y, Cao J. Multifractal Detrended Analysis Method and its Application in Financial Markets. Singapore: Springer; 2018

[113] Barunik J, Aste T, Di Matteo T, Liu R. Understanding the source of multifractality in financial markets. Physica A. 2012;**391**:4234-4251

[114] Yuan Y, Zhuang X-T, Jin X. Measuring multifractality of stock price flucutation using multifractal detrended fluctuation analysis. Physica A. 2009;**388**:2189-2197

[115] Lorenz E. Deterministic nonperiodic flow. Journal of the Atmospheric Sciences. 1963;**20**:130-141

[116] Zelinka I, Chadli M, Davendra D, Senkerik R, Jasek R. An investigation on evolutionary reconstruction of continuous chaotic systems. Mathematical and Computer Modelling. 2013;**57**:2-15

[117] Jokar M, Salarieh H, Alasty A. On the existence of proper stochastic Markov models for statistical reconstruction and prediction of chaotic time series. Chaos, Solitons and Fractals. 2019;**123**:373-382

Chapter 6

Spatial-Temporal Data Analysis in Nonlinear System

Xing He and Minyu Chen

Abstract

Spatial-temporal analysis is at the heart of data mining in Big Data Era. Most mathematical tools are incompetent to deal with spatial-temporal data. This phenomenon has greatly spurred the development of data science, especially in the field of BDA (big data analytics). This chapter proposes random matrix theory (RMT) to handle this problem, which begins by modeling spatial-temporal datasets as sequences, whose term is in the form of a random matrix each. Then, some fundamental RMT principles are briefly discussed, such as asymptotic spectrum laws, transforms, convergence rate, and free probability, in order to extract high-dimensional statistics from the random matrix as the indicators. The statistical properties of these indicators are discussed for a better understanding of the system. Finally, some potential application fields are given.

Keywords: spatial-temporal data, electric power system, data-driven, random matrix theory, situation awareness, big data analyzation

1. Introduction

Electric power system reliability and intelligent management are critical to our daily living. Engineers and academics have recently focused on the use of large-scale phase measuring units (PMUs) to improve wide-area monitoring, protection, and control [1–5].

Most existing algorithms in power grid are model-based, which are built upon mechanism assumptions/simplifications and linear system control theory, with a determined and typically analytic outcome. These models, however, are ineffective for today's power system, which is of ever-increasing complexity and uncertainty [6–10]: 1) Interconnection of nearby utilities may frequently improve overall safety and efficiency, resulting in a huge interconnected system, such as the North American Power Grid, which serves almost 400 million customers throughout the continent [11]; 2) the continuous penetration of cell units (e.g., distributed generations) that are small-size, large-number, distributed-deployment, diverse-behaviors, smart-response, and uncertain-control [12]. 3) physical disciplines (mechanics, magnetism, electric, and electronics) of a system are closely intertwined, especially in a CHP (combined heat and power) system or even IES (integrated energy system) [13]; and 4) the construction of energy foundation for a smart city. Those above characteristics are advantageous to an open, flat, nonlinear, high-uncertainty, and distributed EIoT, as shown in **Figure 1** [14]. For such an EIoT, a

 IntechOpen

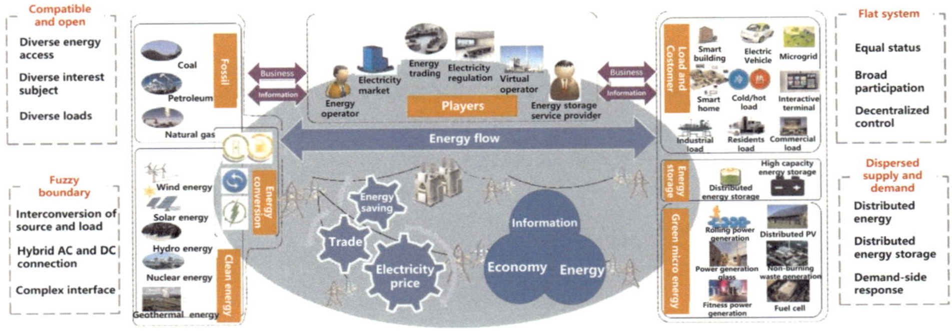

Figure 1.
Diagram of future energy internet of thing: its resource flow, data flow, and participants [14].

precise mechanistic model or even a proper descriptive representation can hardly be formulated, let alone model-based linear mode.

Furthermore, engineering data, such as sampling data in a power system, is not similar to image data. Various sensors, such as phasor measuring units (PMUs), are used to sample data from the grid. The huge dataset is in a high-dimensional vector space and in time series: the temporal variations (T sampling instants) and spatial fluctuations (N grid nodes) are recorded concurrently, and hence it is called spatial-temporal data.

Most mathematical tools are incompetent in this task [15]. Facing the above spatial-temporal data, we can hardly extract statistical information, particularly spatial-temporal correlations; the high-dimensional structure does not match the requirements of most traditional mathematical methods. Also, this task is incompatible with supervised training algorithms such as neural networks, due to the lack of or asymmetry of massive labeled data [16].

Fortunately, random matrix theory (RMT), by unifying time and space through their ratio $c = T/N$, can strictly and mathematically deal with such data. Moreover, linear eigenvalue statistics (LESs) built from data matrices follow Gaussian distributions for very general conditions, and other statistical variables are studied due to the latest breakthroughs in probability on the central limit theorems of those LESs.

2. Spatial–Temporal data analyzation mode, tools, theory, and applications in electric power system

2.1 Big data era, fourth paradigm, and data-driven model

The world's science has altered, as seen in **Figure 2** [17]. Initially, there was only experimental science, followed by theoretical science, which included Newton's Laws, Maxwell's equations, and so on. The theoretical models became too hard to solve analytically for many issues, and people had to start simulating. These simulations have carried us through much of the previous millennia. People nowadays are collecting data through intensive sensors or simulations. The data flood has an impact on experimental, theoretical, and computational science, and several state-of-the-art data technologies and data sciences have converged to provide tremendous promise for data-intensive scientific discovery, the so-called Fourth Paradigm.

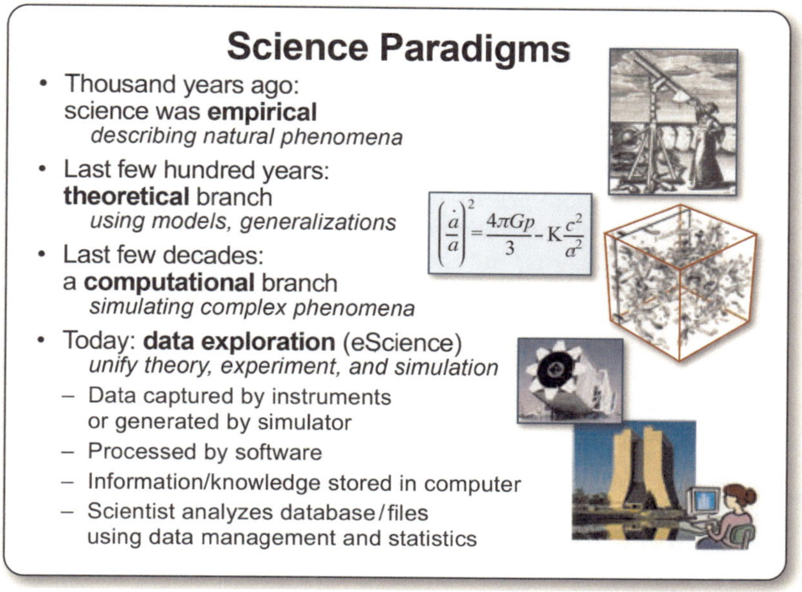

Figure 2.
Science paradigms and fourth paradigm [17].

Data-driven becomes a natural and stressful topic in energy systems, as evidenced by *IEEE TRANSACTIONS ON SMART GRID* special issue "big data analytics for grid modernization" published in 2016 [18]. Data-driven approaches are also characterized as model-free; we no longer rely significantly on physical models, and hence can manage instances where physical parameters are incorrect or even totally unavailable. Data-driven mode enables a quick start to our task, especially for a modern energy system in which the behaviors and discipline of system cell units are strongly coupled.

2.2 Basic of spatial–Temporal data, and high-dimensional information

Spatial–temporal data analysis means that we simultaneously deal with a large number of variables (in N-dimensional spatial space), and each variable ($i = 1, \dots, N$) samples *time series* for a duration (in T-dimensional temporal space). A classical statistic theory treats fixed N only (typically $N < 6$ [3]), e.g., for ABC-dq0 transformation $N = 3$. This fixed small N is called the low-dimensional regime. In practice, we are interested in the case that N can vary arbitrarily in size compared with T (typically $T > 60$, $N > 20$, and $c = N/T > 0$ [15]). This fundamental difference is the *primary motivation* for studying BDA.

Spatial–temporal data mining is expected to contribute some (high-dimensional) information with domain-specific meaning attached as the supplement to DT-based situation awareness (SA). High-dimensional indicators (outputs of high-dimensional statistics) and deep features (outputs of deep learning) are two main types of representation of high-dimensional information.

2.3 Spatial-temporal data utilization architecture and tools

Most mathematical methods struggle to extract information from spatial-temporal data. This phenomenon has accelerated the development of data science, particularly in the field of AI and BDA. We describe a high-focus technique for each field: 1) DL (Deep learning), which is good at massive data modeling in AI [19] and 2) High-dimensional statistics, or more precisely, RMT, which does well in data analytics in BDA. Both tools use a set of (high-dimensional) methodologies for integrated spatial-temporal modeling and analysis, and they have already made profound impacts on many domains. **Figure 3** depicts the architecture of large data mining based on DL and RMT.

2.3.1 Deep learning and its advantages

DL is a cutting-edge data mining algorithm. As demonstrated in **Figure 4**, deep characteristics are learned at some level from comparatively hidden features in the hierarchy [20]. DL uses the enormous data in a non-handcrafted approach to create a

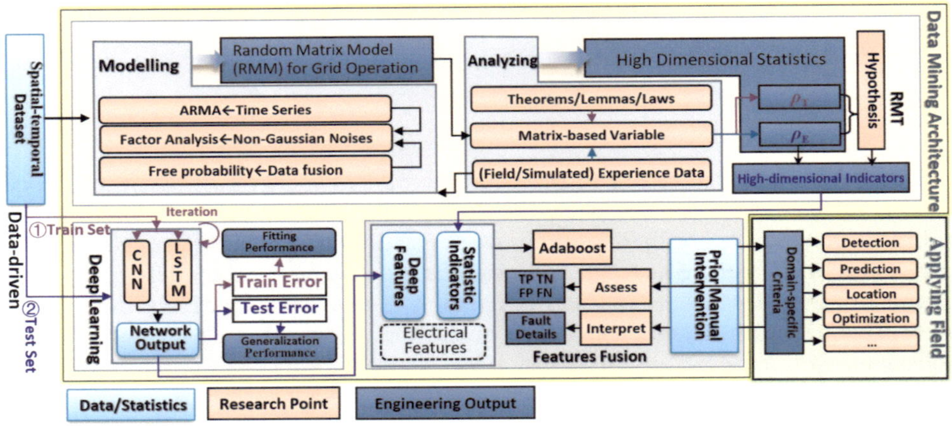

Figure 3.
Architecture of spatial-temporal data utilization.

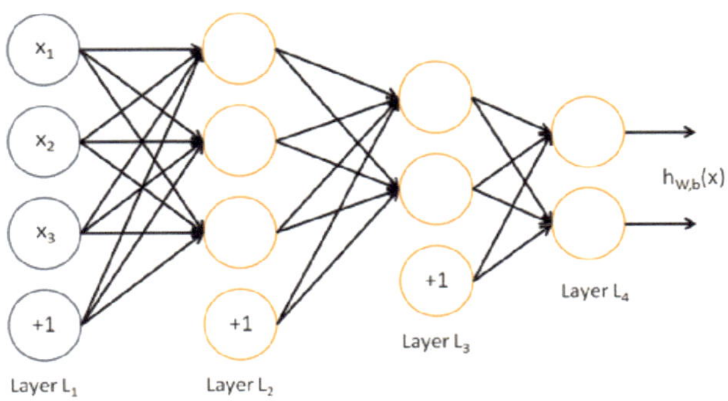

Figure 4.
A typical ANN structure.

deep (nonlinear) network model. A typical ANN (Artificial Neural Network) network is modeled as

$$y = f(x) \triangleq f^L\left(W^L \dots f^2\left(W^2 f^1\left(W^1 x + b^1\right) + b^2\right) \dots + b^L\right) \tag{1}$$

The above DL network model can be built with little prior knowledge relevant to the physical mechanism or causal relationship. As a result, DL may be used in a variety of situations or even systems without major changes. For example, we use CNN (convolutional neural networks) for computer version system modeling [21], LSTM (long short-term memory) for prediction [22], and deep reinforcement learning for strategy optimization [23].

In a complicated system, DL has a competitive edge in terms of possible data use. Furthermore, the test error might be used to quantify the DL model's performance on the generalization task, ensuring its usefulness in a real-world situation.

DL holds a competitive advantage for feasible data utilization in a complex system. In addition, the performance of the DL model on generalization task could be quantitatively evaluated by the test error, ensuring its usefulness in a real-world situation.

2.3.2 Big data analytics and RMT and the advantages

BDA uses spatial-temporal joint analysis to acquire high-dimensional statistics. Matrix-based variables, such as eigenvalue or the matrix variate itself, are likely to provide some insight to BDA [24]. These matrix-based variables are the variables of the $N \times T$ (large-dimensional) spatial-temporal data matrix that have an intrinsic statistical link, whether causal or not. These matrix-based variables are analytically intractable due to their high dimensionality rather than their big size. RMT is inextricably linked to this issue.

RMT understands the joint eigenvalue distribution as the statistic analytics in the asymptotic regime. In particular, by unifying time and space through their ratio c = T/N, BDA is acquired as the functionals of the eigenvalue distributions. For example, the matrix's LES indicators [25] follow Gaussian distributions for very general conditions. Furthermore, many LES-derived variables, whose statistical properties are mostly derivable and provable, are studied due to the latest breakthroughs in high-dimensional probability [15]. In this sense, RMT is rigorous and fundamental in nature. Besides, RMT performs well with only moderate-size (unlabeled) data, which is often true for a domain-specific problem in EIoT.

2.4 Random matrix theory in a nutshell

2.4.1 RMT and its universality principle

Two ensembles, Gaussian unitary ensemble (GUE) and Laguerre unitary ensemble (LUE), are studied first in RMT [10]:

$$\mathbf{\Gamma} = \begin{cases} \frac{1}{2}\left(R + R^H\right), R \in \mathbb{R}^{N \times N}, \text{GUE}; \\ \frac{1}{T} R R^H, R \in \mathbb{R}^{N \times T}, \text{LUE}. \end{cases} \tag{2}$$

where R is i.i.d. standard Gaussian random matrix.

We investigate the rate of convergence of the expected empirical spectral distribution (ESD) of Γ. Let hΓ (x) denotes the true eigenvalue density. Wigner's Semicircle Law and Wishart's M-P Law, respectively, for GUE and LUE, say that

$$h_\Gamma(x) = \begin{cases} \dfrac{1}{2\pi}\sqrt{4-x^2} & ,x \in [-2,2], \text{GUE} \\ \dfrac{1}{2\pi c x}\sqrt{(x-a_1)(x-a_2)} & ,x \in [a_1,a_2], \text{LUE} \end{cases} \tag{3}$$

where $a_1 = (1-\sqrt{c})^2, a_2 = (1+\sqrt{c})^2$

Universality principle enables us to perform hypothesis tests under the assumption that the matrix entries are not Gaussian distributed but use the same test statistics as in Gaussian case. Numerous studies using field data [25, 26] demonstrate that M-P Law is universally valid with moderate matrix sizes, such as tens. This is the very reason why RMT is widely used in engineering.

2.4.2 Linear eigenvalue statistics and its properties

Consider a random matrix $\mathbf{\Gamma} \in R^{N \times T}$, and $\mathbf{M}$ is the covariance matrix $\mathbf{M} = 1/T\mathbf{\Gamma}\mathbf{\Gamma}^H$. The LES τ of $\mathbf{\Gamma}$ is defined in [27].

$$\tau_\varphi = \sum_{i=1}^{N} \varphi(\lambda_i) = \mathrm{Tr}\varphi(\mathbf{M}), \tag{4}$$

Law of Large Numbers tells us that $N^{-1}\tau_\phi$ converges in probability to the limit

$$\lim_{n\to\infty} \frac{1}{N}\sum_{i=1}^{N} \varphi(\lambda_i) = \int \varphi(\lambda)\rho(\lambda)\mathrm{d}\lambda \tag{5}$$

where $\rho(\lambda)$ is the probability density function, which is given in Eq. (4). Therefore, we deduce that

$$\tau_\varphi = \sum_{i=1}^{N} \varphi(\lambda_i) = \mathrm{Tr}\varphi(\mathbf{M}) = N\int \varphi(\lambda)\rho(\lambda)\mathrm{d}\lambda \tag{6}$$

The Central Limit Theorem (CLT) for LES is studied as the natural second step:

$$\sigma^2(\tau_\varphi) = \frac{2}{c\pi^2}\int \int_{-\frac{\pi}{2}<\theta_1,\theta_2<\frac{\pi}{2}} \psi^2(\theta_1,\theta_2)(1 - \sin\theta_1 \sin\theta_2) d\theta_1 d\theta_2 \\ + \frac{\kappa_4}{\pi^2}\left(\int_{-\frac{\pi}{2}}^{\frac{\pi}{2}} \varphi(\zeta(\theta)) \sin\theta d\theta\right)^2 \tag{7}$$

See [25] for details.

2.4.3 LES-based hypothesis testing for random matrix

LES τ, as a positive scalar random variable defined in Eq. (5), is studied instead of the probability distribution of eigenvalues in Eq. (4). It can be viewed as a mathematically rigorous dimensionality reduction—the $N \times T$ random matrix is reduced to a positive scalar random variable.

As $N \to \infty$, the asymptotic limit of LES τ expectation and variance, i.e., $\mathrm{E}(\tau)$ and $\sigma^2(\tau)$, is given, respectively, in Eqs. (7) and (8). These two equations are sufficient to study the scalar random variable τ. Universal principle, as well as engineering experiences, demonstrate that moderate values of N and T are accurate enough for our practical purposes. LES τ is robust against data flaw and insusceptible to noises [10]. All of these statistical properties make LES a good SA indicator.

2.5 High-dimensional situation awareness indicator system and its properties

According to Eq. (5), numerous LESs can be designed from a certain spatial-temporal data Γ. Similarly, other high-dimensional indicators, for instance, statistic indicators, deep features, and electrical features, are calculable as the outputs of data tools according to **Figure 4**. They are tied together to provide an insight into domain-specific SA criteria for detection, prediction, etc. The details about the high-dimensional SA indicator system and its successful application cases can be found in ref. [15].

With these indicators, the high-dimensional indicator system is built; it supplies a multiple view angle to gain insight into the system. Aiming to provide a domain-specific SA task, the test function ϕ plays a role as a flexible filter depending on our task. **Table 1** lists the properties of LES indicators and makes a comparison with classical ones.

Table 1 tells that LES provides a better indicator system in the 4th Paradigm. The relation of the LES indicators to the classical ones, in some sense, is just like that of quantum physics to the classical one. Comparing experimental values with ideal theoretical values, LES conducts SA in a complex system statistically.

High-dimensional (LES) Indicators	Classical indicators
Data-driven	(Mechanism) model based
Supported by data science	Supported by physical laws or experience
Maybe unclearly defined in engineering	Clearly defined
Often probabilistic value	Often determined one
Often in high dimensions	Often in low dimensions
Able to harness the spatial-temporal data flexibly	Only a few data are available
Robust against bad data and insensitive to data selections	Susceptible to data selections (usually a single measurement at a time slice)
Pure statistical procedure	System errors are inevitable
Naturally coupling/decoupling for data block	Coupling/decoupling based on assumptions and simplifications
Random errors can be estimated with the model size (N, T)	Errors accumulation are inevitable and difficult to evaluate

Table 1.
High-dimensional indicator system for EIoT.

In short, RMT supplies us with a data-driven approach to indicator extraction for the informatization of a real system via sampling spatial-temporal data. A cluster of statistical indicators, via a mathematical procedure, is formed as a new epistemology for the system. Some advantages—such as data-driven and model-free mode, theoretic guided, fast in speed, reasonableness, sensitivity, flexibility, and robustness against bad data—have already been shown in our previous work [10, 17].

2.6 LES-based hypothesis testing for random matrix

To study the convergence as a function of N, we study LES instead of the probability distribution of eigenvalues in Eq. (4). For an arbitrary test function with enough smoothness, LES τ (see it as a random variable Y) is a positive scalar random variable defined in Eq. (5). As $N \to \infty$, the asymptotic limit of its expectation, $\mathrm{E}(Y)$, is given in Eq. (6), and the asymptotic limit of its variance, σ_2 (Y), is given in Eq. (7). These two equations are enough to study the scalar random variable Y. This approach can be viewed as a dimensionality reduction—the random data matrix of size $N \times T$ is reduced to a positive scalar random variable Y! This dimension reduction is mathematically rigorous only when $N, T \to \infty$, but $N/T \to c$. Experiences demonstrate, however, that moderate values of N and T are accurate enough for our practical purposes. Moreover, our previous work shows that LES is robust against data errors (e.g., data loss, data out-of-synchronization) and insusceptible to (independent) random noises (not limited to white noises), which is not true to those low dimensional statistics, such as mean and variance of any single variable. All these statistical properties make LES a good matrix-based variable for a hypothesis testing design aiming to provide anomaly detection task.

We formulate the hypothesis testing in terms of the statistical properties of LES. Referring to the Gaussian property and standard scores, the detection is modeled as a binary hypothesis testing: the normal hypothesis H_0 (no anomaly present) and the abnormal one H_1, denoted by:

$$
\begin{aligned}
\mathcal{H}_0 &: \left| \frac{\tau_\varphi - \mathbb{E}(\tau_\varphi)}{\sigma(\tau_\varphi)} \right| < \epsilon, \\
\mathcal{H}_1 &: \left| \frac{\tau_\varphi - \mathbb{E}(\tau_\varphi)}{\sigma(\tau_\varphi)} \right| \geq \epsilon,
\end{aligned}
\tag{8}
$$

where ϵ is a threshold value that needs to be preset—e.g., at a significance level of 0.05, the ϵ should be set at 1.96.

3. Conclusion

This chapter, motivated for the future's electrical grid, studies the nonlinear analysis based on RMT. Three ingredients are discussed in detail: 1) data modeling—modeling the spatial-temporal data as a sequence of random matrices, which are naturally connected to RMT. 2) data analytics—conducting high-dimensional analysis to obtain the statistical indicators. 3) interpretation—interpreting the indicator by studying its properties for a better understanding of the system.

The experimental indicators, which are fully derived from the sampling data, are applicable to various engineering functions. For example, by comparing the LESs with their theoretical prediction, anomaly detection can be implemented.

Future research directions include: (1) Model validation with different implementations of the grid, ranging from statistic, dynamic and real-world systems; (2) Data fusion with a number of random data matrices, using mathematical tools such as free probability; and (3) The use of Gaussian random matrices in replacement for general data matrices that are obtained from the electrical grid. The universality principle of RMT says that this replacement causes negligible errors.

Acknowledgements

This work is supported by the National Natural Science Foundation of China (Grant No. 51907121).

Author details

Xing He* and Minyu Chen
Shanghai Jiao Tong University, Shanghai, China

*Address all correspondence to: hexing_hx@126.com

References

[1] Chakrabarti S, Kyriakides E, Bi T, Cai D, Terzija V. Measurements get together. IEEE Power and Energy Magazine. 2009;7(1):41-49

[2] Luo L, Bei H, Chen J, Sheng G, Jiang X. Partial discharge detection and recognition in random matrix theory paradigm. IEEE Access. 2016;**PP**(99):1-1

[3] Lei C, Qiu RC, Xing H, Ling Z, Liu Y. Massive streaming pmu data model- ing and analytics in smart grid state evaluation based on multiple high-dimensional covariance tests. IEEE Transactions on Big Data. 2018;**4**(1): 2332-7790

[4] Hou W, Ning Z, Lei G, Xu Z. Temporal, functional and spatial big data computing framework for large-scale smart grid. IEEE Transactions on Emerging Topics in Computing. 2017;**PP**(99):1-1

[5] Tu C, Xi H, Shuai Z, Fei J. Big data issues in smart grid – A review. Renewable and Sustainable Energy Reviews. 2017;**79**:1099-1107

[6] Shaker H, Zareipour H, Wood D. A data-driven approach for estimating the power generation of invisible solar sites. IEEE Transactions on Smart Grid. 2016; **7**(5):2466-2476

[7] Motter AE, Myers SA, Anghel M, Nishikawa T. Spontaneous synchrony in power-grid networks. Nature Physics. 2013;**9**(3):191-197

[8] Wang L, Li HW, Wu CT. Stability analysis of an integrated offshore wind and seashore wave farm fed to a power grid using a unified power flow controller. IEEE Transactions on Power Systems. 2013;**28**(3):2211-2221

[9] Xu X, He X, Ai Q, Qiu RC. A correlation analysis method for power systems based on random matrix theory. IEEE Transcations on Smart Grids. 2017; **8**(4):1811-1820

[10] He X, Ai Q, Qiu C, Huang W, Piao L, Liu H. A big data architecture design for smart grids based on random matrix theory. IEEE Transactions on Smart Grid. 2017;**8**(2):674-686

[11] Transmission, Office Of Electric Grid 2030: A National Vision for electricity's Second 100 Years. Washington, DC. Office of Electric Transmission & Distribution. 2003

[12] Yang B, Yu T, Shu H, Dong J, Jiang L. Robust sliding-mode control of wind energy conversion systems for optimal power extraction via nonlinear perturbation observers. Applied Energy. 2018;**210**:711-723

[13] Fu X, Sun H, Guo Q, Pan Z, Xiong W, Wang L. Uncertainty analysis of an integrated energy system based on information theory. Energy. 2017;**122**: 649-662

[14] Guo J. The evolution of power system characteristics and related thinking. In: 2nd "Clean Energy Development and Consumption Symposium". Xi'an, China: Chinese Society for Electrical Engineering; 2019

[15] Qiu R, Antonik P. Smart Grid and Big Data. New York: John Wiley and Sons; 2015

[16] Cheng L, Yu T. A new generation of ai: A review and perspective on machine learning technologies applied to smart energy and electric power systems.

International Journal of Energy Research. 2019;**43**(6):1928-1973

[17] Hey AJ, Tansley S, Tolle KM, et al. The Fourth Paradigm: Data-Intensive Scientific Discovery. Vol. 1. WA: Microsoft Research Redmond; 2009

[18] Hong T, Chen C, Huang J, Lu N, Xie L, and Zareipour H. "Guest editorial big data analytics for grid modernization". IEEE Transactions on Smart Grid. Sept 2016;7(5):2395–2396

[19] Najafabadi MM, Villanustre F, Khoshgoftaar TM, Seliya N, Wald R, Muharemagic E. Deep learning applications and challenges in big data analytics. Journal of Big Data. 2015;**2**(1):1

[20] Ren Y, Zhang L, Suganthan PN. Ensemble classification and regression-recent developments, applicationsand future directions. IEEE Computational Intelligence Magazine. 2016;**11**(1):41-53

[21] Ling Z, Zhang D, Qiu RC, Jin Z, Zhang Y, He X, et al. An accurate and real-time method of self-blast glass insulator location based on faster r-cnn and u-net with aerial images. CSEE Journal of Power and Energy Systems. 2019;**5**(4):474-482

[22] Kong W, Dong ZY, Jia Y, Hill DJ, Xu Y, Zhang Y. Short-term residential load forecasting based on lstm recurrent neural network. IEEE Transactions on Smart Grid. 2017;**10**(1):841-851

[23] Zhang Z, Zhang D, Qiu RC. Deep reinforcement learning for power system applications: An overview. CSEE Journal of Power and Energy Systems. 2019;**6**(1):213-225

[24] Adhikari S. Matrix variate distributions for probabilistic structural dynamics. AIAA Journal. 2007;**45**(7):1748-1762

[25] Shcherbina M. Central limit theorem for linear eigenvalue statistics of the wigner and sample covariance random matrices. Journal of Mathematical Physics, Analysis, Geometry. 2011;**7**(2):176-192.

[26] He X, Qiu RC, Ai Q, Chu L, Xu X, Ling Z. Designing for situation awareness of future power grids: An indicator system based on linear eigenvalue statistics of large random matrices. IEEE Access. 2016;**4**:3557-3568

[27] Lytova A, Pastur L, et al. Central limit theorem for linear eigenvalue statistics of random matrices with independent entries. The Annals of Probability. 2009;**37**(5):1778-1840

Section 3

Analysis of Nonlinear Systems and Nonlinear Control

Chapter 7

Structural Properties and Convergence Approach for Chance-Constrained Optimization of Boundary-Value Elliptic Partial Differential Equation Systems

Kibru Teka, Abebe Geletu and Pu Li

Abstract

This work studies the structural properties and convergence approach of chance-constrained optimization of boundary-value elliptic partial differential equation systems (CCPDEs). The boundary conditions are random input functions deliberated from the boundary of the partial differential equation (PDE) system and in the infinite-dimensional reflexive and separable Banach space. The structural properties of the chance constraints studied in this paper are continuity, closedness, compactness, convexity, and smoothness of probabilistic uniform or pointwise state constrained functions and their parametric approximations. These are open issues even in the finite-dimensional Banach space. Thus, it needs finite-dimensional and smooth parametric approximation representations. We propose a convex approximation approach to nonconvex CCPDE problems. When the approximation parameter goes to zero from the right, the solutions of the relaxation and compression approximations converge asymptotically to the optimal solution of the original CCPDE. Due to the convexity of the problem, a global solution exists for the proposed approximations. Numerical results are provided to demonstrate the plausibility and applicability of the proposed approach.

Keywords: chance-constrained optimization, structural properties, state-constrained boundary-value PDE, probabilistic state constraints

1. Introduction

Partial differential equations (PDEs) are widely used to describe the spatial variations of physical, biological, and social systems as well as processes in mechanical engineering, thermodynamic, chemical engineering, medicine, industrial manufacturing, etc. [1, 2]. Moreover, practical PDE models involve uncertainties arising from imprecise model parameters and the system's operational environment. In real-life applications, external influences have a non-negligible impact and seriously affect system behaviors [2–6]. For example, ambient temperature, wind, and pressure are uncertain external influences that seriously impact system performances.

External input uncertainties will cause output uncertainties in system state variables [4, 7–11]. Such random inputs usually affect the boundary of the system and thus should be compensated by distributed boundary control. Hence, we consider in this study the randomness of the boundary condition of elliptic PDE systems and solve the chance-constrained optimization problems of such systems [12].

This study is an extension of the previous works in [13, 14] in which the randomness from the model parameters of a PDE system was considered but without considering boundary-valued control. In the present study, we consider the randomness from a nonhomogeneous boundary condition of a PDE system, which implies that the required state solution of the chance-constrained optimization of boundary-value elliptic partial differential equation (CCPDE) is a random field [15]. The control input is applied deterministically at the boundary function to compensate for the random disturbances. As a result, the study addresses the issue of chance-constrained optimization of a randomly boundary-valued PDE system.

Mathematically, in this work, we consider a random parameter $\xi \in \Omega$ coupled with a spatial variable $x \in D$ at the boundary condition of the PDE system. We assume that the uncertainty is under a given probabilistic measure Pr of the complete probability space (Ω, Σ, Pr) where Σ is a sigma-Algebra in the Borel set Ω. This study analyzes the properties of infinite-dimensional optimization problems in the reflexive and separable Bochner space with the elliptic PDE system as equality constraint and its probabilistic state constraints as inequality constraints. In general, for CCPDE problems, significant difficulties arise from chance constraints. Specifically, the main structural properties such as continuity, compactness, convexity, and differentiability of the probabilistic state-constraint functions are difficult to analyze. In addition, solving chance-constraints problems is generally not a trivial task.

Therefore, our investigation first focuses on the theoretical analysis of the main structural properties of the probability pointwise state-constrained functions in the CCPDE. The presence of uncertainties on the nonhomogeneous and nonlinear Dirichlet boundary conditions impacts the required state solutions. It is necessary to investigate the optimality conditions to the existence and uniqueness of the solution to the CCPDE problem. Subsequently, since such CCPDE problems are generally difficult to solve directly and also potentially nonsmooth [6, 16], this work proposes smoothing approximation methods to address this difficulty [2, 13].

The numerical computation for solving the CCPDE problem needs a finite-dimensional representation of the infinite-dimensional space through a discretization coupled with an appropriate sampling of the random variables by the multilevel Mont-Carlo method (ML-MCM). Since the resulting finite-dimensional chance-constrained optimization problem is generally nonsmooth, nonconvex, and difficult to solve directly, we use the recently proposed inner-outer approximation approach [6] for the solution of the CCPDE problem. Several structural properties of the inner-outer approximation-based CCPDE are also analyzed in this study. In the previous work, [13] the convexity of the outer approximation was investigated. In this study, we address the convexity issue of the inner approximations to guarantee the optimal global solution of the CCPDE.

2. Problem definition

We consider chance-constrained optimization of a boundary-value elliptic PDE system (CCPDE),

$$CCPDE: \quad \min_u E[J(y,u,\xi)] := \left\{ E\left[\|y - y_d\|^2_{H^1_g(D)} \right] + \frac{\rho}{2} \left[\|u\|^2 \right]_{L^2(D)} \right\} \tag{1}$$

subject to :

$$-\nabla.(\kappa(x)\nabla y) = f(x) \, \text{in} \, D \times \Omega, \tag{2}$$

$$y|_{\partial D} = g(x,u,\xi), \quad \xi \in \Omega, x \in \partial D, u \in U \tag{3}$$

$$Pr\{y_{min}(x) \le y(x,u,\xi) \le y_{max}(x)\} \ge \alpha, x \in D, \tag{4}$$

$$u_{min} \le u(x) \le u_{max}, u \in U \tag{5}$$

where $D \subset \mathbb{R}^n$ is a given bounded convex open spatial domain with Lipschitz boundary ∂D and $n \ge 2$, ρ is a given regularization parameter, D_c is a given compact subset of the closure $\overline{D}$ of D. $\nabla.$ and ∇ represent the divergence and gradient operator w.r.t. x in the weak sense of Sobolev spaces, respectively. The state function $y(x,u,\xi) : \overline{D} \times U \times \Omega \mapsto \mathbb{R}$ is a random continuous function in $H^1_g(D) = \{\Gamma \in W^{1,2}(D) \mid \Gamma(x) = g(x), x \in \partial D\}$ with $H^1_g(D)$ being a closed subspace of the Sobolev space $H^1(D) = W^{1,2}(D)$ for any $\xi \in \Omega$. $y_d \in H^1(D)$ is a given function describing the desired profile of the state and is assumed twice differentiable w.r.t. $x \in D$.[1] With $< h, g >_{H^1_g(D)}$ and $< h, g >_{H^1(D)}$, we denote the related standard scalar product (see [1–5, 7, 8, 17, 18] for more details on Sobolev spaces).

The triple (Ω, Σ, Pr) represents a complete probability space, with a set of all possible outcomes $\Omega \subset \mathbb{R}^p$, with σ-Algebra $\Sigma \subset 2^\Omega$ and probability measure $Pr(\cdot) : \Sigma \mapsto [0,1]$ and $Pr(\Omega) = 1$. The parameter ξ represents uncorrelated input random vector variables distributed homogeneously acting on the system through the boundary ∂D. Such disturbances are position-dependent random parameters and distributed inside or outside of the boundary of the spatial domain ∂D. In general, such infinite-dimensional random parameters can be treated by a dimensional-reduction method using the Karhunen-Loeve (KL) expansion (see [19]) or a finite-dimensional representation using a discretization method [3, 4].

The input data $g(x,u,\xi)$ might vary randomly from one point of the boundary domain ∂D to another point and thus their uncertainty should be described in terms of random fields, which can be dealt with a sampled covariance from multilevel Monte Carlo method (MLMCM) [2, 9, 20]. The expected value $E[\cdot]$ is taken with respect to the probability space and the probability measure possesses the Radon-Nikodym derivative ϕ w.r.t. the Lebesgue measure μ, i.e., $dPr(\xi) = \phi(\xi)d\mu(\xi)$. Moreover, we suppress the measure μ and write simply $dPr(\xi) = \phi(\xi)d\xi$. The random variable ξ is assumed to have a continuous probability density function $\phi(\xi)$ with Ω being its sample space of the support set. For each $u \in U$, due to the random variable ξ, the solution y of the boundary values (2)–(3) is a stochastic linear boundary value state function indicated by $y(u,\xi;x)$.

---

[1] The space $H^1(D)$ is a Hilbert space with norm $\|\cdot\|_{H^1(D)}$. The Sobolev space $H^1(D)$ is the completion of $C^1(\overline{D})$ w.r.t. $\|\cdot\|_{H^1(D)}$.

After the solution of the PDE system, Eq. (4) expressed by $P(.,x) = Pr\{y(x,u,\xi) \leq y_{max}(x)\} \geq \alpha, \forall x \in D$ defines a single pointwise probability state constraint, for each $x \in D$, to be satisfied with a given reliability level α, where $\alpha \in (0,1]$. It should be noted that chance constraints for the PDE system considered in this study can be expressed in the following two forms:

1. Single chance constraints

$$Pr\{y_{min}(x) \leq y(x,u,\xi) \leq y_{max}(x)\} \geq \alpha, \forall x \in D, \tag{6}$$

2. Joint chance constraint

$$Pr\{y_{min} \leq y(x_i,u,\xi) \leq y_{max}, \forall x_i \in D\} \geq \alpha, \tag{7}$$

The first one describes the chance constraints imposed on individual points in D (i.e., pointwise chance constraints), while the second one requires the satisfaction of the constraints at all points with a probability level. In this study, only the form of single pointwise constraints is considered. A joint CCPDE is mathematically complex and needs further studies.

The right-hand side of Eq. (2) is assumed to be a function in $L^2(D)$. Due to (3), y depends on u and ξ, and therefore, we need to analyze the existence and uniqueness of a weak solution. In Section 3, we will prove these by verifying the continuous-bilinear and coercivity form in the Bochner space with the associated expectation of the norm $E(\| \cdot \|_{H^2(D)})$ based on the Lax-Milgram theorem.

Specifically in this study, (2) and (3) lead to a boundary value of an elliptic PDE system with nonhomogeneous Dirichlet boundary condition $y|_{\partial D} = g(x,u,\xi), \xi \in \Omega$ and $u \in U$. u is a control variable bounded by u_{min} and u_{max} by (5). Since the output y is constrained, we have to find an optimal control profile w.r.t. $x \in D$, in the admissible set (U_{adm}) by variational analysis.

Now, we define a separable and reflexive Bochner space by mapping $\mathcal{W} := L(\Omega; W(D))$ from the Borel space to the Sobolev space $W(D)$

$$\begin{aligned} H(\Omega; W(D)) = \{ & v : \Omega \to W(D) : v \text{ is measurable}, \|v\|_{\mathcal{W}} \\ & = \int_{\Omega} \|v(\cdot,\xi)\|^2_{W(D)} \phi(\xi) d\xi < +\infty \} \end{aligned} \tag{8}$$

From the PDE system defined in (2)–(3), the related function spaces of individual inputs are defined as follows:

$$\mathcal{H} := L^2\left(\Omega; H^1_g(D)\right), \quad \mathcal{L} := L^2(D), \mathcal{G} := L^2(\Omega, H^2(D)), \mathcal{B} := L^2\left(\Omega, H^{1/2}(D)\right), \mathcal{K} := L^\infty(D). \tag{9}$$

Since the spaces defined above are separable, the weak measurability for the random PDE system is equivalent to the strong measurability (see [[21], Section 3.5 Cor. 2]).

In addition, we define scalar products in $\mathcal{L}$ and $\mathcal{H}$ spaces, respectively,

$$\langle a,b \rangle_{\mathcal{L}} = \int_{\Omega} \int_{D} a(x,\xi) b(x,\xi) dx \phi(\xi) d\xi, \quad \|a\|^2_{\mathcal{L}} = <a,a>_{\mathcal{L}}, \tag{10}$$

$$\langle a,b\rangle_{\mathcal{H}} = \int_\Omega\int_D a(x,\xi)b(x,\xi)dx\phi(\xi)d\xi + \int_\Omega\int_{\partial D} (\nabla a(x,\xi))^t \nabla b(x,\xi)dx\phi(\xi)d\xi, \quad (11)$$

$$\|a\|^2_{\mathcal{H}} = <a,a>_{\mathcal{H}} = <a,a>_{\mathcal{L}} + <\nabla a, \nabla a>_{\mathcal{L}}, \quad (12)$$

for $a, b \in \mathcal{L}$ and $a, b \in \mathcal{H}$, respectively. In addition, in (9),

$$\mathcal{B} = \left\{v : \Omega \to W(\partial D) : v \text{ is measurable}, L_2\left(\Omega \times H^{-1/2}(\partial D)\right) \to \mathbb{R}\right\},$$

$$\|v\|_{\mathcal{B}} = \left(\int_\Omega \|v\|^2_{H^{-1/2}(\partial D)}\phi(\xi)d\xi\right)^{1/2}$$

$$= \left[\|v\|^2_{L_2(\Omega\times\partial D)} + \sum\nolimits_{|m|=[s]}\int_\Omega\int_{\partial D} \frac{|\partial^m v(x_1) - \partial^m v(x_2)|^2}{|x_1 - x_2|^{n+2+2(s-[s])}} d\sigma\phi(\xi)d\xi\right]^{1/2}, \quad (13)$$

$$\|v\|^2_{\mathcal{B}} = \left[\|v\|^2_{L_2(\Omega\times\partial D))} + \int_\Omega\int_{\partial D} \frac{(|v(x_1) - v(x_2))^2|}{|x_1 - x_2|^{\frac{s+1}{2}}} d\sigma\phi(\xi)d\xi\right] < \infty, s = 1/2,$$

is a norm of trace function in the boundary space $\mathcal{H}$ of the boundary value with a compact embedding from $\mathcal{B}$ where $d\sigma$ is the surface measure at the boundary of D [22]. Finally, the space of the model parametric coefficient $\mathcal{K} = \{v : L^\infty(D) \to \mathbb{R} :$ v is measurable $: \|v\|_{L^\infty(D)} = maxsup_{n\to\infty}|v_n|^2 < \infty\}$. Thus, $v \in \mathcal{H}$, it implied that $v(\cdot,\xi) \in H^1_g(D)$ and $E[\|v(\cdot,\xi)\|^2_{H^1_g(D)}] < +\infty$.

The probability density ϕ is assumed to be Lebesgue measurable and almost everywhere positive on Ω. Hence, the spaces $\mathcal{L}$, $\mathcal{G}$, and $\mathscr{H}$ are a reflexive Bochner space, e.g., Hilbert spaces using the standard equivalence classes. Note also that $\mathcal{G}, \mathcal{H}, \mathcal{K}, \mathcal{B}$ are dense subspaces of $\mathcal{L}$ in the topology of $\mathcal{L}$.

The variable $u \in L^2(D)$ is a decision variable that belongs to the set of admissible decisions

$$U_{adm} := \left\{u(x) \in L^2(D) \mid u_a \le u(x) \le u_b\right\}, \forall x \in D, \text{for } u_a \ge u_{min} \text{ and } u_b \le u_{max}, \quad (14)$$

where $u_a, u_b \in L^2(D)$ are given functions with $u_a \le u_b$. Observe that equalities and inequalities of functions in the Lebesgue space $L^2(D)$ and corresponding Sobolev spaces are valid only almost everywhere on D. The term almost everywhere (a.e.) will be suppressed in this study assuming almost surely (a.s.) without any confusions arising. Note that U_{adm} is a nonempty, convex, closed, and bounded subset of $L^2(D)$.

In the elliptic PDE system (2)–(3), the random parameters in the boundary condition in $\mathcal{B}$[2] represent the effect of external and internal disturbances such as ambient temperature, pressure, and wind; also, there is a factor of imprecise model parameters, while those in the *forcing term f* are nonrandom input function. For the sake of simplicity of presentation, the coefficient and forcing term (κ, f) respectively are

[2] The boundary of the elliptic operator needs to be C^1 smooth for ensuring $x \mapsto y(u,x,\xi) \in H^{1/2}(\partial D)$ (see: [7]).

considered as nonrandom input functions in this study. Moreover, the forcing term is continuous w.r.t. $x \in D$ a.s. and $f(x) \in \mathcal{L}$.[3].

As a result, pointwise probabilistic state constraint $P(.,x) = Pr\{y_{min} \leq y \leq y_{max}\} \geq \alpha$, overall the spatial region $x \in D a.s.$, is conservative (worse-case) if, its reliability level $\alpha = 1$, with no chance of constraints violation. The internal function $y_{min} \leq y \leq y_{max}, \forall x \in D$ cannot be computed deterministically. Hence, the expression in (4) defines a chance (probabilistic) constraint by stipulating the satisfaction of the inequality constraint on $y \in \mathcal{H}$ with a given probability value of a reliability level $\alpha \in [0.95, 1)$. Moreover, (4) represents a pointwise chance constraint, i.e., the constraint on the state variable is required to hold with the same reliability level α at each individual point of $x \in D$.

The required random state solution $y(x, u, \xi)$ is a function in the infinite-dimensional space $\mathcal{H}$, so that, the infinite number of probabilistic constraints make sense whenever in the equivalence class y is w.r.t. x a continuous element, which is ensured by Sobolev embedding theorems in $L^2(\Omega, H^2(D) \cap H^1_g(D))$, excellent properties of the inhomogeneity term f, and the convexity of D. From the embedding theory of Sobolev space, one can use a more general setting in $\mathcal{H}^p = W^{1,p}$ with $p > d$ and sufficient regular D. Thus, the convexity of D can be relaxed. We give here only one opportunity, where it works. It is essential for our approximation approach (see, e.g., theorem 3.5 continuity of y, that space for y can be continuously embedded in the space of continuous functions).

For instance, at some critical spatial locations $x \in D_c \subset D$, the reliable level $\alpha \in [0.95, 1)$ can be chosen, as a result, this study focuses on the solution of CCPDE with pointwise constraints but considers reliable level independent of x for simplicity of representation, and it is not trivial to directly extend our inner and outer approximation concept in [6] to joint and uniform chance constraints for the infinite number of $x \in D$. Therefore, solving the CCPDE problem is not a trivial task since there is no simple equivalent deterministic representation. Also, there is no closed-loop analytic representation for the probabilistic state constraint in the equation expressed in (4). The structural properties of (4) are not yet analyzed properly, generally unknown, nondifferentiable, and nonconvex.

3. Existence of the solution of the PDE system

In this paper, we need to solve the weak variational form of the random PDE system with nonhomogeneous Dirichlet boundary value of the elliptic PDE system as we defined in equations expressed (2) and (3), the control is applied on the boundary of polygonal spatial domain D and the control $u \in L^2(D)$,

$$-\nabla \cdot (\kappa(x) \nabla y(x, \xi) = f(x), \text{ on } D \times \Omega \text{ a.s.}, \tag{15}$$

$$y(x, u, \xi)|_{x \in \partial D} = g(x, u, \xi), \quad \xi \in \Omega \text{ a.s.} \tag{16}$$

For every test function $v \in \mathcal{H}$, we can apply integration by part,

---

[3] The density function $\phi(\xi) = \prod_{i=1}^{\infty} \phi_i(\xi_i)$ is infinite-dimensional probability density function, where $\xi_i \in \Omega$ is distributed homogeneously through the boundary of the spatial domain ∂D. It has a finite-dimensional representation from KL expansion [19].

$$E\left[\int_D \kappa(x,\cdot)\nabla y(x,\cdot)\nabla v(x,\cdot)dx\right] = E\left[\int_D f(x)v(x,\cdot)dx\right] + E\left[\int_{\partial D} g(u,x,\cdot)v(x,\cdot)ds\right], \forall v \in \mathcal{H}, \tag{17}$$

which is the weak form of the PDE system (15)–(16) with $f(x)$ in the Hilbert space $\mathcal{L}$ and solution $y(.) \in \mathcal{H}$. The relevant functions of the inputs are in separable and reflexive Bochner space. Since $f(x) \in \mathcal{L}$, $\kappa \in \mathcal{K}$, and $g \in \mathcal{B}$, the spatial domain D is convex, the well-known *shift statements* (see, for instance, [[23], Th. 3.30]; i.e., higher regularity of f is shifted to higher regularity of y) imply that $x \mapsto y(u, x, \xi) \in C(\overline{D})$. Since the continuity of $x \mapsto y(u, x, \xi)$ is required only on the subset D, the convexity of D is not necessary whenever the critical spatial domain $D_c \subset intD$, $intD$ is the interior of space D. However, to guarantee the well-posedness of the weak form, our investigation is based on the following standard assumptions.

Assumption 3.1. **(A1.1)** The domain D is convex, the set $D_c \subset \overline{D}$ is compact and $y_c \in C(D) \cap H_g^1(D), y_d \in H_g^1(D) \cap H^2(D)$.

(A1.2) The coefficient $\kappa(\cdot) \in \mathcal{K}$ is positive and bounded such that

$$0 < \kappa_{min} \leq \kappa(x) \leq \kappa_{max}, x \in D \text{ a.s.}, \tag{18}$$

where $\kappa_{min}, \kappa_{max}$ are finite constants.

(A2.1) For each $u \in L^2(D)$, the random forcing term $u \mapsto f(u, \cdot) : L^2(D) \to \mathcal{L}$ is continuous.

(A2.2) For each $u \in L^2(D)$, the random forcing term $u \mapsto f(u, \cdot) : L^2(D) \to \mathcal{L}$ is continuously Fréchet differentiable.

(A3) The forcing term has a Taylor expansion form m $f(x, u) = u(x) + \sum_{n=0}^{\infty} f^n(x_0)/(n!)) * (x - x_0)^n$, where $u \in L^2(D)$ and $f_0 \in \mathcal{L}$.

(A4) For each $u \in L^2(\partial D)$, the random forcing term $u \mapsto g(u, \cdot, \cdot) : L^2(\partial D) \to \mathcal{L}$ is continuous. For each $u \in L^2(\partial D)$, the random forcing term $u \mapsto g(u, \cdot, \cdot) : L^2(\partial D) \to \mathcal{L}$ is continuously Fréchet differentiable. and g is linear w.r.t. u.

(A5) The random variables $\xi^T = \left(\xi_1, \ldots, \xi_p\right)$ are independently, identically distributed with a continuous joint multivariate probability density function $\phi(\xi) = \Pi_{i=1}^p \phi_i(\xi_i)$ and the set $\Omega = \Pi_{i=1}^p \Omega_i$, where $\Omega_i \subset \mathbb{R}, i = 1, \ldots, p$, such that

$$f(x, u) = u(x) + \sum_{n=0}^{\infty} f^{(n)}(x_0)/(n!) * (x - x_0)^n \tag{19}$$

$$g = u(x) + g_0(\xi)\left(\frac{\partial y_0(x)}{\partial x_0}\right) + \sum_{n=1}^{n} g_k(\xi)\frac{\partial y(x)}{\partial x_k} = u(x) + \sum_{k=1}^{N} g_k(\xi)\frac{\partial y_k}{\partial x_k} \tag{20}$$

with $u, a_k \in L^2(D), k = 0, 1, 2, \ldots, n$..

In the assumption A5, f and g are commonly given as a series, which is called *finite-dimensional noise* representation (20) (see [4, 5, 20]) for the boundary condition g. In fact, for numerical computations, it is essential to reduce the dimension of the uncertainties in g from KL dimension reduction method.

3.1 Solution of the random PDE with nonhomogeneous boundary control

From the equations expressed in (15) and (16), we have to show that the Lax-Milgram theorem of the continuous-bilinear and coercivity property for every test function in $v \in \mathcal{H}$,

$$-(\nabla \cdot (\kappa_0 \nabla y) + y_{\partial D})v(x,\xi) = fv(x,\xi) + gv(x,\xi). \tag{21}$$

This implies for $v \in \mathcal{H}$,

$$\begin{aligned} &-\int_{\Omega}\int_{D}(\nabla \cdot (\kappa_0 \nabla y + y_{\partial D}))v(x,\xi)dx\phi(\xi)d\xi \\ &= \int_{\Omega}\int_{D} fv(x,\xi)dx\phi(\xi)d\xi + \int_{\Omega}\int_{\partial D} gv(x,\xi)d\sigma\phi(\xi)d\xi. \end{aligned} \tag{22}$$

It implies the following integration functions of the expectation,

$$-E\left[\int_{D}\nabla \cdot (\kappa_0 \nabla y + y_{\partial D}))v(x,\xi)dx\right] = E\left[\int_{D} fv(x,\xi)dx\right] + E\left[\int_{\partial D} gv(x,\xi)d\sigma\right], \forall v \in \mathcal{H}. \tag{23}$$

The Sobolev space plays several roles in the study of stochastic PDE system [7]. The space $L^2(D) = H^0(D)$ equivalence class of real-valued Lebesgue measure and square integrable function defined on the spatial domain D. Let $H^1 = H^1(D)$ denote the vector subspace of H defined by $H^1 = \{v \in H^1 : \nabla v_{xi} \in H^1\}$ $i = 1, 2, 3, \dots, n$ the equipped with the norm

$$\|v\|^2_{H^1} = \|v\|^2_{L^2(D)} + \|\nabla_{x_i} v\|^2_{L^2(\partial D)}. \tag{24}$$

Space $H^1(D)$ is Hilbert space, and it is known as the Sobolev space of order 1. Let $Domain(D)$ denote the $C^\infty(D)$ with a compact support. The closure of $Domain(D)$ is norm topology of $H^1(D)$, the required random solution is in $H^1_g(D)$ is subspace of $H^1(D)$. The dual of H^1_g is H^{-1}, is the dual space of continuous linear function on H^1_g, both of Sobolev space H^1_g and H^{-1} are space of functional distributions in the sense of Schwartz and have nonunique representations, and the functional at boundary has a polynomial approximation in Eq. (20), with $g_k(\xi)$ and $\frac{\partial y}{\partial x_k}(x)$ for $k = 1, 2, 3. \dots, n$. Where the derivative of $y_k(x)$ is understood in the sense of distribution. Generally, the input random boundary conditions,

$$g \in L^2\left(\Omega; H^{1/2}(\partial D)\right) \equiv \mathcal{B}, \tag{25}$$

that defined by above expansion of orthonormal function the adjoint is in $L^2(\Omega; H^{-1/2}(\partial D)$ for the defined $\xi \in \Omega$. Since there is nonzero and nonlinear random function $g \neq 0$ at the Dirichlet boundary condition. The boundary g is approximated by Fourier transform, and one can define Sobolev space one can define Sobolev space H^s for all real numbers s, $s < 0$ these are genuine distributions as characteristic function, for $s = 0$, we have $H^0 = L^2(D)$ for $s > 0$ these are the regular function spaces contained in H^1, for example, y belongs to $H_g^1(D)$ see [21, 24]. There is the trace operator $\Gamma : L^2(\Omega; H_g^1(D)) \mapsto L^2(\Omega; H^{1/2}(\partial D), \Gamma(y) = y = g$ for $x \in \partial D$. The space $H^m \subset H^{m-k}$ for $k > 0$ and the injection is compact. The trace function loses its interior smoothness and may be a distribution on the boundary as $\frac{\partial y}{\partial n}$ belongs to $H^{-1/2}(\partial D))$, n is a normal to the test function v, i.e., $\frac{\partial y}{\partial v} \in H^{-1/2}(\partial D))$. We shall define the inner product space (.,.) or $\prec ., . \succ$ by the double integral function on the set D and Ω, there is the duality pairing between H_1 and H_1^* at the trace operator. In each case of boundary-value condition of bilinear-continuity form $a(y, v)$ and coercive (elliptic) form, we can formulate the following equation:

$$a(y,v) - (f,v) - (g,v) = \int_\Omega \int_D v(Ly - f)dx\phi(\xi)d\xi + \sum [a_{ik}\partial_k y - g]vds \tag{26}$$

that vanishes the right-hand side of second integral, and L and a_{ik} are the linear and differential operators, respectively [7]. The weak PDE system gives

$$-E\left[\int_D \nabla \cdot (\kappa_0 \nabla y)v(x,\xi)dx\right] = E\left[\int_D fv(x,\xi)dx\right] + E\left[\int_{\partial D}(y-g)v(x,\xi)d\sigma\right], v \in \mathcal{H}. \tag{27}$$

Definition 3.2. The system of elliptic PDE in Eqs. (2) and (3) has a weak solution $y \in L^2(\Omega; H_g^1(D))$, if there exists a measurable random variable y w.r.t. ξ defined in Ω such that $E[a(y,\vartheta)] = E[l(y,v)] = E[(f,\vartheta) + E[(g,\vartheta)]$ for all $\vartheta \in \mathcal{H} = L^2(\Omega; H_g^1(D))$. The operator a satisfies the continuous bilinear and coercive form.

Definition 3.3. The system of elliptic PDE said to be stable in $L^2(\Omega; H_g^1(D))$, if it has a weak solution $y \in L^2(\Omega; H_g^1(D))$ and the forcing term $f \in \mathcal{L}$ expressed in Eq. (9) and the boundary input $g \in L_2(\Omega, H^{1/2}(\partial D)$. The solution y is continuous depending on the random parameter, i.e., $E[y] = E[y(f,g)]$. The subspace of all function form $L^2(\Omega, H^l(D)$ whose generalized derivatives up to order l exist and belong to $L^2(\Omega, H^l(D)$. The space $H^l(D) = W^l{}_2(D)$ is called Sobolev space order l..

Theorem 3.4. ***Lax-Milgram Theorem:*** *Let $\kappa(.) \in L^\infty(D)$ be a functional, and there exists a constant $\kappa_{min} > 0$ and $\kappa_{max} > \kappa(x) > \kappa_{min}$ almost surly and the test function $v \in \mathcal{H} = L^2(\Omega \times H_g^1(D)), g \in L_2(\Omega \times H^{1/2}(\partial D))$ and $f(x) \in (H^1(D))^*$. The operators a and l defined in Eq. (25) hold continuous-bilinearity and coercivity. Thus, the variational problem defined in Eq. (1) and (2) has unique solution $y \in L_2(\Omega, H_g^1(D))$ for all $\xi \in \Omega$.*

Proof. The elliptic PDEs in Eqs. (15) and (16) have weak solution if there exists a test function $v \in \mathcal{H}$, this boundary condition g moves to left-hand side

$$\begin{aligned} -\nabla(\kappa(.)\nabla y)v &= f(x)v \,\text{in}\, D; \\ (y - g(.,\xi,u))v &= 0 \,\text{on}\, U \times \partial D \times \Omega; \\ \Rightarrow \int_\Omega \int_D &- \nabla\kappa(.)\nabla y.vdx\Phi(\xi)d\xi \\ &= \int_\Omega \int_D fvdx\Phi(\xi)d\xi + \int_\Omega \int_{\partial tD}(g-y)(.,\xi,u).vd\sigma\Phi(\xi)d\xi. \end{aligned} \tag{28}$$

Where $d\sigma = \overleftarrow{n}.\overleftarrow{ds}$ is the surface measure at the boundary of D from integration by part,

$$\Rightarrow \int_{\Omega}\int_{D}\kappa(.)\nabla y.\nabla v dx\Phi(\xi)d\xi - \int_{\Omega}\int_{\partial D}\frac{\partial y}{\partial n}(\xi,x)v(\xi,x).d\sigma\Phi(\xi)d\xi$$
$$= \int_{\Omega}\int_{D} fv dx\Phi(\xi)d\xi + \int_{\Omega}\int_{\partial D}(g-y).v d\sigma\Phi(\xi)d\xi \quad (29)$$

Since $y.v(\xi,x) = 0$ for $x \in \partial D$ because the unit normal vector is perpendicular to the boundary ∂D. Let

$$a(y,v) = \int_{\Omega}\int_{D}\kappa(.)\nabla y.\nabla v dx\Phi(\xi)d\xi;$$
$$l(y,v) = \int_{\Omega}\int_{D} fv dx\Phi(\xi)d\xi + \int_{\Omega}\int_{\partial D} g(.,\xi,u).v d\sigma\Phi(\xi)d\xi; \quad (30)$$
$$\Rightarrow a(y,v) = l(y,v).$$

We need to show that continuous-bilinear and coercivity form. These two properties are sufficient for the existence and uniqueness of weak solution [7, 8]. The input random functions are in a reflexive and separable Bochner space. Thus, weak solution is the same as the classical solution, the weak measurability is also similar to the strong measurability seen in [24].

$$\Rightarrow |a(y,v)| = |\int_{\Omega}\int_{D}\kappa(x)\nabla y.\nabla v dx\Phi(\xi)d\xi| \leq |k(x)|\int_{\Omega}\int_{D}|\nabla y.\nabla v dx\Phi(\xi)d\xi|$$
$$\leq \|k(.)\|_{L^{\infty}(D}\|\nabla y\|_{L^2(\Omega\times D)}\|\nabla v\|_{L^2(\Omega\times D)} \quad (31)$$
$$\leq \|\kappa(.)\|_{L^{\infty}(D}\|y\|_{L^2\Omega\times H^1{}_g(D)}\|v\|_{L^2(\Omega\times D)}.$$

Hence, the operator a is continuous bilinear. The same case for $l(y,v)$ this is integral of duality product between a mapping in Hölder theorem

$$|l(y,v)| = \left|\int_{\Omega}\int_{D} fv dx\Phi(\xi)d\xi + \int_{\Omega}\int_{\partial D} g.v d\sigma\Phi(\xi)d\xi\right|$$
$$\leq \left|\int_{\Omega}\int_{D} fv dx\Phi(\xi)d\xi\right| + \left|\int_{\Omega}\int_{\partial D} g.v d\sigma\Phi(\xi)d\xi\right| \quad (32)$$
$$\leq \|f\|_{(L^2(D)}\|v\|_{(L_2(\Omega\times D)} + \|g\|_{(L^2(\Omega\times\partial D)}\|v\|_{(L_2(\Omega\times\partial D)}$$
$$\leq \|f\|_{(H^1(D)}\|v\|_{(L_2(\Omega\times D)} + \|g\|_{(L_2(\Omega,H^{1/2}(\partial D)))^*}\|v\|_{(L_2(\Omega,\partial D)}.$$

Therefore, both a and l are continuous bilinear form. To see coercivity, $\forall v \in V$ Fubini theorem implies that $|a(v,v)| = \int_{\Omega}\int_{D}|\kappa(x)\nabla v.\nabla v dx\Phi(\xi)d\xi| \geq \kappa_{min}\int_{\Omega}\int_{D}|\nabla v.\nabla v dx\Phi(\xi)d\xi|$ because κ_{min} is bounded from below a.s., and independent of random ξ, $\kappa_{min}(x)$ is lower bounded as well. Thus, $|a(v,v)| \geq \kappa_{min}\int_{\Omega}\int_{D}\|\nabla v\|_V^2 dx\Phi(\xi)d\xi \geq k_{min}/C$ $\|v\|_V^2$ Where $C = c^2$ by Poincare-Friedrichs inequality $|a(v,v)| \geq \kappa_{min}/c^2\|v\|_{\mathcal{H}}^2 \geq \kappa_{min}/c^2\|v\|^2{}_V$ the same for l. The required numerical solution is obtained from stochastic finite difference method (SFDM) or finite element method (SFEM), for the indexed $x \in D$ the numerical solution

$$y(x,u,\xi) = A^{-1}\left(f_{ij}/(\kappa)(u) + g_{ij}(u,\xi)\forall x \in D a.s. \right. \quad (33)$$

The $y(x,u,\xi)$ have a linear property w.r.t. u, ξ jointly. This case the matrix A obtained from the discretization operator a is positive definite [22]. Therefore, the weak solution $y \in L^2(\Omega \times H^1{}_g(D))$ exists and is unique. We will analyze the continuously dependent of the solution $y \in L^2(\Omega \times H^1{}_g(D))$ on f, g, u and $k, \forall \xi \in \Omega$. □

Theorem 3.5. *Suppose the coefficient operator* $\kappa = \sum_{ij}\kappa_{ij}x^i(\omega)x^j(\omega)$ *and* $\kappa_{ij}(x) \in L^\infty(D)$ *is nonrandom, is independent of* ξ *seen in [7] and deterministic coefficient function of* $\kappa(.)$ *is indexed by* $x \in D$. *There exists a lower bound number* $\kappa_{min} > 0$ *such that* $\kappa_{min} \le \kappa(x) \le \kappa_{max}$ *a.s. and* $\sum_{ij}\kappa_{ij}x^ix^j \ge \kappa_{min}x^2$ *for* $x \in D$ *subset of* $\mathbb{R}^n$. *Thus elliptic PDE system is* L^2 *stable in the sense of distribution in Definition (3.2) and Definition (3.3). There exist* $c > 0$, *c is independent of f, g for all fixed* $x \in D$, *and c is dependent only* κ, *such that*

$$E(\|y(f,g)\|^2)_{H^1(D)} \le c\left\{E(\|f\|^2)_{H^1(D)} + E(\|g\|^2)_{H^{-1/2}(\partial D))}\right\} \quad (34)$$

$\forall f \in H^1(D)$ *and* $g \in L^2\left(\Omega, H^{\frac{-1}{2}}(\partial D)\right), E(\|f\|^2) = \int_\Omega \|f\|^2\Phi(\xi)d\xi = \|f\|^2\int_\Omega\Phi(\xi)d\xi = \|f\|^2$. *This theorem is proved and extended in the work [25].*

Remark: For dimension $n \le 3$, the map of the solution is continuous embedding from $L^2(\Omega \times H^1(D)) \to L^2(\Omega \times (H^1(D))^* \cap L^2(\Omega \times H^{-1/2}(\partial D))$ is fulfilled and the mapping is continuous and linear $E\left[\|y\|_{H^1{}_g(D)}\right] \le c\left\{\left[\|f\|^*_{H^1(D)}\right] + E\left[\|g\|_{H^{1/2}(\partial D)}\right]\right\}$ for each ξ a.s., see the prove in [7].

Theorem 3.6. *Let U and V be Hilbert spaces. Then, the linear mapping* $L : U \to V$ *is an isomorphic mapping if and only if the associated form* $a : U \times V \mapsto \mathbb{R}$ *satisfies*

i. Continuity, there exists $C > 0$ such that $|a(u,v)| \ge C\|u\|_U\|v\|_V$ for all $u \in U$

ii. Inf-sup condition, i.e., there exists $c > 0$ such that $sup\left(\frac{a(u,v)}{\|v\|_V}\right) \ge c\|u\|_U, \forall u \in U$.

iii. *For every* $v \in V$, *there exists* $u \in U$ *with* $a(u,v) \ne 0$ *and if we assume continuity and inf- sup conditions above, then* $L : U \to \{v \in V : a(u,v) = 0, \forall u \in U\} \subset V$ *is an isomorphism. Thus, the equation* $sup\left(\frac{a(u,v)}{\|v\|_V}\right) \ge c\|u\|_U$ *is equivalent to* $\|Lu\|_V \ge c\|u\|_U, \forall u \in U$. *It follows the equivalent formulation* $inf_{\ u \in U}\left\{sup_{v \in V}\left(\frac{a(u,v)}{\|v\|_V\|U\|_U}\right)\right\} \ge c > 0$.

Proof. We need to show that injective and surjective map $L : U \to V$. The equivalence of continuity of $L : U \to V$, $L(u_1) = Lu_1 = L(u_2) = Lu_2, u_1, u_2 \in U$

$$\Rightarrow a(u_1,v) = a(u_2,v), \forall v \in V \quad (35)$$

$\Rightarrow a(u_1 - u_2, v) = 0$ implies $u_1 - u_2 = 0$. For the surjective $\forall f \in L(u), L(u)$ from the image of the pre image u. There exists a unique $u = L^{-1}(f)$. Thus, $c\|u\|_U \le sup\left(\frac{a(u,v)}{\|v\|_V}\right) = sup\left(\frac{a(f,v)}{\|v\|_V}\right) = \|f^*\|$. □

3.2 Reduced optimization of PDE with random data

The problems of CCPDEs are not properly studied. Moreover, the pointwise or uniform probabilistic state-constrained function is a nonsmooth, nonconvex, intractable, and infinite-dimensional state constraints. The following assumptions are needed for reducing dimension and variability for analyzing the structural properties of CCPDE.

Assumption 3.7. Assume that the functionals

1. The $\kappa(x) \in \mathcal{K}$ of Eq. (9), the $g(x,.,.) \in \mathcal{B}$, the $f(.) \in H^*$, the H^* is the dual in Sobolev space of $\mathcal{H}$ defined by Eq. (9) and D_c is subset of convex set D have Lipschitz smooth boundary ∂D.

2. The control functional $u(.) \in \mathcal{L}$ is compact and convex, the $\mathcal{L}$ is subset of the admissible set U_{adm}.

3. The objective functional E[J(y(u))] is mapping from $L_2(\Omega, H^2(D))) \times L_2(\Omega, H^{1/2}(\partial D))) \to \mathbb{R}$ is weakly sequentially lower semicontinuous (wsls) and bounded below by $\frac{\varrho}{2}\|u\|^2_{L^2(D)}$.

4. The $E(J(y(u)))$ is a convex w.r.t. u.

5. The f w.s.l.s w.r.t. $x \in D$ and g are w.s.l.s and convex w.r.t (u, ξ) simultaneously.

The optimization problem of

$$\min_{u(.) \in \mathcal{L}} E[J(y(x,\xi), u(.))] \tag{36}$$

subject to:

$$P(u;x) = p_x\{u\} = Pr\{y(x,u,\xi) \leq y_{max}\} \geq \alpha, x \in D \tag{37}$$

the random variable $\xi \in \Omega$ have a log-concave density function $\phi(\xi)$. If the objective function in Eq. (36) is convex, then the chance-constrained programming is a convex optimization problem and has a globally optimal solution.

Proof. The functional J is a convex function from the convexity property of norm. The expectation E is the integral of the convex function, is convex. Moreover, the solution of the PDE system y is a linear w.r.t (x, ξ). The internal functional $\gamma = y - y_{max}$ is a quasiconcave from proposition (3.4), then $P_x(u)$ is a convex w.r.t. u. Suppose the constrained $P = \{u \in U/P(u;x) > \alpha\}$ is a convex set [26]. Hence, the composition functional $E(J(y(u)))$ is the composition of convex and lower-continuous function. Therefore, $E(J(y(u)))$ is a convex function and the set $\mathcal{B} = L^2(\Omega, H^{1/2}(\partial D) \subset P \cap U_{adm}$, convex intersection of convex set is convex. □

The optimization problem of CCPDE reduced to the following programming form and the solution y is continuous dependent on the parameters f and g in theorem (3.5),

$$\min_{u(.) \in P \cap U_{adm}} q(u(.)) = E[J(y(f(x), (g(x,\xi), u(.))], \tag{38}$$

where the probability function in Eq. (37) expressed $P(.,x) = Pr\{\gamma = y(f(x), g(\xi), u(.)) - y_{max} \leq 0, \forall x \in D\}$, the set $P = \{u \in U/P(x; u) > \alpha\}$ this optimization problem admits unique optimal solution. The random variable $\xi \in \Omega$ has a log-concave density function $\phi(\xi)$. If the objective function in Eq. (47) is a convex, then the chance-constrained programming is a convex optimization problem and has a global optimal solution.

3.3 The structural property of probabilistic constrained function

Most of the recent research works on the chance-constrained optimization problems do not include a probabilistic state constraints. In this study, we have deliberated a pointwise probabilistic state constraints, which are expressed in Eq. (41). The structural properties of the probabilistic state-constrained function are not properly analyzed. The continuity, differentiability, compactness, and convexity properties of the state constraints are important for guaranteeing the optimality criteria. The function of state constrained looks like,

$$P(u;x) := p_x\{u\} := Pr\{y((x,\xi),u)) \leq y_{max}, \forall x \in D\} \geq \alpha. \tag{39}$$

The internal part of the probability function is

$$\gamma(u,x,\xi) = y((x,\xi),u)) - y_{max} \leq 0, \forall x \in D, \tag{40}$$

which is a continuous differentiable function from the equation expressed (33), it does not imply the continuity and differentiability of the probability function in Eq. (41). The following propositions guaranteed the structural properties for the probabilistic uniform constrained functions, and these are a strong form of a pointwise state constrained,

$$P(u;x) := p_x\{u\} := Pr\{y((x,\xi),u)) = y_{max},, \forall x \in D\} = 0. \tag{41}$$

it holds the measure zero property

$$P(u;x) := p_x\{u\} := Pr\{y((x,\xi),u)) = y_{max},, \forall x \in D\} = 0. \tag{42}$$

Proposition 3.8. *Assume that $\gamma(x,.,\xi)$ are Borel measurable w.r.t $\xi \in \Omega$ for all $u \in U$ and for all $x \in D$ and $\gamma(x,.,\xi)$ are weakly sequentially lower semicontinuous (wsls) or weakly sequentially upper semicontinuous (wsus) vice versa for $x \in D$ and $\xi \in \Omega$. Then, $P(u;x) = Pr(\gamma(x,u,\xi) \leq 0)$ is wsls or wsus vice versa and $\gamma(x,u,\xi) = y - y_{max} \leq 0, \forall x \in D$. The same case, for fixed $x \in D$ the function $\mathcal{P}(u) = \{u \in U_{adm} : P(.,x) \geq \alpha\}$ is wsls or wsus vice versa.*

Proof. From a given assumption, $\mathcal{P}$ is well defined by Borel measurability of $\gamma(x,u,.)$ w.r.t. $\xi \in \Omega$ in the second argument. Fix an arbitrary $\hat{u}$ and let $u_n \in U_{adm}$, $u_n \rightarrow \hat{u}$, arbitrary weakly convergent sequences in U_{adm} and any arbitrary fixed $x \in D$ and let $x_n \in D$, $x_n \rightarrow \hat{x}$. Denote by u_{n_l} a subsequence such that $liminf_{n\rightarrow\infty}\mathcal{P}(u_n) = lim_{l\rightarrow\infty}\mathcal{P}(u_{n_l})$. The variable of the decision $u = u(x)$ linearly continuous dependent on $x \in D$..

Define the sets $P = \{\xi \in \Omega : \gamma(x,u,\xi) \leq 0, \forall \xi \in \Omega\}$ and $P_n = \{\xi \in \Omega : \gamma(x,u_n,\xi) \leq 0 \forall \xi \in \Omega\}, n \geq n_o \in \mathcal{N}$. Since, by γ being wsls in the first argument, we have

$$liminf_{n\to\infty}\gamma(x,u_n,\xi) \geq lim_{l\to\infty}\gamma\left(x,u_{n_l},\xi\right) \geq \gamma(x,\hat{u},\xi), \forall\xi\in\Omega/P. \tag{43}$$

Consequently, $\gamma(x,u_n,\xi) < 0, \forall\xi\in\Omega/P$ and all $n \geq n_0, \xi\in\Omega$.
Denoting by

$$h(\xi) = \begin{cases} 1 & \text{if } \xi\in P \\ 0 & \text{if } \xi\notin P \end{cases} \tag{44}$$

the characteristic function of a set U_{adm}, this entails that $h(P_n(\xi)) \to 0$ as $n \to \infty, \forall\xi\in\Omega/P$. By the Lebesgue dominance convergence theorem, $\int_\Omega h[\gamma(x,u_n,\xi)\phi(\xi)d\xi \to 0$ for all $\xi\in\Omega/P$.
On the other way $E[h[\gamma(x,u_n,\xi)] \leq E[h[\gamma(x,u,\xi)] = 1, \forall\xi\in P$.
Therefore,

$$\begin{aligned} lim\,\mathcal{P}\left(u_{n_l}\right) &= lim_{l\to\infty}Pr\left[\gamma\left(x,u_{n_l},\xi\right)\leq 0\right] = lim_{l\to\infty}\int_\Omega h\left(\gamma\left(x,u_{n_l},\xi\right)\right)\phi(\xi)d\xi \\ &= lim_{l\to\infty}\int_{\Omega/P} h\left[\gamma(x,u_n,\xi)\phi[\xi]d\xi + lim_{l\to\infty}\int_P h[\gamma(x,u_n,\xi)\phi(\xi)d\xi\right. \\ &\geq lim_{l\to\infty}\,inf\int_{\Omega/P} h\left[\gamma(x,u_n,\xi)\phi[\xi]d\xi + lim_{l\to\infty}\,inf\int_P h[\gamma(x,u_n,\xi)\phi(\xi)d\xi\right. \\ &\geq lim_{l\to\infty}\,inf\int_P h[\gamma(x,u_n,\xi)\phi(\xi)d\xi \end{aligned} \tag{45}$$

$\geq lim_{l\to\infty}\,inf\int_P\phi(\xi)d\xi = Pr[\xi\in P] = Pr[\xi\in P:\gamma(x,.,\xi)\leq 0] = \mathcal{P}(\hat{u})$. Thus, the function $\mathcal{P}$ is wsls from the equation in above $liminf_{n\to\infty}\mathcal{P}(u_n) = lim_{l\to\infty}\mathcal{P}\left(u_{n_l}\right) \geq \mathcal{P}(\hat{u})$. For wsusc property, related propositions also proved in work of [11, 16]. □

Proposition 3.9. *Assume that D is compact subset of* $\mathbb{R}^n$, *if* γ *is wsus then* $P(u,x)$ *is wsus at* $\forall u\in U$ *satisfying* $pr(\gamma^*(u,\xi)=0)=0$, *this is said to be measure zero property where* $\gamma^* = inf\,(\gamma(x,u,\xi))$ *defined in proposition (3.8).*

The prove is the similar to the proof of (3.8), has been proved in [16]. Let $x\in D$ be fixed. We need to show the convexity property of probability function $u\to p(u,\xi)$ needs convex property with respect to (u,ξ) along the continuous probability density function $\phi(\xi)$. From the solution of PDE system, we have a continuous y in Bochner space and $y(x,\xi,u) = \left((A^{-1})/(k)\right)\left(j_H(f(x)+kg(u,.,.))(x.\xi)\right)\forall x\in D, \forall\xi\in\Omega$. Thus, $\gamma(x,.,.) = \left((A^{-1})/(k)\right)\left(j_H(f(x)+k(g(u,.,.))(x,u,\xi)-y_{max}\right)\leq 0$ and $Pr\left\{\left((A^{-1})\right.\right.$ $\left.\left.\left(j_H(f(x)+(k)g(u,.,.))(x.\xi)-(k)y_{max}\right)\right)\leq 0\right\}\geq\alpha$ is a convex w.r.t. (u,ξ) jointly from the linearity of the internal function for any fixed $x\in D$. Therefore, it is a sufficient in the finite-dimensional optimization, continuity, and linearity of y guarantee for continuity and convexity of $P(u;x)$ [26, 27]. Our aim is to extend it to infinite-dimensional case.

Remark: Assume that $\gamma(u,x,\xi) = \left(A^{-1}/(k)\right)\left(j_H f(x)\right) + g(x,.,.) - y_{max}$ is a concave w.r.t.$\xi\in\Omega$, $\forall u\in U$ and $\forall x\in D$, for each u there exists random vector $\xi\in\Omega$ such that $\gamma\leq 0, \forall x\in D$ then, ξ has a density $\phi(\xi)$ distributed continuously and

$$Pr\{\gamma^*(u,\xi)=0\} = 0 = Pr\left\{\left(A^{-1}/(k)\right)\left(j_H f(x)\right) + g(u,.,.)(x,\xi) = y_{max}\right\} = 0, \forall x\in D, \tag{46}$$

where, $\gamma^* = inf\,(\gamma(x,u,\xi)) = inf_u\{(A^{-1}/(k))\,(j_H f(x)) + g(u,.,.)(x,\xi) - y_{max}, \forall x \in D\}$.

Proposition 3.10. *Let U be Banach space and D be arbitrary index set. Let the n-dimensional random vector $\xi \in \Omega$ have a log-concave density, i.e., density of its logarithm is possibly extended valued concave function. Assume that the function $\gamma(x,u,\xi) = (A^{-1}/(k))\,(j_H(f(x)) + g(u,.,.)(x,\xi)) - y_{max}$ is quasi-concave for all x in D, then the set $M = \{u \in U/P(u;x) > \alpha\}$ is a convex set.*

Proof. From structural property of wslsc on the proposition (3.8) pick up $\gamma^* = inf_u(\gamma(x,u,\xi))$ and we defined $P(u;x) := Pr(\gamma^*(x,u,\xi) \le 0)$ for all u in Banach space $U_{adm} = M \subset U$ to fix arbitrary elements in U, $u_1, u_2 \in U$ and quasi concavity of γ^* such that ξ_1 and ξ_2 in Ω and $(u_1,\xi_1), (u_2,\xi_2) \in U \times \Omega$ and $\lambda \in [0,1]$ there exists indexed $x \in D$ such that $\gamma^*(\lambda(u_1,\xi_1) + (1-\lambda)(u_2,\xi_2)) \ge min\,\{(\gamma^*(u_1,\xi_1), \gamma^*(u_2,\xi_2)\}$ and the density ξ having log-concavity density shown in Prekopa ([11, 28] Theorem [4.2.1]) for the given log-concave distribution. The $pr(\xi \in \lambda P + (1-\lambda)q) \ge pr(\xi \in p)^{\lambda} Pr(\xi \in q)^{1-\lambda})$ for p and q are convex subset of Ω. We need to show the convexity of set M in u_1 and u_2 in M and $\lambda \in (0,1)$ we define a map $H : U \to \mathbb{R}$ u in U observe that $H(u_1)$ and $H(u_2)$ are the convex set an immediate consequence of the quasi concave on γ^*. The sets $\lambda H(u_1 + (1-\lambda)H(u_2)) \subset H(\lambda u_1 + (1-\lambda)u_2)$. Thus, H is a concave set (γ is a quasi-concave). In other words, let $\xi_1 \in H(u_1)$ and $\xi_2 \in H(u_2)$, $\xi \in P(\lambda u_1 + (1-\lambda)u_2) = pr[\xi \in H(\lambda u_1 + (1-\lambda)u_2)] \ge pr\left[\xi \in H(u_1]^{\lambda} * pr[\xi \in H(u_2)]\right]^{(1-\lambda)} = \alpha^{\lambda}\alpha^{(1-\lambda)} = \alpha$, for reliable level α. The $\xi = \lambda\xi_1 + (1-\lambda)\xi_2 \in \Omega$, $\gamma^*(u_1,\xi_2), \gamma^*(u_2,\xi_2) \le 0$ obtaining from the quasi concavity of γ^* we have $\gamma^*(\lambda u_1 + (1-\lambda)u_2, \xi) = \gamma^*(\lambda u_1, \xi_1) + (1-\lambda)(u_2, \xi 2)) \ge min\,(\gamma^*(u_1, \xi 1), \gamma^*(u_2, \xi 2)) \ge 0$. Finally, the result is proved the convexity of chance-constrained, $\lambda u_1 + (1-\lambda)u_2 \in M$. Hence, M is a convex set. □

Proposition 3.11. *Let $M = U$ be separable Banach space defined in proposition of the convexity (3.10) and M^* is a dual space of M, assume that the $\gamma(x,.,.) \le 0$ is a weakly sequentially upper continuous (w.s.u.s) for each $u \in M$ and $x \in D$. Then, it satisfies the following three conditions and the three statements are equivalent [11, 16].*

i. *The $\gamma(x,u,.)$ is a Borel measurable function w.r.t. $\xi \in \Omega$ $\forall u \in M$ and $x \in D$.*

ii. *The set $M = \{u \in M : \mathcal{P}(u;x) \ge \alpha\}$ is weakly closed.*

iii. *The $\mathcal{P}(u;x)$ is weakly sequentially upper continuous (wsus).*

Proof. From the assumption, the function $\gamma(x,u,.) : \mathbb{R}^n \to \mathbb{R}$ upper semicontinuous for each $u \in U$ and each $x \in D$. Consequently, the sets $\{\xi \in \Omega : \gamma(x,u,\xi) \le 0, \forall x \in D\}$ are closed, which implies the Borel measurability w.r.t. ξ (i). The proposition (3.8) explains the continuity property with measure zero property in (42), justifies to talk about probability of events as in (i) and (iii). It is an immediate consequences of (ii). Hence, in order to (i) prove (ii) let $\mathcal{P}(u)$ in $\mathbb{R}$ be arbitrary and consider weakly convergent sequences $u_n \to u$ has a convergent subsequence $u_{n_l} \in M$ with $u_n \in M$ for all n, it is from Arzela-Ascoli, every sequence of a given family of real-valued continuous functions defined on a closed and bounded interval has a uniformly convergent subsequence. We have to show that u in M we define $H(u) = \{\xi \in \Omega : \gamma(x,u,\xi) \le 0\}$ from $u_n \in M$, and $P\{\xi/H(u_n) \ge \alpha\}$. Boundness of u_n by weak convergence implies that there is some closed ball B with sufficiently large radius such that u in B for all n. From separability of U^*, the weak topology B metrizable w.r.t. ξ for fixed x. Regarding the finite-dimensional case, it has been proved in work [16].

Lemma 3.12. *All the assumptions of Proposition (3.11) are true, there are constants* $\varepsilon > 0, \sigma > 0$, *such that with d referring to the Hausdorff distance from the point on* U_{adm} *to the set in probabilistic set* $P_x(u;\xi)$ *such that* $d(\in U_{adm}, \{u \in \mathcal{P} : P(u,x) \geq \wp\} \leq \sigma max\{\log(\wp) - \log(P_x(u,x)), 0\}, \forall \wp \in [\alpha - \varepsilon, \alpha + \varepsilon]$.

For an infinite-dimensional problem of CCPDE, we have proved it in the previous work [14, 25]. This lemma has been verified for the finite-dimensional problem without the PDE system in [16]. It needs Arzela-Ascoli theorem, this CCPDE constrained function also holds bounded and continuous property [16, 21]. The Lax-Milgram theorem is sufficient for the bounded and continuous property from the continuous bilinear and coercivity form; see in the theorem (3.4).

4. Approximation for CCPDE by inner and outer functions

The approximation of the probability constrained function has been analyzed in the previous work [6, 14]. This proposed approach is a smooth parametric approximation for the nonsmooth and intractable probability function. This study will analyze some open issues related to the topological and structural properties of the CCPDE. Some of the issues are continuity, differentiability, closedness, compactness, and convexity of the inner and outer approximation functions in the infinite-dimensional Bochner space. They are significant for assuring the optimality criteria for the existence and uniqueness of the optimal control. Furthermore, the convergence approach and the numerical results are studied properly.

The smooth parametric inner and outer approximations are analyzed in the work [6]. This section briefly analyzes the parametric functions to define a smooth approximations to the optimal control of the boundary-value CCPDE. The optimal control of the CCPDEs is approximated by the family of the sequence of solutions of the inner-outer approximation problems, when the approximation parameter τ_k tends to zero as k goes to infinity. In case of convexity and norm convergence of the approximate solutions to a solution of CCPDE can be proved and to some extended the structural analysis of the proposed approach for guaranteeing the optimality criterion of CCPDE. For this purpose, we employ and extend our recent work [6, 13, 14, 25], where the inner-outer approximation methods, for finite-dimensional and smooth CCPDE, were proposed to solve the reduced CCPDE problem.

We consider the parametric Geletu-Hofmann function

$$\theta(\tau, s) = \frac{1 + m_1\tau}{1 + m_2\tau \exp\left(-\frac{s}{\tau}\right)}, \quad for \quad \tau \in (0,1), \ s \in \mathbb{R} \tag{47}$$

to approximates a nonsmooth problem of the CCPDE in the work [13]. If we fix the parameters $m_1 = 0, m_2 = 1/\tau$, the parametric function $\theta(\tau, s)$ is the same as the sigmoid function. Thus, we can approximate the probabilistic constraint by sigmoid function in terms of $\theta(\tau, s)$. Unfortunately, the sigmoid function is not bound to the probabilistic constrained function from the below, and the computation is a too slowing in comparison with $\theta(\tau, s)$. The advantage of theta is bound to the probabilistic function from above and below. Thus, the probabilistic constraint of CCPDE is bounded by the family of smooth functions of inner $E[\theta(\tau, -\gamma(u,\xi;t,x)])$ and outer $1 - E[\theta(\tau, \gamma(u,\xi;t,x)])$ approximations from the lower and upper bound, respectively. We have a piecewise continuous function,

$$h(s) = \begin{cases} 1 & \text{if } s \geq 0 \\ 0 & \text{if } s < 0. \end{cases} \tag{48}$$

Multiplied by the discontinuous function, with jump nonlinear function and an infinitely differentiable function $E[h(s)]$, where $s = \gamma(x, u, \xi)$ is defined above in Eq. (48). Then, the product is infinitely differentiable, and the difference of approximation on the parametric functions in the region $0 < x < \varepsilon$ that is tiny for small positive number epsilon but the probabilistic constraint has measure zero property defined in Eq. (42). This makes the approximation functions a nice function property. The m_1 and m_2 are positive constants, the parametric family each property of θ function is written and proved in [6, 13] together with its continuity, differentiability, and convexity of its approximation functions. The following well-known identities are obtained

$$p(u; x) = E[h(-\gamma(u, \xi; x))] = 1 - E[h(\gamma(u, \xi; x))], \tag{49}$$

despite the appealing expected value $E[h(\gamma(u, \xi; x))] = \int_{\Omega} h(\gamma(u, \xi; x))\phi(\xi)d\xi$ representation of probability functions in (49), the missing smoothness of the unit jump function does not provide computational advantages. Nevertheless, the function h provides an insight for the construction of a smoothing approximation for the probability function p and the internal function $\{\theta(\tau, \gamma), \tau \in (0, 1)\}$ of functions $\theta :$ $(0, 1) \times \mathbb{R} \rightarrow \mathbb{R}_+$ possessing the following *strict monotonicity* and *uniform limit* properties.

Assumption 4.1. Suppose there is a parametric family of a functions $\theta(\tau, s)$, which possess the following properties such as monotonic (strictly increasing) uniform limit properties

i. There is a constant C with $1 < C < \infty$ such that $C > \theta(\tau, s) > h(s), \forall s \in \mathbb{R}$ and $\forall \tau \in (0, 1)$.

ii. The parametric $\theta(\tau, s)$ is strictly increasing in all $s \in \mathbb{R}$ and $\tau \in (0, 1)$.

iii. The parametric $\theta(\tau, s)$ is continuous and infinitely differentiable smooth function.

iv. The infimum $inf_{\tau \in (0,1)}(\theta(\tau, s)) = h(s), \forall s$.

v. The $limsup_{\tau \rightarrow 0+}(\theta(\tau, s)) - h(s) = 0, \tau \in (0, 1)$ and $s \in (-\infty, 0) \cup (0, \infty)$.

vi. The parametric $\theta(\tau_k, s)$ uniform convergence and $lim_{k \rightarrow \infty} sup \frac{\partial \theta(\tau_k, s)}{\partial s} = 0$, for $\tau_k \rightarrow 0$ as $k \rightarrow \infty$.

Thus, from the above boundedness property of jump function in above assumption we have,

$$1 - \theta(\tau, s) \leq h(-s) \leq \theta(\tau, -s), \tag{50}$$

It follows that, $E(1 - \theta(\tau, s)) \leq E(h(-s)) \leq E(\theta(\tau, -s))$, from property of expectation. Hence,

$p(u;x) := Pr\{y(u,\xi,x) \le y_{max}\} \ge \alpha, \forall x \in D \Rightarrow Pr\{\xi \in \Omega : \gamma(x,u,\xi) \le 0\} = E(h(s))\big|_{s=\gamma(x,u,\xi) \le 0} \ge \alpha, \forall x \in D.$
Now, based on the parametric function θ, the following functions are defined

$$\varphi(\tau,u;t,x) := E[\theta(\tau, -\gamma(u,\xi;t,x)]), \tag{51}$$

$$\psi(\tau,u;t,x) := 1 - E[\theta(\tau,\gamma(u,\xi;t,x))], \tau \in (0,1). \tag{52}$$

Under the measure zero property and smoothness properties of $\gamma(u,\xi;t,x)$, the functions $\psi(\tau,u;t,x)$ and $\varphi(\tau,u;t,x)$ can be shown to be smoothing approximations of $1 - p(u;t,x)$ and $p(u;t,x)$, respectively (see Geletu et al. [6]). Moreover, the following convergence properties

$$\inf_{\tau \in (0,1)} \varphi(\tau,u;t,x) = p(u;t,x); \tag{53}$$

$$\sup_{\tau \in (0,1)} (1 - \psi(\tau,u;t,x)) = p(u;t,x); \tag{54}$$

and the Lebesgue dominance convergence properties analyze the almost indeed convergence properties of the inner and outer function sequence. The detail convergence approach of the outer approach to CCPDE has been studied in the work [6, 13]. However, the convergence analysis for the inner approximation to the optimal solution of CCPDE of the smooth function IA_τ has not been properly analyzed in the previous work [13] because the study needs the convex approximations and subdifferentiability for probabilistic constraints as the convergence of stationary points of IA_τ is very relevant for the existence of the optimal solution.

For several chance constraints $p_i(u) \ge \alpha, i = 1, 2, \ldots, m$, the regularity is given by $\{u \in U | p_i(u) \ge \alpha, i = 1, 2, \ldots, m\} = cl\{u \in U | p_i(u) > \alpha, i = i = 1, 2, \ldots, m\}$ [29]. For continuously differentiable probability functions p_i, a sufficient condition for the validity of the regularity assumption is the satisfaction of the Mangasarian-Fromowitz constraint qualification (MFCQ) at the active points, [16, 29] Proposition 3.7.

The respective feasible sets of inner and outer approximations are defined as follows:

$$\mathcal{M}(\tau) := \{u \in U \mid \psi(\tau,u;x) \le 1 - \alpha, (x) \in D\}; \tag{55}$$

$$\mathcal{S}(\tau) := \{u \in U \mid \varphi(\tau,u;x) \ge \alpha, (x) \in D\}. \tag{56}$$

As a consequence of the properties of the functions $\psi(\tau,u;x)$ and $\varphi(\tau,u;x)$, we have the following relations among the feasible sets of CCPDE, IA_τ, and OA_τ. That is,

$$M(\tau) \subset \mathcal{P} \subset S(\tau), \text{for} \tau \in (0,1). \tag{57}$$

The tightness property of the relaxation of IA and compression of OA sequentially analyzed in the previous work [29], if $0 < \tau_2 \le \tau_1 < 1$, then.

$$\mathcal{M}(\tau_2,x) \subset \mathcal{M}(\tau_1) \subset \mathcal{P}(x) \subset \mathcal{S}(\tau_1,x) \subset \mathcal{S}(\tau_2). \tag{58}$$

It leads $lim_{\tau \to 0+} S(\tau) = \cap_{\tau \in (0,1)} S(\tau) = P$ and $lim_{\tau \to 0+} M(\tau) = Cl\{\cup_{\tau \in (0,1)} M(\tau)\} = P$. Both $M(\tau)$ and $S(\tau)$ are closed sets as $\tau \to 0+$; see the property in proposition 5.1 and 5.2 in the next section.

The new monotonic properties of the objective function $E(J(T(u,.)))$ with respect to τ are a continuous function on $u \in U_{adm}$. If $0 < \tau_1 < \tau_2 < 1$ with $inf\ [\varnothing] = \infty$,

$$\begin{aligned} inf_{\ u \in M(\tau_2)}[E(J(T[u]))] \geq inf_{\ u \in M(\tau_1)}[E(J(T[u]))] \geq inf_{\ u \in P}[E(J(T[u]))] \\ \geq inf_{\ u \in S(\tau_1)}[E(J(T[u]))] \geq inf_{\ u \in S(\tau_2)}[E(J(T[u]))], \forall \tau \in (0,1). \end{aligned} \tag{59}$$

It implies the following compact form conditions of nondegenerate fuzzy optimality,

$$\begin{aligned} inf_{\ u \in M(\tau_2)}[E(J(T[u])) + \mu_i(\alpha - (1 - \psi_i(u_i))] \geq inf_{\ u \in M(\tau_1)}[E(J(T[u])) + \mu_i(\alpha - (1 - \psi_i(u_i))] \\ \geq inf_{\ u \in P}[E(J(T[u])) + \mu_i(\alpha - \mathcal{P}_i(u_i)] \leq inf_{\ u \in S(\tau_1)}[E(J(T[u])) + \mu_i(\alpha - \varphi_i(u_i)] \\ \geq inf_{\ u \in S(\tau_2)}[E(J(T[u])) + \mu_i(\alpha - \varphi_i(u_i)], \forall \tau \in (0,1). \end{aligned} \tag{60}$$

The complementary property of the probability function $1 - p(u;x) \leq \psi(\tau, u; x) \leq 1 - \alpha$ and $\varphi(\tau, u; x) \geq p(x) \geq \alpha$, for $x \in D$ hold true. Now, using the parametric functions $\psi(\tau, \cdot)$ and $\varphi(\tau, \cdot)$, we define the following problems with the same objective function q in the equation given (38) as in CCPDE.

$$\begin{array}{llll} & \min\limits_u q(u) \quad (IA_\tau) & & \min\limits_u q(u) \quad (OA_\tau) \\ \text{s.t.} & & \text{s.t.} & \\ & \psi(\tau, u; x) \leq 1 - \alpha, x \in D_c & & \varphi(\tau, u; x) \geq \alpha, \\ & u \in U, & & u \in U. \end{array} \tag{61}$$

The feasible set of CCPDE defined in the above set P in Eq. (41) has a property that satisfies Mangasarian-Fromowitz constraint qualification (MFCQ). It is an important prerequisite for applying the necessary optimality criteria in nonlinear optimization. The MFCQ is a condition for the regularity of a permissible point. The MFCQ is in one of point x, and if this point is a local minimum, then the Karush-Kuhn-Tucker conditions are met at this point. If the MFCQ is valid, it is easy to check whether a given point is optimal or not; see the work [13]. From Section 6 proposition (3.10), it follows that the convexity of inner approximation is convex conservative, ψ is concave function or all fixed x in D and $\xi \in \Omega$. The convexity of φ analyzed in the work [13]. The Slater condition is a sufficient condition for strong duality to hold for a convex optimization problem named after Slater; for further analysis, please see the work in [13].

The necessary optimality condition is not valid for expressing local optimal solution in this technique, but can be shown to hold true for generalized stationary points of Fritz–John types [27, 29]. This essentially requires the uniform convergence of partial gradients of inner and outer functions $\varphi_i(\tau, u, \xi)$ and $1 - \psi_i(\tau, u, \xi)$ of $\nabla\varphi_i(\tau, u, \xi) = 0$ and $\nabla(1 - \psi_i(\tau, u, \xi)) = 0$, for $\tau_k \to 0+$, as $k \to \infty$ on bounded subsets W of U_{adm} [29]. In addition, each strict local minimum of CCPDE is a cluster point of local minima of the inner approximation problems IA_τ under the satisfaction of tightness or the outer approximation problems OA_τ; see ([29], Proposition 3.3–3.7). The following sections are more about the structural properties of approximation functions such as closedness, convergence, and differentiability of IA_τ and OA_τ; those properties are not properly analyzed.

5. Closedness, convergence, and differentiability of IA_τ and OA_τ

5.1 Closedness property of M set of IA_τ and S set of OA_τ

The nice properties of the parametric function are given in assumption (4.1); closedness property is defined by the distance from a particular point on the P to set $MorS$ (this distance is close to zero). It is called Hausdorff distance (point to set distance). The specific point of P is closed under the arbitrary sequence of the set M_n and S_n. This property is an essential property for the compactness of the specified M and S. In this case, compactness is boundedness, closedness, and any convergent sequence has a convergent subsequence with the same limit point. The boundness and subsequential convergence property have been clear from (4.1) and tightness property.

Let $u \in U_{adm}$ be separable Banach space and $U_{adm}{}^*$ is dual space of U_{adm}, assume that the $\gamma(x,.,.)$ are the continuous differentiable functions as shown in proportions (3.8) and (3.9) for the continuity property of $P_x(u)$ for any fixed $x \in D$. Then, it satisfies the compactness of set P, which is a feasible set of CCPDE and the conditions as shown (proposition 3.7 in the work [11]). The $\gamma(x,u,.)$ is a Borel measurable w.r.t. $\xi \in \Omega$, thus, $P = \{u \in U_{adm} : P(u) \geq \alpha\}$ is weakly closed in the proposition (3.11), it is from the property of the wssc of P. Now, the closedness property of M and S proved in the following proposition.

Proposition 5.1. *Assume that the $\gamma(u,.,x)$ are Borel measurable for all $u \in U$ and $x \in D$ and $\gamma(.,x,\xi)$ are weakly sequentially semicontinuous (wss) for $x \in D$ and $\xi \in \Omega$. Then $\varphi(\gamma(x,u,\xi),\tau)$ and $\psi(\gamma(x,u,\xi),\tau)$ are wssc and $\gamma(x,u,\xi) = y - y_{max} \leq 0, \forall x \in D$.*

Proof. This is extension of the proposition 3.1 and 3.2. The prove is directly related with proofs of those propositions. Observe first, that ρ is well defined by Borel measurability of γ in the second argument. This γ is wssc, then the parametric function $\theta(\gamma,\tau)$ is wssc. It satisfies the six properties of monotonicity, smoothness, boundedness, and uniform convergence property are expressed in Assumption 4.1. The inner function and outer functions are continuous.

Proposition 5.2. *The function $1-\psi(u,x)$ of M of IA_τ and $\varphi(u,x)$ of OA_τ are continuous from the continuity property $\theta(\tau,\gamma)$. The set M of IA_τ and the set S of OA_τ are closed if the following Hausdorff distance holds true, and there are constants $\forall\varepsilon > 0, \sigma > 0$, such that with d referring to the Hausdorff distance $d(u,\{u \in U_{adm} | \varphi(u,x) \geq \alpha\}) \leq \sigma max\{\log(\alpha) - \log(\varphi(u,x)), 0\}, \forall\alpha \in [1-\varepsilon, 1]$. The same case for IA, there are constants $\forall\varepsilon > 0, \sigma > 0$, such that with d referring to the Hausdorff distance $d(u,\{u \in U_{adm} | 1-\psi(u,x) \geq \alpha\}) \leq \sigma max\{\log(\alpha) - \log(1-\psi), 0\}, \forall\alpha \in [1-\varepsilon, 1]$.*

Proof. The proof directly follows from the nice property of expectation in the given assumptions (4.1, i–vi). Moreover, the proposition (3.11), the inequality $E(h(\gamma)) \geq \alpha$ is bounded by the two parametric functions $\varphi(u,x)$, and $1-\psi(u,x)$. This is a direct extension of the lemma (3.12), we have shown that, for arbitrary $u_1, u_2 \in U_{adm}$ and $\lambda \in [0,1]$ the inequality $\varphi(\lambda u_1 + (1-\lambda)u_2) \leq \lambda\varphi(u_1) + (1-\lambda)\varphi(u_2)$ and $(1-\psi(\lambda u_1 + (1-\lambda)u_2)) \geq \lambda(1-\psi)(u_1) + (1-\lambda)(1-\psi(u_2))$ hold true from convexity property in (3.10), so $\psi(u)$ is a concave function. This means that $log\varphi$, $\log(1-\psi)$ are the log-convex functions and, hence, the inequality $\mu(\varphi(u)) \geq \alpha$ and $\mu(\psi(u)) \leq 1-\alpha$ is equivalent with $\hat{P}(u) \leq -\log(\alpha)$, where $\hat{P}(u) := -logP$ is a log-convex function. The Proposition (3.11), P(u) is wsus, the expectation of P(u) is $\hat{P}$, is wsls from continuity property in proposition (3.8) and (3.9). The function given in proportion (3.10) is a convex function. From Robison-Ursescu theorem ([16], Lemma 4), the continuous convex function is closed. Therefore, the probabilistic constrained function P(u) is closed.

We have proved that the sets M and S are defined in the following section, are the convex feasible sets. It has a closed and convex graph. Consider an arbitrary sequence $(u_n, t_n) \to (\hat{u}, \hat{t})$ with $u_n \in M$ of IA and $(u_n, t_n) \to (\hat{u}, \hat{t})$ with $u_n \in S$ of OA. Then, $u_n \in U_{adm}$ and, hence,$u \in U_{adm}$ by closedness of U_{adm} see in Proposition (3.11).

Moreover, $\hat{\varphi}(u_n) \le t_n$ and $\hat{\psi}(u_n) \le t_n$. Since $\hat{\varphi}$ and $\hat{\psi}$ are wsls, we derive that $\hat{\varphi}(\hat{u}) \le liminf_{n\to\infty}\hat{\varphi}(u_n) \le liminf_{n\to\infty}t_n = \hat{t}$ and $\hat{\psi}(\hat{u}) \le liminf_{n\to\infty}\hat{\psi}(u_n) \le liminf_{n\to\infty}t_n = \hat{t}$. So, $\hat{t} \in M(\hat{u})$ and $\hat{t} \in S(\hat{u})$ implying that the graph of M is closed.

The same case for OA, we have that $\hat{\psi}(u_1) \le t_1$ and $\hat{\psi}(u_2) \le t_2$. Then, convexity of $hat\varphi$ yields that $\hat{\psi}(\lambda u_1 + (1-\lambda)u_2) \le \lambda t_1 + (1-\lambda)t_2$. In other words, $\lambda t_1 + (1-\lambda)t_2 \in S(\lambda u_1 + (1-\lambda)u_2)$, proving that the graph of S is also convex. Further analysis on the convexity of IA is in the following section.

For the properties of closedness of IA and OA in the infinite-dimensional space, we need further analysis of Robinson- Ursescu theorem; see [16]. It is depend on the Hausdorff distance between the point of the set and the set of probabilistic constrained function expressed in Eq. (41). Thus, closedness and boundedness are sufficient for compactness in the finite-dimensional case.

Lemma 5.3. *Under the assumptions of Proposition 3.7, there are constants* $\varepsilon > 0, \sigma > 0$, *such that with d referring to the Hausdorff distance* $d(u \in M, u \in U_{adm} | \psi(u,x) \le 1-\tau) \le \sigma max\{\log(\tau) - \log(\psi(u,x)), 0\} \forall \tau \in [\alpha - \varepsilon, \alpha + \varepsilon]$ *for IA and* $d(u \in S, u \in U_{adm} | \varphi(u,x) \ge \tau) \le \sigma max\{\log(\tau) - \log(\varphi(u,x)), 0\} \forall \tau \in [\alpha - \varepsilon, \alpha + \varepsilon]$.

The prove of this lemma is analyzed in [16].

5.2 Convergence of the stationary points and differentiability of IA_τ and OA_τ

The continuity and continuous differentiability of the functions $\psi(s,\tau), \varphi(s,\tau)$ are directly from the continuity and continuous differentiability of the parametric function $\theta(\tau, s)$. The expectation of a continuous function is also continuous. Additionally, the expectation of a continuously differentiable function is continuously differentiable.

The associated probability functions $P_x(u) \ge \alpha, \forall x \in D$, in the chance constraints, are allowed to be lower semicontinuous or continuous in Proposition (3.8). Also, it cannot be differentiable. Hence, the characterization of optimality properties of CCPDE calls for generalized subdifferentiation such as Fréschet, Clarke, and Mordukhovich subderivative and implicit formula of gradient computations on the reflexive and separable Bochner space. It is analyzed in the previous submitted work in [25]. This concept is applicable to lower semicontinuous (lsc) functions. Furthermore, the epigraph of lsc function is closed everywhere. These subdifferentials assure the optimality criteria for the CCPDE. Also, the $P_x(u) \ge \alpha, x \in D$ are Lipschitzian functions, assuring the explicit formula for the Clarke subgradient under special conditions.

For each $u \in U_{adm}$, there is some neighborhood $W \subset U_{adm}$ and some measurable function $q : \mathbb{R}^n \to \mathbb{R}+$ such that, for continuously partially differentiable up to order b on $u_o \in U_{adm}$ for $\xi \in \Omega$ almost surely (a.s.),

$$sup_{u_j \in W} max \sum_{j=1}^{n} {}_{m_j} \left| \frac{\partial \sum l_{i} \gamma}{\partial u_1^{m_1} \partial u_2^{m_2} \partial u_3^{m_3} \dots \partial u_j^{m_j}} \right| \le q(\xi); \tag{62}$$

$$E(q) = \int_{\Omega} q(\xi)\phi(\xi)d\xi < \infty, \tag{63}$$

for $b = \sum_{j=1}^{n} l_j \geq \sum_{i=1}^{j} m_i$. Suppose each property above holds true. Then, for the parametric function $\theta = \mathcal{H}$, the functions $\varphi(\gamma, \tau)$ and $\psi(\gamma, \tau)$ are continuously partially differentiable w.r.t. u up to order b on U_{adm}, for all $\tau \in (0, 1)$. The higher derivatives

$$\frac{\partial^{\sum_{j=1}^{n} l_j}(1 - \psi(\mathcal{H}(\tau, u))}{\partial u_1^{m_1} \partial u_2^{m_2} \partial u_3^{m_3} \ldots . \partial u_j^{m_j}} = \int_{\Omega} \frac{\partial^{\sum_{j=1}^{n} l_j} \psi(\gamma, \tau, u)\phi(\xi)d\xi}{\partial u_1^{m_1} \partial u_2^{m_2} \partial u_3^{m_3} \ldots . \partial u_j^{m_j}} \phi(\xi)d\xi; \tag{64}$$

for the case for outer approximation,

$$\frac{\partial^{\sum_{j=1}^{n} l_j} \varphi(\mathcal{H}(\tau, u)}{\partial u_1^{m_1} \partial u_2^{m_2} \partial u_3^{m_3} \ldots . \partial u_j^{m_j}} = \int_{\Omega} \frac{\partial^{\sum_{j=1}^{n} l_j} \varphi(\gamma, \tau, u)\phi(\xi)d\xi}{\partial u_1^{m_1} \partial u_2^{m_2} \partial u_3^{m_3} \ldots . \partial u_j^{m_j}} \phi(\xi)d\xi; \tag{65}$$

for $\sum_{j=1}^{n} l_j \geq \sum_{i=1}^{j} m_i$. Further analysis of convergence approach and uniform convergence gradient of the approximation functions are seen in [29, 30] of proposition 4.2. All the above properties ensure the application of Lebesgue's dominating convergence theorem [29] for interchanging the integration and differentiation operations. Observe the chain rule yields powers $E(J(T(u))$ till order $l = k$ in the upper estimation of the integrand. The differentiability up to order k of the lower and upper approximating functions $E(\mathcal{H}(\tau, \gamma))$ and $E(\mathcal{H}(\tau, -\gamma))$ converges to the corresponding order derivative of $p(x, u) \forall t \in Ia.s$. It depends mainly on the existence of the corresponding derivatives of the internal function γ w.r.t.u and the finite expectation of these derivatives, uniformly on some neighborhood of $W = B(u, r)$, whenever the family $\mathcal{H}$ is suitably chosen. Thus, it needs the newly generalized derivative in the neighborhood set of W, such as Fréchet and the weak derivatives (Clarke and Mordukhovich derivatives) for computing gradient, stationary points, and their conveniences; see in the previous work [25].

Assumptions on the properties P_1 *to* P_5, together with additional properties in above γ, it is smooth function w.r.t. u. Because the random state solution T is smooth w.r.t u. The differential operator A computed from coercivity properties of a and l, A is invertible and $\frac{\partial T}{\partial u} = A$.

Assumption 5.4. Assume that $u \in \mathcal{P}$, is a neighborhood of the set M and super-set S in Eq. (55). This set $\mathcal{P}$ is a connected neighborhood super set S of the OA_τ and inside the interior set M of IA_τ, the following conditions hold true for the approximation smooth functions.

i. The support of the probability density function (pdf) $supp\{\phi\}$ is compact.

ii. The pdf is continuous on some set $\mathcal{P}$.

iii. The internal function γ and $\nabla_u(\gamma) = A$ are continuous on the set $\mathcal{P} \times supp\{\phi\}$. The functions $\nabla\gamma$ and $\phi(\xi)$ vanish over $\mathcal{P} \times \Omega/supp\{\phi\}$.

iv. The $\nabla\gamma$ is continuous at $\Gamma(u)$ the boundary of S and M, and $\nabla\gamma \neq 0$ for any $\xi \in \Gamma(u)$ in boundary of M and S.

Theorem 5.5. *If the assumption (5.4) holds true for any family of the parametric function* $\{\theta(s_i(u),\tau)\}|_{s_i=\gamma_i}$ *for any* $i\in I$*, the probabilistic function* $p(u)$ *has a continuous differentiable or a continuous gradient in some open ball W around point* $u\in U_{adm}$ *and the gradient of inner and outer approximation functions respectively converges uniformly to the gradient of* $p(u)$ *on W for any decreasing sequence* $\{\tau_k\}_{k\in N}\in(0,1)$ *for any* $i\in I$ *such that*

$$lim_{\,k\to\infty}\, inf_{\,u\in W}\nabla_u(1-\psi_i(\tau_k,u)) = \nabla_u p_i(u); \tag{66}$$

$$lim_{\,k\to\infty} sup_{u\in W}\nabla_u\varphi_i(\tau_k,u) = \nabla_u p_i(u); \tag{67}$$

for a particular $\tau\to 0+$, all of the gradient of the functions is zero. Thus, as $k\to\infty$,

$$lim_{\,\tau_k\to 0+}\, inf_{\,u\in W}\nabla_u(1-\psi_i(\tau_k,u)) = \nabla_u p_i(u) = 0; \tag{68}$$

$$lim_{\,\tau_k\to 0+} sup_{u\in W}\nabla_u\varphi_i(\tau_k,u) = \nabla_u p_i(u) = 0. \tag{69}$$

This theorem has been proved in the previous submitted work [25]. Suppose each property above assumption (5.4) holds true. Then, for the appropriate parametric function $\theta\equiv\mathcal{H}$ the functions $\varphi(\gamma,\tau)$ and $\psi(\gamma,\tau)$ are continuously partial differentiable w.r.t. u up to order b on U_{adm}, for all $\tau\in(0,1)$. The higher derivatives,

$$\partial^{\sum_{j=1}^{n} l_j} E\Big(\mathcal{H}(\tau,\gamma)/\partial u_1\partial u_2\partial u_3.\dots\partial u_j = \int_\Omega \partial^{\sum_{j=1}^{n} l_j}\psi(\gamma,\tau)/\partial u_1\partial u_2\partial u_3.\dots\partial u_j\phi(\xi)d\xi. \tag{70}$$

And the case for outer approximation

$$\partial^{\sum_{j=1}^{n} l_j} E\Big(\mathcal{H}(\tau,-\gamma)/\partial u_1\partial u_2\partial u_3.\dots\partial u_j = \int_\Omega \partial^{\sum_{j=1}^{n} l_j}\varphi(\gamma,\tau,u)/\partial u_1\partial u_2\partial u_3.\dots\partial u_j\phi(\xi)d\xi. \tag{71}$$

Further analysis of convergence approach and uniform convergence gradient of the approximation functions are seen in [29] of proposition 4.2.; thus, the equation given in (70) and (70) are equivalent.

6. Convexity property of inner and outer approximations

We notice that the coefficient parameter is either constant or depends on the $x\in D$, and the convexity issues depend on the following three cases.

Case 1 κ is constant dependent of x.

Then $(u,\xi)\mapsto y(u,x,\xi): U\times\Omega\to\mathbb{R}$ is linear by see Eq. (16), hence, jointly convex w.r.t. (u,ξ), for each $x\in D$ (see Theorem 5.4 of the continuous dependent of the solution on the random parameter.

Case 2 κ is a positive independent x but depends on random variable, and concave function.

Case 3 κ depends on x, positive and nonlinear; we have to use linear approximation of order 1 such as Taylor approximation of order one for k.

For the above two cases, $\kappa\in\mathcal{K}$, the elliptic PDE system defined in Eqs. (3)–(4) becomes

$$-\Delta y(x,\xi) = \frac{1}{\kappa} f(u,x), \text{for } x \in D, \tag{72}$$

$$y(x,u,\xi) = g(x,u,\xi), \text{for} x \in \partial D, \quad \xi \in \Omega \text{a.s.} \tag{73}$$

For all three cases, the κ is independent or depends on x we can solve the PDE system by stochastic finite difference method (SFDM). It follows $(\mathcal{A}(\kappa))^{-1} = \frac{1}{\kappa}\mathcal{A}^{-1}$ and thus

$$y(u,x,\xi) = \frac{1}{\kappa}\mathcal{A}^{-1}\left[j_{\mathcal{H}}(f(\cdot)) + kg(\cdot,\xi))\right](x) \leq y_{max}. \tag{74}$$

It is equivalent to

$$\left[\mathcal{A}^{-1} j_{\mathcal{H}}\left(f_{ij}(\cdot)\right) + kg_{ij}(\cdot,\xi))\right](x) - \kappa y_{max}(x) \leq 0. \tag{75}$$

The expression on the left-hand side of the last inequality is jointly convex w.r.t. (u,ξ) because of the linearity of f w.r.t. (u,ξ), the concavity of κ and the nonnegativity of y_{max}. Since

$$Pr\{y(u,x,\xi) \leq y_{max}, \forall x \in D\} = Pr\{\mathcal{A}^{-1}\left[j_{\mathcal{H}}(f(x)) + kg(\cdot,\xi))\right] - \kappa y_{max}(x) \leq 0, \forall x \in D\}. \tag{76}$$

is valid, in Proposition 4.2 [13] yields the following result with measure zero property. Hence, for any arbitrary random function $g \in \mathcal{B}$, the internal part of the probabilistic function is a linear, it implies the quasi-concavity/quasi-convexity of γ. Hence, the probabilistic uniform-constrained function is a convex; see proposition (3.10). For any general case, the convexity of the proposed approach is proved in the following proposition.

Proposition 6.1. *Let* $\gamma(x,u,\xi) = y - y_{maxd} \leq 0, \forall \xi \in \Omega, u \in U_{adm}$ *with a measure zero property see in the proposition 4.2. and* $x \in D$. *If* $\gamma(x,u,\xi)$ *is quasi convex or quasi concave, then the approximation functions* $\psi_i(\tau,u,\xi) = 1 - E(\theta(\gamma(x,u,\xi) \leq 1 - \alpha$ *and* $\varphi_i(\tau,u,\xi) = E(\theta(\tau, -\gamma(x,u,\xi)) \geq \alpha$ *are convex.*

Proof. The outer approximation convexity is proved in [13, 29]. Let $m(s) = e^{-s/\tau}$ is a quasi-convex function for all $s \in \mathbb{R}$ and $\tau \in [0,1]$. Then, the sum and the constant multiple of quasi-convex functions are a quasi-convex. Also, based on this statement $l(s) = 1 + m_2\tau e^{-s/\tau}$ is a quasi-convex function. The reciprocal of $l(s)$ which is

$$(l(s))^{-1} = \left(1 + m_2\tau e^{-s/\tau}\right)^{-1}, \tag{77}$$

is said to be reciprocally quasi-concave [31]. It is from

$$\frac{\partial\left(l(s)^{-1}\right)}{\partial s} = m_2\tau e^{-s/\tau}\left[1 + m_2\tau e^{-s/\tau}\right]^{-2}, \tag{78}$$

$\frac{\partial(l(s))^{-1}}{\partial s} = 0$, as $\tau \to 0+$. The second derivative with respect to s is negative at the stationary point $\tau \to 0+$. To check convexity property through the second-order derivative test

$$\frac{\partial^2\left(l(s)^{-1}\right)}{\partial s^2} = -\left(m_2\tau e^{-s/\tau}\right)\left[1 + m_2\tau e^{-s/\tau}\right]^{-2}\left(1 - 2m_2\tau e^{-s/\tau}\right) < 0, \tag{79}$$

not valid for $0 < m_2 \leq m_1/(1+m_1)$ and $2m_2\tau e^{-s/\tau} > 1$. The function l is a strictly concave function for every $\tau \in (0,1]$. Every concave function is a quasi-concave. Therefore, $(l(s))^{-1} = \left(1+m_2\tau e^{-s/\tau}\right)^{-1}$ is a concave function. Also, it holds the quasi-concavity property. It follows that $\theta(\tau,s) = (1+m_1\tau).^* \left(l(s)^{-1}\right)$ is a constant multiple of concave function and a concave function. The negative of concave function is a convex, it implies that $-\theta(\tau,s) = -(1+m_1\tau).^* l(s))$ is a convex function. Since, the integral of the convex function is convex, $-E(\theta(\tau,s) = -\int_\Omega (1+m_1\tau.^* l(s))\phi(\xi)d\xi$ is a convex function. Furthermore, $\psi(\tau,s) = 1 - E(\theta(\tau,s)) \geq \alpha$ is a convex function, based on the convexity property, which stipulates that the sum of convex functions is convex. Therefore, the inner approximation is generally a convex function. □

Theorem 6.2. *The properties of the functions* $E(\theta(\tau,s)) = 1 - \psi(\tau,s) \geq \alpha$ *and* $\varphi(\tau,u,\xi) = E(\theta(\tau,-s)) \geq \alpha$ *are convex in the proposition (6.1). The regularization parameter* $\rho \geq 0$*, the objective function* $E(J(y(.)))$ *is convex in Eq. (38), If* $P(u,x)$ *is a continuous and convex function w.r.t. u, then the* IA_τ *and the* OA_τ *have the unique optimal solutions as* $\tau \rightarrow 0+$. *The optimal solutions of the approximations converge to the optimal solution of CCPDE.*

This theorem has been proved in the work [13].

7. Numerical implementation and case study

This work considers the stationary boundary value elliptic PDEs, where the randomness comes from the boundary conditions of the PDE system. It analyzes the observed temperature variation and distribution processes widely used in biological applications, such as hyperthermia treatment of cancer tissue in human body organs. This work is an extension of the previous works Kibru et al. [13, 14]. The optimal heating of an enclosed, thermally well-insulated, spatial domain $D_c \subset D$ to the desired temperature y_d is given. The heat injection is elected through a distributed stationary heat source $(y(u^*) = T(u^*))$ [20, 32].

The heat source is assumed to be highly affected by uncertainties, e.g., due to inaccuracies arising from heating devices and inlet heating processes. The boundary condition is supposed to be nonhomogeneous and nonlinear depending on random parameters so that the thermal conductivity is spatially constant, which is not precisely known or position-dependent $x \in D$. Furthermore, despite the specified overall desired temperature y_d is deterministic. The required state solution is $y(x,u^*,\xi)$ is random, and it should be kept below a maximum allowed value y_{max} with a high-reliability level $\alpha \leq 0.95$ in a given subset Dc of D. More practical applications are studied in the work [14, 33].

The numerical random state solution y is obtained from the stochastic finite difference method (SFDM [34]). We have discretized 10,000 points from the x_1 and x_2 axes with the mesh generation. After the solution $y(x = (x_1,x_2),u,\xi)$ obtained from the infinite-dimensional space D of the PDEs, the optimization problem is reduced to finite-dimensional reduced $CCPDE_{red}$. The nonsmooth analysis is relevant for solving this nondifferentiable CCPDE where the random fluctuation comes from the system's boundary. We have developed a smooth parametric approximation called IA and OA in equations.

The variables $\xi \in \Omega$ are identical independently distributed (iid) as standard normal distributed random variables. The samples for the random variable are generated by using the multilevel Monte Carlo method (ML-MCM) sampling approach. Subsequently, the PDE system is solved through the SFDM using a MATLAB implementation at each step of the optimization algorithm. After discretization, the inner and outer approximation problems are solved using the MATLAB Optimization function **fmincon.m**, each decreasing values of $\tau = 10^{-k}, k = 1, \dots, 4$.

In our previous work, we have considered practical applications on hyperthermia treatment (HT) and planning as a case study Kibru et al. [14]. The HT and planning are used as an accompanying strategy in modern clinical cancer therapy [20]. Hyperthermia treatment consists of the heating of tumor tissue in order to subdue or eradicate the growth of tumor cells from a given organ. The hyperthermia treatment procedure aims to heat the tumor tissue in the human body to a given temperature without causing damage to healthy surrounding sensitive tissue due to overheating.

The heating is usually done through multiple electromagnetic (EM) sources, where each EM source generates an electric field $G(x)$ of the heat capacity c and density ρ the phases and amplitude p. As a result, the electric fields facilitate a net power deposition on the tumor region given by [20] (e.g., $Q(x) = \frac{\sigma(\xi)}{2}|G(x)|^2$, where $G(x) = \sum_{j=1}^{N} p_j G_j x$ is a linear superposition of the individual fields and $\sigma(x)$ the electric conductivity. In general, the phases and the power Q, corresponding to each individual antenna, are not known in advance.

Figure 1 displays the surface of the state solution $y(x, u^*, \xi)$ and the adjoint operator P for the optimality criteria of the infinite-dimensional CCPDE problem where

$$S^*(y - y_d) + \rho(u - u^*)) \geq 0, \forall u \in U_{adm}, \tag{80}$$

where $S : U_{adm} \mapsto \mathcal{H}$ is a control to state map,

$$y^* = S(u^*), \tag{81}$$

is the optimal state of CCPDE, since CCPDE is a convex optimization problem w.r.t. u in the proposition expressed (3.10), and

$$P = S^*(y - y_d) = S^*(S(u) - y_d) \tag{82}$$

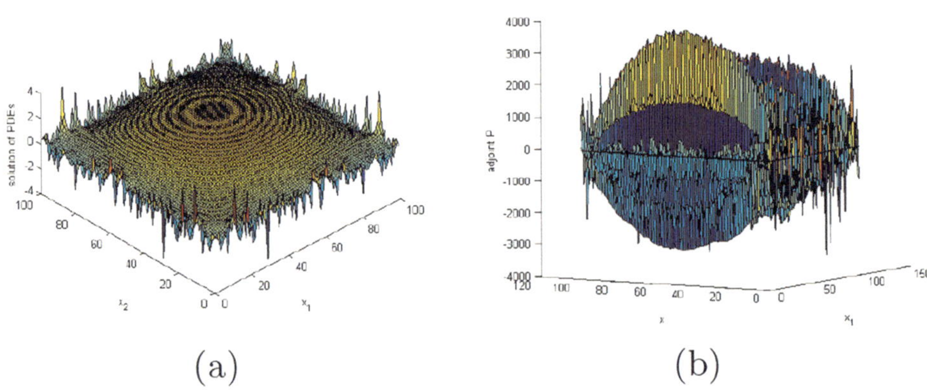

Figure 1.
The solution of PDEs and adjoint P. (a) State y. (b) Adjoint P.

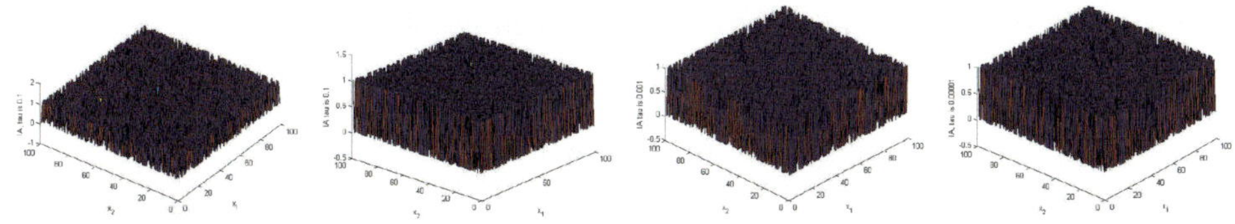

(a) (b) (c) (d)

Figure 2.
The control IA_τ. (a) IA, tau=0.1 (b) IA, tau=0.01 (c) IA, tau=0.001 (d) IA, tau=0.0001.

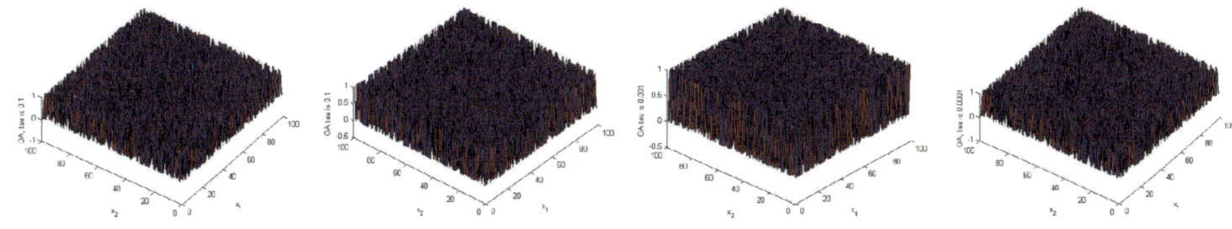

(a) (b) (c) (d)

Figure 3.
The control OA_τ. (a) OA, tau=0.1 (b) OA, tau=0.01 (c) OA, tau=0.001 (d) OA, tau=0.0001.

are displayed **Figure 1** in (a) and (b), respectively. Hence, the state and adjoint are depending on the random variable. So, they are the random state and adjoint functions. The solution of boundary value PDEs is displayed in (a) and (b) from SFDM in this **Figure 1**.

The controls obtained from IA and OA are displayed in **Figures 2** and **3** with different values of $\tau = 10^k$ for $k = 1, \dots, 4$. the optimal controls from the optimization approach of the IA and OA are displayed in (a) to (d) respectively.

The error between the IA and OA is displayed in (a–d) **Figure 4**, when τ is reduced, that mean $\tau \to 0+$ the error is near to zero. This shows that the controls of inner and outer approximation are equal at $\tau = 0.001$.

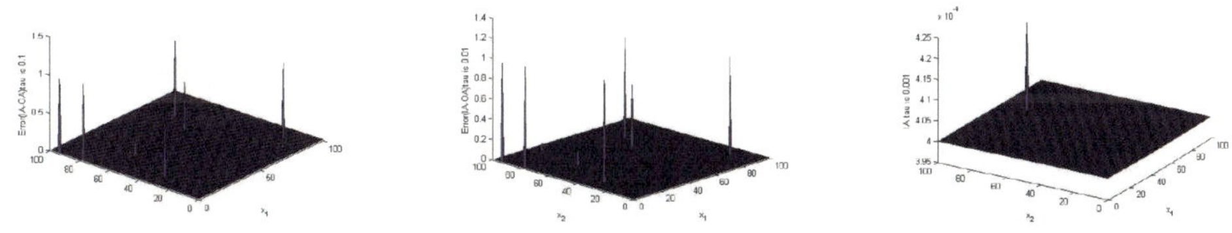

(a) (b) (c)

(d) (e)

Figure 4.
The errors of IA_τ, OA_τ, Obj. (a) Error of IA and OA, tau=0.1; (b) Error of IA and OA, tau=0.01; (c) Error of IA and OA, tau=0.001; (d) Error of IA and OA, tau=0.0001; (e) Error of IA and OA, objective of OA and IA.

Finally, the **Figure 4e** shows the objective functions of $J_{IA}(\tau)$ and $J_{OA}(\tau)$ when $\tau \rightarrow 0+$, the $IA \equiv OA, \tau = 10^{-4}$. The error is zero when tau is less than 0.001 in **Figure 4a**.

Example: To solve the following CCPDE, where ξ are iid $\mu = 0$ and $\sigma = 1$:

$$-\nabla(\kappa(x)\nabla y) = \Delta(y), \kappa(x) = 1, f(x) = 0, g(u, \xi, x) = (u(x) - y)\xi, y_{max} = 2.999; \quad (83)$$

$$yd = 2 - 2.^*(x1.^*(x1 - 1) + (x2.^*(x2 - 1))), u_{min} = 0.5, u_{max} = 4.5; \quad (84)$$

$$\rho = 10^{-3}, \alpha = 0.95; \quad (85)$$

$$D = [0, 1] \times [0, 1]. \quad (86)$$

Acknowledgements

This work is financially supported by KAAD and DFG.

Author details

Kibru Teka*, Abebe Geletu and Pu Li
Process Optimization Group, Technical University of Ilmenau, Ilmenau, Germany

*Address all correspondence to: kibru-teka.nida@tu-ilmenau.de

References

[1] Robert AA, Fournier JJF. Sobolev Spaces. In: Pure and Applied Mathematics. Second ed. Vol. 140. Amsterdam: Acadamic press (Elsevier); 2003

[2] Babusk I, Nobile F, Tempone R. A Stochastic collocation method for elliptic partial differential equations with random input data. SIAM Review. 2010; **52**(2):317-355

[3] Babuska I, Chelbun J. Effects of uncertainities in the domain on the solution of Neumann boundary value problems in two spatial dimensions. Mathematics of Computation. 2001;**71**: 1339-1370

[4] Babuska I, Tempone R, Zourakis GE. Galerkin finite element approximations of stochastic elliptic partial differential equations. SIAM Journal on Numerical Analysis. 2004;**24**:800-825

[5] Babuska I, Tempone R, Zourakis GE. Solving elliptic boundary value problems with uncertain coefficients by the finite element method - the stochastic formulation. Computer Methods in Applied Mechanics and Engineering. 2005;**194**:1251-1294

[6] Geletu A, Hoffmann A, Klöppel M, Li P. An inner-outer approximation approach to chance constrained optimization. SIAM Journal on Optimization SIAM Journal on Optimization. 2017;**27**(3):1834-1857

[7] Ahmed NU. Stablity class of stochastic distributed parameter systems with random boundary conditions. Journal of Mathematical Analysis and Applications. 1983;**92**:274-298

[8] Alla A, Volkwein S. Asymptotic stability of POD based model preditive control for a semilinear parabolic PDE. Advances in Computational Mathematics. 2015;**41**:1073-1102

[9] Ghanem RG, Kreuger RM. Numerical solution of spectral Stochastic finite element system. Computer Methods in Applied Mechanics and Engineering. 1996;**168**(289–303):1996

[10] Ghanem R. Ingredients for a general perpose Stochastic finite element formulation. Computer Methods in Applied Mechanics and Engineering. 1999;**168**(19–34):1999

[11] Pre'kopa A. Stochastic programming, Text book on stochastic optimization. In: Mathematics and Its Applications (MAIA, volume 324). Dordreicht: Kluwer; 1995. ISBN 978-90-481-4552-2 ISBN 978-94-017-3087-7

[12] Casas E, Töltzsch F. First-and second-order optimality conditions for a class of optimal control problems with quasilinear ellipitc equations. SIAM Journal on Control and Optimization. 2009;**49**(2):688-718

[13] Geletu A, Hoffmann A, Schmidt P, Li P. Chance constrained Optimization of ellptic PDE system with smoothing convex approximations. International Journal of Systems Science. 2018;**26**:70 1991 Mathematics subject classification

[14] Teka K, Geletu A, Li P. A Computation Approach to Chance Constrained Optimization of Boundary-Value Parabolic Partial Differential Equation Systems. In: IFAC World Congress 2020: Control and Athomatica. Berlin, Germany: Paper Plaza; 2020

[15] Kibru TN, Geletu A, Li P. A computation approach to chance constrained optimization of

boundary-value parabolic partial differential equation systems. In: Publication in IFAC World Congress 2020: Control and Athomatica. 2020

[16] Farshbaf-Shaker MH, Henrion R, Hömberg D. Properties of chance constraints in infinite dimessional with an Application to PDE consrained optimization. Set Valued Variational Analysis, Mathematics subject classification. 2018;**26**:821-841

[17] Altmüller N, Grüne L. Distributed and boundary model predictive control for the heat equation. GAMM-Mitteilungen. 2012;**35**(2):131-145

[18] Brezis H. Functional Analysis, Sobolev Spaces and Partial Differential Equations. USA: Book of springer in Applied Mathematics; 2011

[19] Karhunen K. Über lineare Methoden in der Wahrscheinlichkeitsrechnung. Annales Academiæ Scientiarum Fennicæ Series A. I. 1947;**37**(3–79):18

[20] Deuflhard P, Schiela A, Martin W. Mathematical Cancer Therapy Planning in Deep Regional Hyperthermia. Berlin; ZIB-Report: Konrad-Zuse-Zentrum für Informationstechnik Berlin; 2011. pp. 11-39

[21] Hille E, Plillips RS. Functional Analysis and semi-Groups. Vol. 31. Providence, Rhode Island: American Mathematical Society Colloquium Publications; 2018). International Journal of Systems Science, Book of Functional Analysis 2000 Mathematics Subject Classification. P. 47-02, sec. 46-02, 47Dxx

[22] Tröeltzsh F. Optimal control of partial differential equations theory, methods and applications. Rhode Island: American Mathematical society providence; 2005). 2000 MSC

[23] Grossman C, Roos HG, Stynes M. Numerical treatment of PDEs. In: Springer Verlag, Berlin, Heidelberg. New York: Mathematics subject classification; 2007

[24] Kouri DP, Heinkenschloss M, Ridzal D, B: G., Waanders VB. A trudt region Algorithm with adaptive stochastic collocation for PDE optimization under uncertainity. SIAM Journal on Scientific Computing. 2013;**35**(4):A1847-A1879

[25] Kibru TN, Geletu A, Li P. New development on the properties of generalized derivatives and smoothing approximation approach for the chance constrained optimization of boundary value elliptic PDE systems. Submitted for publication in AMS; 2022

[26] Nemiroroviski A, Shapiro A. Convex approximation of chance constraint programs. SIAM Journal on Optimization. 2006;**17**:969-996

[27] Zowe J, Kurcyuusz S. Regularity and stablity for the mathematical programming in Banach space. Applied Mathematics and Optimization. WiJrzburg, Am Hubland, West Germany. 1979;**5**:49-62

[28] van Ackooij W, Aleksovsca I, Munoz-Zuniga M. (Sub) diffrentiability of Probablity Function with Elliptical Distributions, Mathematics Subject Classification. WiJrzburg, Am Hubland, West Germany: Springer Science; 2010

[29] Geletu A, Hoffmann A, Li P. Analytic Approximation and Differentiablity of Joint Chance Constraints. Tu-Ilmenau, Germany: Taylor and Francis; 2019 AMS classification, 90c15; 90c31; 26A42: 57R12: 58B10

[30] Hantoute A, Henrion R, Pérez-Aros P. Subdifferential characterization of

probability functions under Gaussian distribution. Germany, part of Springer Nature and Mathematical Optimization Society: Springer-Verlag GmbH; 2018. p. 2018

[31] Merkele M. Reciprocally convex functions. ELSEVIER. Journal of Mathematical Analysis and Applications. 2004;**293**(1):210-218

[32] Dubljevic S, Christofides PD. Predictive output feedback control of parabolic partial differential equations (PDEs). Industrial and Engineering Chemistry Research. 2006;**45**:8421-8429

[33] Dubljevic S, El-Fara NH, Mhaskar P, Christofides PD. Predictive control of parabolic PDEs with state and control constraints. International Journal of Robust and Nonlinear Control. 2006;**16**: 749-772

[34] Eiermann M, Ernst OG, Ullmann E. Computational aspects of the stochastic finite element method. Computing and Visualization in Science. 2007;**10**(1):3-15

Chapter 8

Isochronous Oscillations of Nonlinear Systems

*Jean Akande, Kolawolé Kêgnidé Damien Adjaï,
Marcellin Nonti and Marc Delphin Monsia*

Abstract

Real-world systems, such as physical and living systems, are generally subject to vibrations that can affect their long-term integrity and safety. Thus, the determination of the law that governs the evolution of the oscillatory quantity has become a major topic in modern engineering design. The process often leads to solving nonlinear differential equations. However, one can admit that the main objective of the theory of differential equations to obtain explicit solutions is far from being carried out. If we know how to solve linear systems, the case of systems of nonlinear differential equations is not in general solved. Isochronous nonlinear systems have therefore received particular attention. This chapter is devoted to presenting some recent developments and advances in the theory of isochronous oscillations of nonlinear systems. The harmonic oscillator as a prototype of isochronous systems is investigated to state some useful definitions (section 2), and the existence of second-order isochronous nonlinear systems having explicit elementary first integrals with an exact sinusoidal solution and higher-order autonomous nonlinear systems that reproduce the dynamics of the harmonic oscillator is proven (section 3). Finally, higher-order nonautonomous nonlinear systems that can exhibit isochronous oscillations are shown (section 4), and a conclusion for the chapter is presented.

Keywords: nonlinear dynamic systems, Hamiltonian systems, higher-order autonomous and nonautonomous equations, isochronous oscillations

1. Introduction

The enormous literature generated by the qualitative theory of dynamic systems suggests that all questions about nonlinear systems are well studied and answered. Far from it, we can see that there are many interesting questions that are not fully resolved. We must note that a dynamic system is a time-dependent system that can undergo regular or chaotic processes. The dynamics of such systems are often described by finite-difference equations (discrete dynamic systems) or differential equations (continuous dynamic systems). Since many problems in engineering physics, biology, and applied mathematics are formulated in terms of differential equations, continuous dynamic systems have been the subject of an intensive investigation in the literature. In particular, planar polynomial dynamic systems given by [1–4]

$$\dot{x} = P(x, y), \quad \dot{y} = Q(x, y), \tag{1}$$

where overdot indicates differentiation with respect to time, and P and Q are polynomials in x and y, are widely investigated from the perspective of the existence of isochronous centers, limit cycles, and elementary first integrals. One question that has been the object of special attention is the determination of the maximum number of limit cycles of the polynomial system (1), motivated by the second part of the Hilbert 16th problem [1]. Another important question is the determination of polynomial and rational first integrals that ensure the complete integrability of polynomial dynamic systems of type (1) inspired by the work of Darboux [2–4]. When the first integrals do not explicitly depend on the time, the system is said, in the case of autonomous systems, conservative system and can exhibit periodic solution. A prototype of dynamic systems that can experience conservative oscillations is the harmonic oscillator described by [5–7]

$$\ddot{x} + x = 0, \tag{2}$$

such that

$$x(t) = \cos t. \tag{3}$$

The harmonic oscillator (2) is characterized by a fixed constant period $T = 2\pi$. Therefore, the system (2) is said to be an isochronous dynamic system in contrast to dynamic systems exhibiting amplitude-dependent frequency oscillations known as nonlinear systems. Nonlinear systems differ from linear systems that exhibit amplitude-independent frequency. A typical example of a nonlinear dynamical system is given by the well-known cubic Duffing equation [5–8]

$$\ddot{x} + a_1\dot{x} + a_2 x + a_3 x^3 = 0, \tag{4}$$

which can exhibit $a_1 = 0$, an amplitude-dependent period, and experience softening and hardening phenomena under a periodic forcing function, where $a_1 > 0$, $a_2 > 0$, and a_3 are constants. Linear systems, such as Eq. (2), cannot exhibit softening and hardening, leading in general to fatigue and failure of material systems [9, 10]. Consequently, the problem of finding dynamic systems, more precisely nonlinear dynamic systems since real-world systems are nonlinear systems, preserving the feature of amplitude-independent frequency, has become a vital question for modern engineering design. Thus, the design and identification of nonlinear isochronous systems have generated a major and attractive research field in the theory of dynamic systems such that the well-established qualitative theory of dynamic systems has been widely applied to identify isochronous centers or systems that can exhibit amplitude-independent periods. In this way, theorems for the existence of a center and an isochronous center are established, particularly for the system (1), where $P(x, y)$ and $Q(x, y)$ are not necessarily polynomials in x and y [5, 6, 11–17]. Additionally, a multitude of approximation methods for periodic solutions has been developed in the literature on the basis of the predictions of the qualitative theory of differential equations and numerical results, while no exact explicit solutions are known. However, many of these studies are not mathematically consistent, as shown by the recent developments and advances in the theory of differential equations due to the considerable contribution of Monsia and coworkers. Consider as an example of illustration the unusual Lienard equation

$$\ddot{x} - \frac{ax}{\sqrt{\mu^2 - x^2}}\dot{x} = 0, \tag{5}$$

investigated by Akande et al. [18], where a and μ are constants. The authors [18] showed that Eq. (5) has the exact isochronous harmonic solution

$$x(t) = \mu \sin\left[-a(t + K)\right], \tag{6}$$

where $a < 0$, $\mu > 0$, $a \neq -\mu$ and K is an arbitrary constant, while Eq. (5) obviously does not satisfy the classical existence theorems for a center for the Lienard equation of the form

$$\ddot{x} + f(x)\dot{x} + g(x) = 0, \tag{7}$$

where $f(x)$ and $g(x)$ are functions of x [5, 11, 14, 16, 17]. The inadequacy of the mentioned theorems can also be shown by considering the following exceptional quadratic Lienard-type [19]

$$\ddot{x} + \frac{u'(x)}{u(x)}\dot{x}^2 = 0, \tag{8}$$

where $u(x)$ is a function of x and prime means differentiation with respect to x. The authors [19] proved that Eq. (8) can exhibit, for example, that when $u(x) = (\mu^2 - x^2)^{-\frac{1}{2}}$, the exact isochronous harmonic solution

$$x(t) = \mu \sin\left[b(t + K_1)\right], \tag{9}$$

where $\mu > 0$, $b > 0$, and K_1 are arbitrary parameters such that $b \neq \mu$. Eq. (8) belongs to the general class of Lienard-type equation

$$\ddot{x} + \vartheta(x)\dot{x}^2 + g(x) = 0, \tag{10}$$

where $\vartheta(x)$ is a function of x. Eq. (10) can be generalized in the form

$$\ddot{x} + h(x, \dot{x})\dot{x} + g(x) = 0, \tag{11}$$

where $h(x, \dot{x})$ is a function of x and $\dot{x}$. Obviously, Eq. (8), where

$$u(x) = \left(\mu^2 - x^2\right)^{-\frac{1}{2}}, \tag{12}$$

does not satisfy the classical theorems for the existence of at least one periodic solution [5, 6] or for the existence of an isochronous center, as stated in Refs. [12, 16, 17]. Other counterexamples of classical existence theorems can be seen in Refs. [20–27]. If some progress has been made with the work of Calogero and coworkers [28], it will be very difficult to say the same thing concerning the dynamic systems represented by nonlinear differential equations having an exact elementary function solution, more precisely an exact explicit isochronous sinusoidal solution before the contribution of Monsia and his group (see Refs. [29–31] and References therein). The work of Monsia and his group revealed not only the inadequacy of the qualitative theory of dynamic systems to predict the effective behavior of nonlinear systems but also showed the existence of many autonomous and nonautonomous nonlinear dynamic systems with an exact explicit isochronous sinusoidal solution of a second and high order. The present chapter aims to contribute to these recent developments and advances in identifying and generating second-order and higher-order autonomous and nonautonomous nonlinear dynamic systems with an exact isochronous sinusoidal solution. To do so, we study the harmonic oscillator considered as the prototype of isochronous systems (section 2), the isochronous oscillations of higher-order autonomous nonlinear systems (section 3), and the isochronous oscillations of higher-order nonautonomous nonlinear systems (section 4). Finally, we present a conclusion for the chapter.

2. Harmonic oscillator

The equation of the harmonic oscillator (2) can be rewritten as a dynamical system in the form

$$\dot{x} = y, \quad \dot{y} = -x, \tag{13}$$

such that the integral curves are given by

$$\frac{dy}{dx} = -\frac{x}{y}. \tag{14}$$

By separation of variables and integration, we have

$$H(x, y) = \frac{1}{2}y^2 + \frac{1}{2}x^2, \tag{15}$$

where H is a constant of integration known as the Hamiltonian or

$$H(x, \dot{x}) = \frac{1}{2}\dot{x}^2 + \frac{1}{2}x^2, \tag{16}$$

so that Eq. (2) is said to be a Hamiltonian system. When $H(x, \dot{x}) = \frac{1}{2}$, the formula

$$x(t) = \cos(t + \varphi), \tag{17}$$

such that

$$\dot{x}(t) = -\sin(t + \varphi), \tag{18}$$

where φ is an arbitrary constant, verifies the first integral (16). Thus, Eq. (17) is the general solution of the harmonic oscillator (2), which exhibits periodic oscillations of period $T = 2\pi$, independent of the oscillation amplitude, as shown in **Figure 1**.

Such oscillations are said to be isochronous. Since all solutions given by Eq (17) are periodic with a fixed constant period T, the harmonic oscillator is called an isochronous system. Therefore we can state the following definitions.

Definition 1: *A system exhibits isochronous oscillations if the period T is independent of amplitude.*

Definition 2: *If the periodic general solution with a fixed constant period T of a system (S) verifies $H(x, \dot{x}) = c$ where c is a constant, then such a system (S) of differential equations corresponding to the Hamiltonian H is an isochronous system.*

On the basis of these definitions, we can investigate the isochronicity of nonlinear systems below.

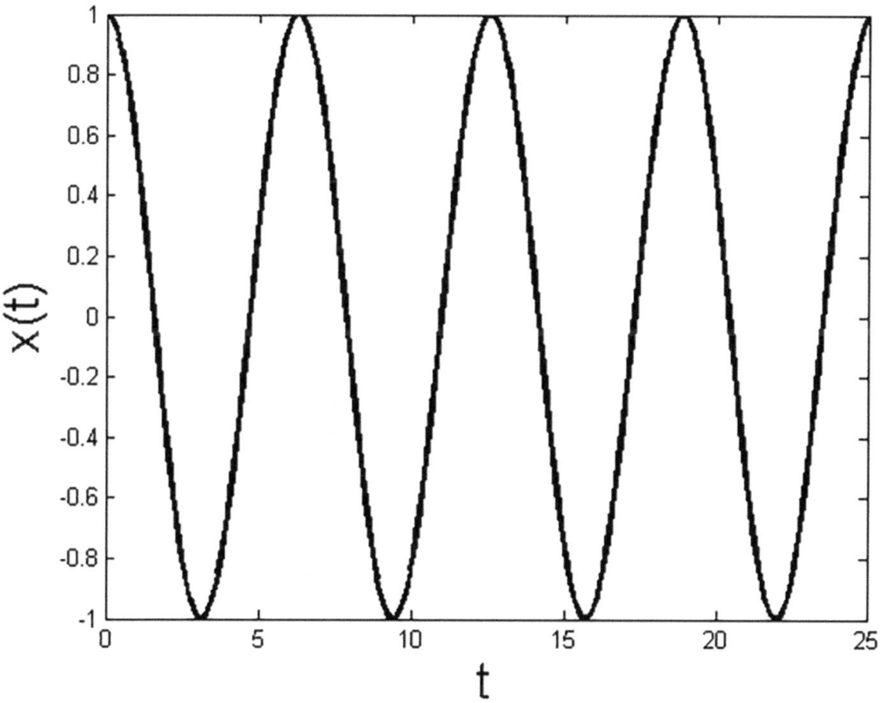

Figure 1.
Typical behavior of solution (17) when $\varphi = 0$.

3. Autonomous nonlinear systems

Recently, Monsia and coworkers introduced a new class of first integrals in the literature [29, 30, 32]. This type of class of first integrals contains $(n+1)$ first integrals $H(x, \dot{x})$ such that $H(x, \dot{x}) = c$ when $x(t) = \cos(t+\varphi)$, where $n \geq 0$ is an integer. The corresponding $(n+1)$ second-order autonomous nonlinear differential equations admit the exact sinusoidal general solution $\cos(t+\varphi)$. In this part, we consider such classes of first integrals to secure isochronous oscillations of autonomous nonlinear systems.

3.1 Isochronous nonlinear systems

Consider a second-order autonomous equation

$$E(x, \dot{x}, \ddot{x}) = 0. \tag{19}$$

Thus, we have the following results.
Theorem 1.1: Assume that

$$H_1(x, \dot{x}) = b = \dot{x}^2 \sum_{\ell=0}^{n} x^{2\ell} + \dot{x}^3 + \dot{x}(x^2 - 1) + x^{2n+2}, \tag{20}$$

is a class of $(n+1)$ first integrals of Eq. (19), where b is a constant and $n \geq 0$ is an integer. Then, Eq. (19) takes the form

$$\frac{d}{dt}H_1(x, \dot{x}) = \ddot{x}\left(2\dot{x}\sum_{\ell=0}^{n} x^{2\ell} + 3\dot{x}^2 + x^2 - 1\right) + 2\dot{x}\left[\dot{x}^2\sum_{\ell=0}^{n} \ell x^{2\ell-1} + x\dot{x} + (n+1)x^{2n+1}\right] = 0, \tag{21}$$

with the exact sinusoidal general solution

$$x(t) = cos\,(t+\varphi) \tag{22}$$

where φ is an arbitrary constant.

Proof. Differentiating with respect to time, the first integral (20) immediately yields Eq. (21). To prove that formula (22) is a solution of Eq. (21), it suffices to prove that Eq. (22) verifies Eq. (20). However, it is also possible to give direct proof by substituting Eq. (22) into Eq. (21). From Eq. (22),

$$\dot{x}(t) = -\sin(t+\varphi), \tag{23}$$

and

$$\ddot{x}(t) = -\cos(t+\varphi), \tag{24}$$

Inserting Eqs. (22)–(24) and the trigonometric equation

$$\cos^2(t+\varphi) + \sin^2(t+\varphi) = 1, \tag{25}$$

into Eq. (21), leads to

$$
\begin{aligned}
&\ddot{x}\left(2\dot{x}\sum_{\ell=0}^{n}x^{2\ell}+3\dot{x}^2+x^2-1\right)+2\dot{x}\left[\dot{x}^2\sum_{\ell=0}^{n}\ell x^{2\ell-1}+x\dot{x}+(n+1)x^{2n+1}\right]\\
&=-\cos(t+\varphi)\left[-2\sin(t+\varphi)\sum_{\ell=0}^{n}\cos^{2\ell}(t+\varphi)+3\sin^2(t+\varphi)+\cos^2(t+\varphi)-1\right]\\
&\quad-2\sin(t+\varphi)\left[\sin^2(t+\varphi)\sum_{\ell=0}^{n}\ell\cos^{2\ell-1}(t+\varphi)-\cos(t+\varphi)\sin(t+\varphi)\right.\\
&\quad\left.+(n+1)\cos^{2n+1}(t+\varphi)\right]\\
&=2\sin(t+\varphi)\sum_{\ell=0}^{n}\cos^{2\ell+1}(t+\varphi)-3\cos(t+\varphi)\sin^2(t+\varphi)-\cos^3(t+\varphi)+\cos(t+\varphi)\\
&\quad-2\sin^3(t+\varphi)\sum_{\ell=0}^{n}\ell\cos^{2\ell-1}(t+\varphi)+2\cos(t+\varphi)\sin^2(t+\varphi)\\
&\quad-2(n+1)\sin(t+\varphi)\cos^{2n+1}(t+\varphi)\\
&=2\sin(t+\varphi)\sum_{\ell=0}^{n}\cos^{2\ell+1}(t+\varphi)-\cos(t+\varphi)\sin^2(t+\varphi)-\cos^3(t+\varphi)+\cos(t+\varphi)\\
&\quad-2\sin(t+\varphi)\left[1-\cos^2(t+\varphi)\right]\sum_{\ell=0}^{n}\ell\cos^{2\ell-1}(t+\varphi)-2(n+1)\sin(t+\varphi)\cos^{2n+1}(t+\varphi)\\
&=2\sin(t+\varphi)\sum_{\ell=0}^{n}\cos^{2\ell+1}(t+\varphi)-\cos(t+\varphi)\left[\sin^2(t+\varphi)+\cos^2(t+\varphi)-1\right]\\
&\quad-2\sin(t+\varphi)\sum_{\ell=0}^{n}\ell\cos^{2\ell-1}(t+\varphi)+2\sin(t+\varphi)\sum_{\ell=0}^{n}\ell\cos^{2\ell+1}(t+\varphi)\\
&\quad-2(n+1)\sin(t+\varphi)\cos^{2n+1}(t+\varphi)\\
&=\sin(t+\varphi)\sum_{\ell=0}^{n}(2\ell+2)\cos^{2\ell+1}(t+\varphi)-\sin(t+\varphi)\sum_{\ell=0}^{n}2\ell\cos^{2\ell-1}(t+\varphi)\\
&\quad-2(n+1)\sin(t+\varphi)\cos^{2n+1}(t+\varphi)\\
&=\sin(t+\varphi)\left[\sum_{\ell=0}^{n-1}(2\ell+2)\cos^{2\ell+1}(t+\varphi)-\sum_{\ell=0}^{n}2\ell\cos^{2\ell-1}(t+\varphi)\right]\\
&=\sin(t+\varphi)\left[2\cos(t+\varphi)+4\cos^3(t+\varphi)+6\cos^5(t+\varphi)+\dots+2n\cos^{2n-1}(t+\varphi)\right.\\
&\quad\left.-\sum_{\ell=0}^{n}2\ell\cos^{2\ell-1}(t+\varphi)\right]\\
&=\sin(t+\varphi)\left[\sum_{\ell=0}^{n}2\ell\cos^{2\ell-1}(t+\varphi)-\sum_{\ell=0}^{n}2\ell\cos^{2\ell-1}(t+\varphi)\right]\\
&=0,
\end{aligned}
\tag{26}
$$

such that Theorem 1.1 is proved.

Remark 1. *The* $(n+1)$ *first integrals given by Eq.* (20) *are the* $(n+1)$ *Hamiltonians of the* $(n+1)$ *equations given by the class of Eq.* (21). *Theorem 1.1 shows that* $H_1(x,\dot{x})$ *is a time-independent constant. One can check that* $H_1(x,\dot{x})=1$ *under* $x(t)=\cos(t+\varphi)$. *Therefore, the* $(n+1)$ *systems of differential Eq.* (21) *are isochronous and exactly reproduce the dynamics of the harmonic oscillator. These results are impossible to predict by the qualitative theory of dynamic systems, mainly by the classical existence theorems [5, 6]. Indeed, the class of Eq.* (21) *can be rewritten as*

$$\ddot{x}+\frac{2\left[\dot{x}^2\sum_{\ell=0}^{n}\ell x^{2\ell-1}+x\dot{x}+(n+1)x^{2n+1}\right]}{3\dot{x}^2+x^2-1+2\dot{x}\sum_{\ell=0}^{n}x^{2\ell}}\dot{x}=0 \tag{27}$$

Eq. (27) has the form of the mixed Lienard-type differential Eq. (11), where

$$h(x,\dot{x})=\frac{2\left[\dot{x}^2\sum_{\ell=0}^{n}\ell x^{2\ell-1}+x\dot{x}+(n+1)x^{2n+1}\right]}{3\dot{x}^2+x^2-1+2\dot{x}\sum_{\ell=0}^{n}x^{2\ell}}, \tag{28}$$

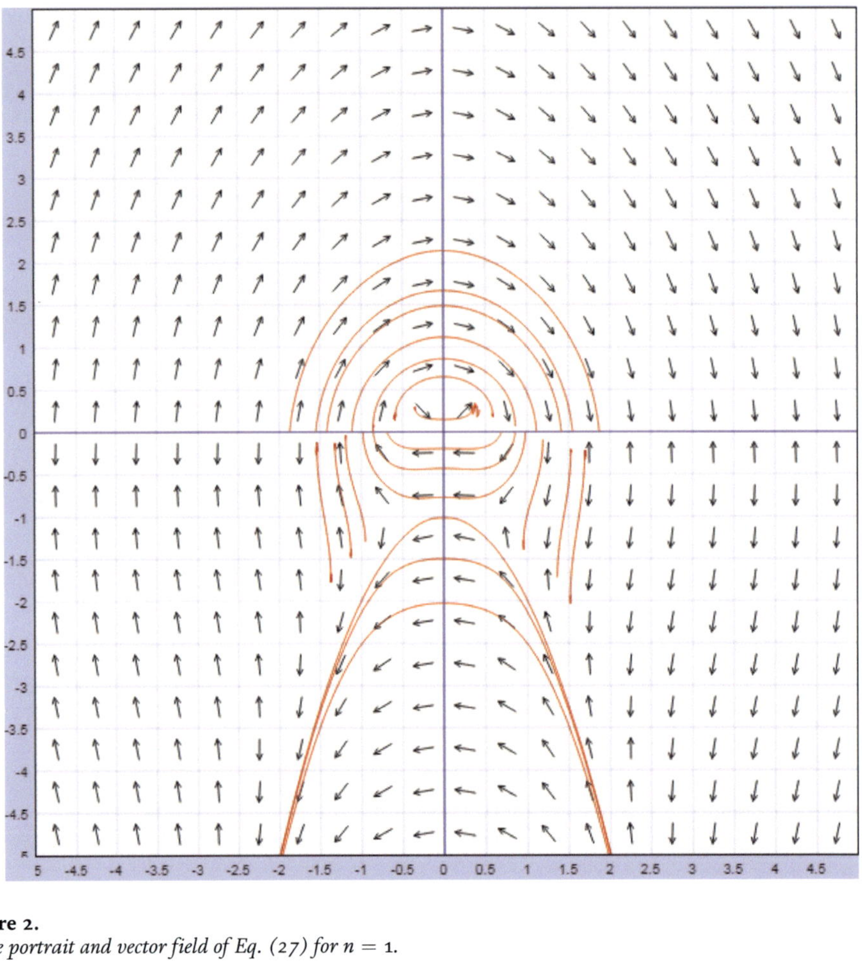

Figure 2.
Phase portrait and vector field of Eq. (27) for $n=1$.

and

$$g(x) = 0. \tag{29}$$

Since $h(0, 0) = 0$ is not negative, $g(x) = 0$ for $x \neq 0$ and $g(x)$ is not odd, then Eq. (21) does not satisfy the classical theorems for the existence of at least one periodic solution (see Theorem 11.2 of ([5], p. 387) and the Lienard-Levinson-Smith theorem of [6]) in contrast to Theorem 1.1. As an example of illustration, let $n = 0$. Thus, Eq. (27) becomes

$$\ddot{x} + \frac{2x(1+\dot{x})}{3\dot{x}^2 + 2\dot{x} + x^2 - 1}\dot{x} = 0 \tag{30}$$

The phase portrait and vector field of Eq. (27) are shown in **Figures 2** and **3** for $n = 1$ and $n = 2$. Consider now the class of the first-order differential equation

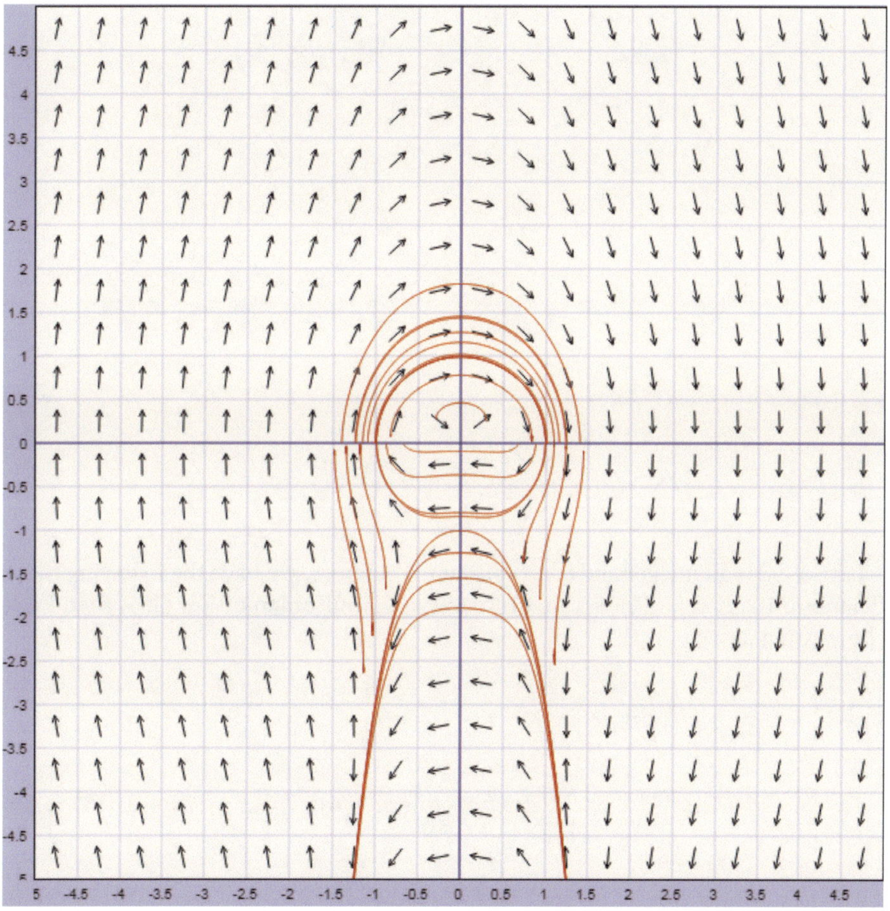

Figure 3.
Phase portrait and vector field of Eq. (27) for n = 2.

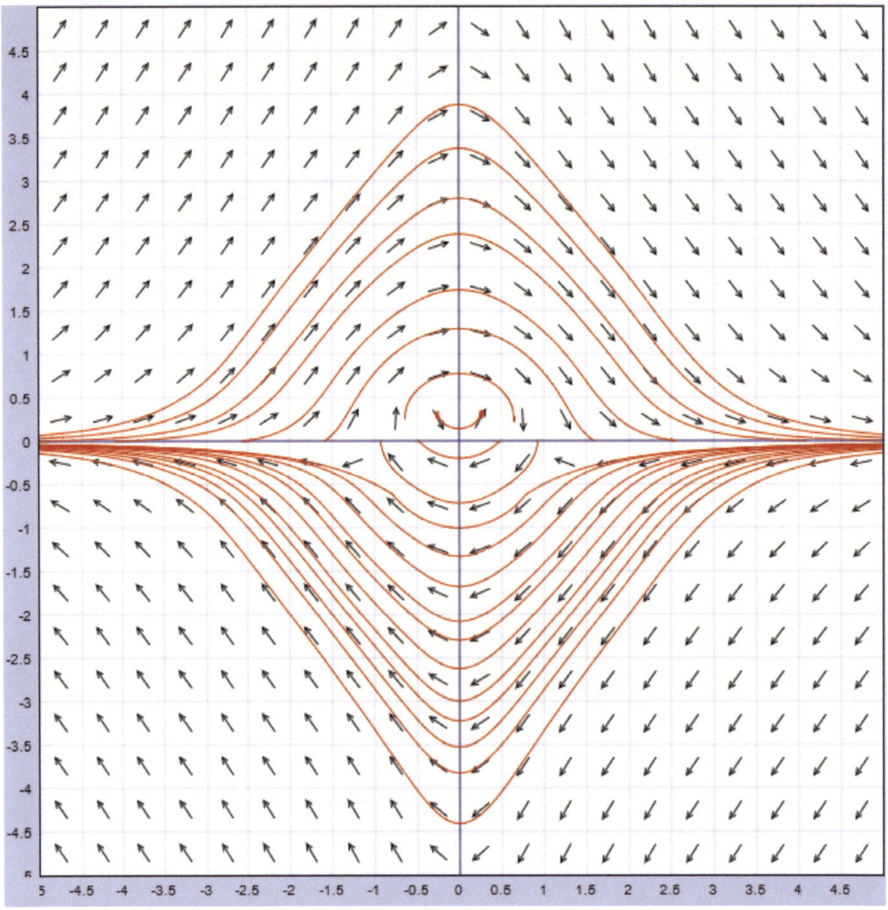

Figure 4.
Phase portrait and vector field of Eq. (36).

$$H_2(x, \dot{x}) = b = \dot{x}^2 \sum_{\ell=0}^{n} x^{2\ell} + \dot{x}^3 \left(1 + \sum_{\ell=0}^{n} x^{2\ell+2}\right) + \dot{x}\left(x^{2n+4} - 1\right) + x^{2n+2}, \tag{31}$$

where $n \geq 0$ is an integer and b is a constant. Therefore, we have the following theorem.

Theorem 1.2: If Eq. (31) is a first integral or Hamiltonian of Eq. (19), then Eq. (19) can be written as

$$\ddot{x}\left[3\dot{x}^2\left(1 + \sum_{\ell=0}^{n} x^{2\ell+2}\right) + 2\dot{x}\sum_{\ell=0}^{n} x^{2\ell} + x^{2n+4} - 1\right]$$
$$+\left\{\dot{x}^3 \sum_{\ell=0}^{n}(2\ell+2)x^{2\ell+1} + \dot{x}^2 \sum_{\ell=0}^{n} 2\ell x^{2\ell-1} + \left[(2n+4)x^{2n+3}\right]\dot{x} + (2n+2)x^{2n+1}\right\}\dot{x} = 0, \tag{32}$$

with the exact general solution

$$x(t) = \cos(t + \varphi). \tag{33}$$

Proof. By differentiation with respect to time, Eq. (31) immediately leads to Eq. (32). It suffices to show that Eq. (33) verifies Eq. (31) to prove that formula (33) is a solution of Eq. (32). Substituting Eqs. (22)–(25) into Eq. (31) yields

$$\begin{aligned}
H_2(x,\dot{x}) &= \sin^2(t+\varphi)\sum_{\ell=0}^{n}\cos^{2\ell}(t+\varphi) - \sin^3(t+\varphi)\left[1+\sum_{\ell=0}^{n}\cos^{2\ell+2}(t+\varphi)\right] \\
&\quad - \sin(t+\varphi)\left[\cos^{2n+4}(t+\varphi)-1\right] + \cos^{2n+2}(t+\varphi) \\
&= \left[1-\cos^2(t+\varphi)\right]\sum_{\ell=0}^{n}\cos^{2\ell}(t+\varphi) - \sin(t+\varphi)\left[1-\cos^2(t+\varphi)\right]\left[1+\sum_{\ell=0}^{n}\cos^{2\ell+2}(t+\varphi)\right] \\
&\quad - \sin(t+\varphi)\left[\cos^{2n+4}(t+\varphi)-1\right] + \cos^{2n+2}(t+\varphi) \\
&= \sum_{\ell=0}^{n}\cos^{2\ell}(t+\varphi) - \sum_{\ell=0}^{n}\cos^{2\ell+2}(t+\varphi) - \sin(t+\varphi) \\
&\quad - \sin(t+\varphi)\sum_{\ell=0}^{n}\cos^{2\ell+2}(t+\varphi) + \sin(t+\varphi)\cos^2(t+\varphi) \\
&\quad + \sin(t+\varphi)\sum_{\ell=0}^{n}\cos^{2\ell+4}(t+\varphi) - \sin(t+\varphi)\cos^{2n+4}(t+\varphi) + \sin(t+\varphi) \\
&\quad + \cos^{2n+2}(t+\varphi) \\
&= \sin(t+\varphi)\left[\cos^2(t+\varphi) + \sum_{\ell=0}^{n}\cos^{2\ell+4}(t+\varphi) - \sum_{\ell=0}^{n}\cos^{2\ell+2}(t+\varphi)\right. \\
&\quad \left. - \cos^{2n+4}(t+\varphi)\right] + \sum_{\ell=0}^{n}\cos^{2\ell}(t+\varphi) - \sum_{\ell=0}^{n}\cos^{2\ell+2}(t+\varphi) + \cos^{2n+2}(t+\varphi) \\
&= \sin(t+\varphi)\left[\cos^2(t+\varphi) + \cos^{2n+4}(t+\varphi) + \sum_{\ell=0}^{n-1}\cos^{2\ell+4}(t+\varphi) - \cos^2(t+\varphi)\right. \\
&\quad \left. - \sum_{\ell=1}^{n}\cos^{2\ell+2}(t+\varphi) - \cos^{2n+4}(t+\varphi)\right] + \sum_{\ell=0}^{n}\cos^{2\ell}(t+\varphi) \\
&\quad - \sum_{\ell=0}^{n-1}\cos^{2\ell+2}(t+\varphi) - \cos^{2n+2}(t+\varphi) + \cos^{2n+2}(t+\varphi) \\
&= \sin(t+\varphi)\left[\sum_{\ell=0}^{n-1}\cos^{2\ell+4}(t+\varphi) - \sum_{\ell=1}^{n}\cos^{2\ell+2}(t+\varphi)\right] \\
&\quad + \left[1+\sum_{\ell=1}^{n}\cos^{2\ell}(t+\varphi) - \sum_{\ell=0}^{n-1}\cos^{2\ell+2}(t+\varphi)\right] \\
&= \sin(t+\varphi)\left[\cos^4(t+\varphi) + \cos^6(t+\varphi) + \ldots + \cos^{2n+2}(t+\varphi)\right. \\
&\quad \left. - \sum_{\ell=1}^{n}\cos^{2\ell+2}(t+\varphi)\right] + 1 + \left[\sum_{\ell=1}^{n}\cos^{2\ell}(t+\varphi)\right. \\
&\quad \left. - \left(\cos^2(t+\varphi) + \cos^4(t+\varphi) + \ldots + \cos^{2n}(t+\varphi)\right)\right] \\
&= 1 + \sin(t+\varphi)\left[\sum_{\ell=1}^{n}\cos^{2\ell+2}(t+\varphi) - \sum_{\ell=1}^{n}\cos^{2\ell+2}(t+\varphi)\right] \\
&\quad + \left[\sum_{\ell=1}^{n}\cos^{2\ell}(t+\varphi) - \sum_{\ell=1}^{n}\cos^{2\ell}(t+\varphi)\right] \\
&= 1,
\end{aligned} \tag{34}$$

proving Theorem 1.2.

Remark 2. *Eq. (32) takes the form*

$$\ddot{x}+\frac{\left\{\dot{x}^3\sum_{\ell=0}^{n}(2\ell+2)x^{2\ell+1}+\dot{x}^2\sum_{\ell=0}^{n}2\ell x^{2\ell-1}+\left[(2n+4)x^{2n+3}\right]\dot{x}+(2n+2)x^{2n+1}\right\}}{\left[3\dot{x}^2\left(1+\sum_{\ell=0}^{n}x^{2\ell+2}\right)+2\dot{x}\sum_{\ell=0}^{n}x^{2\ell}+x^{2n+4}-1\right]}\dot{x}=0, \tag{35}$$

Obviously, Eq. (35) has the form of Eq. (27), so the classical existence theorems cannot predict its general solution (33). As previously stated, Eq. (35) contains $(n+1)$ nonlinear isochronous Hamiltonian oscillators. One can verify that $H_2(x,\dot{x})=1$, when $x(t)=\cos(t+\varphi)$. As an example of Eq. (35), put $n=0$. Then, Eq. (35) becomes

$$\ddot{x}+\frac{2x\dot{x}^3+(4x^3+2x)\dot{x}}{3\dot{x}^2(1+x^2)+2\dot{x}+x^4+x^2-1}\dot{x}=0. \tag{36}$$

The phase diagram and vector field of Eq. (36) are shown in **Figure 4**. **Figures 5** and **6** show the phase portrait and vector field of Eq. (35) for $n=1$ and $n=2$.

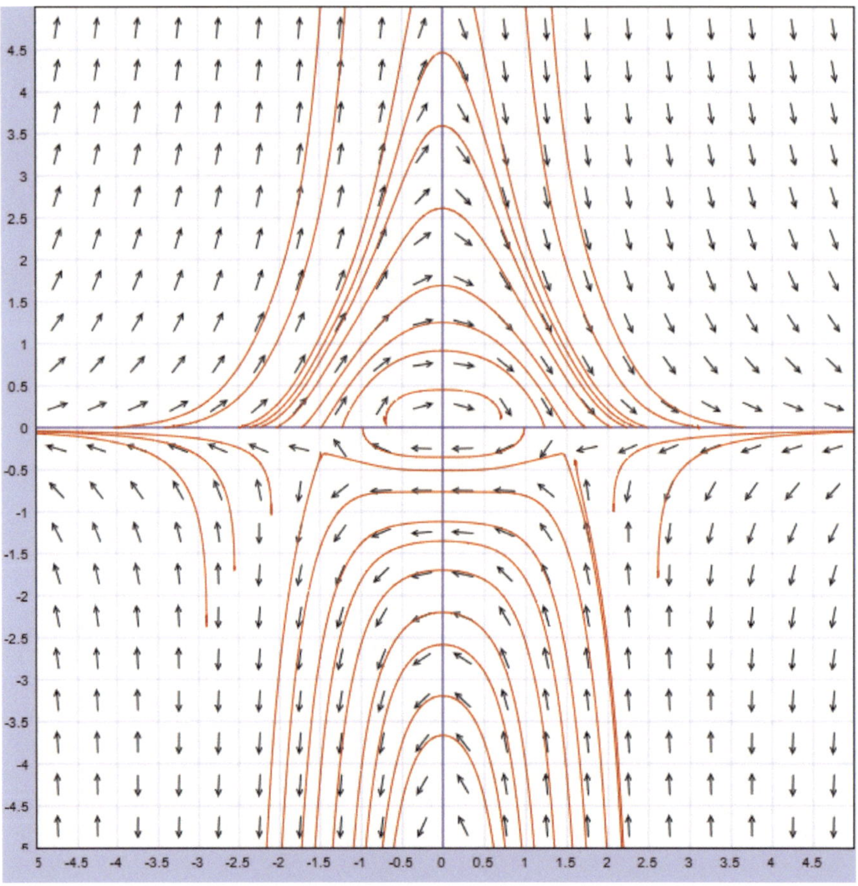

Figure 5.
Phase portrait and vector field of Eq. (35) for $n = 1$.

3.2 Higher-order nonlinear equation

Nonlinear systems have been extensively investigated from the perspective of chaotic behavior. Chaos in higher-order systems has been widely studied in the literature since they are subject in general to a dramatic change in their qualitative behavior under a small change in initial conditions. Consequently, the determination of exact explicit solutions has been less explored in the literature. It follows the high importance of finding higher-order systems that admit a general solution with a regular predictable behavior when the initial conditions change. In this regard, higher-order systems having a sinusoidal general solution such as the harmonic oscillator cannot, in an analytic way, exhibit chaotic behavior. We, therefore, focus on these systems in this part. It is obvious that [30, 32], if

$$H(x, \dot{x}) = b, \tag{37}$$

where b is a constant, and

$$x(t) = \cos(t + \varphi) \tag{38}$$

then

$$\frac{d^m}{dt^m}[H(x, \dot{x}) - b] = 0, \tag{39}$$

where $m \geq 0$ is an integer, with the exact solution (38). Indeed

$$\frac{d^m}{dt^m}[H(x, \dot{x}) - b] = \frac{d^{m-1}}{dt^{m-1}}\left\{\frac{d}{dt}[H(x, \dot{x}) - b]\right\} = 0. \tag{40}$$

Therefore, the following theorems have been proven.

Theorem 1.3: Consider the Hamiltonian (20). Then, the equation

$$\frac{d^m}{dt^m}\left[\dot{x}^2\sum_{\ell=0}^{n}x^{2\ell} + \dot{x}^3 + \dot{x}(x^2 - 1) + x^{2n+2} - 1\right] = 0, \tag{41}$$

where $b = +1$, has the general solution

$$x(t) = \cos(t + \varphi). \tag{42}$$

Remark 3. *Eq. (41) is a class of $(n + 1)$ nonlinear $(m + 1)$th order autonomous systems that can exhibit isochronous oscillations.*

Theorem 1.4: Consider the Hamiltonian or first integral (31), where $b = 1$. Then, equation

$$\frac{d^m}{dt^m}\left[\dot{x}^2\sum_{\ell=0}^{n}x^{2\ell} + \dot{x}^3\left(1 + \sum_{\ell=0}^{n}x^{2\ell+2}\right) + \dot{x}(x^{2n+4} - 1) + x^{2n+2}\right] = 0, \tag{43}$$

possesses the general and harmonic isochronous solution

$$x(t) = \cos(t + \varphi). \tag{44}$$

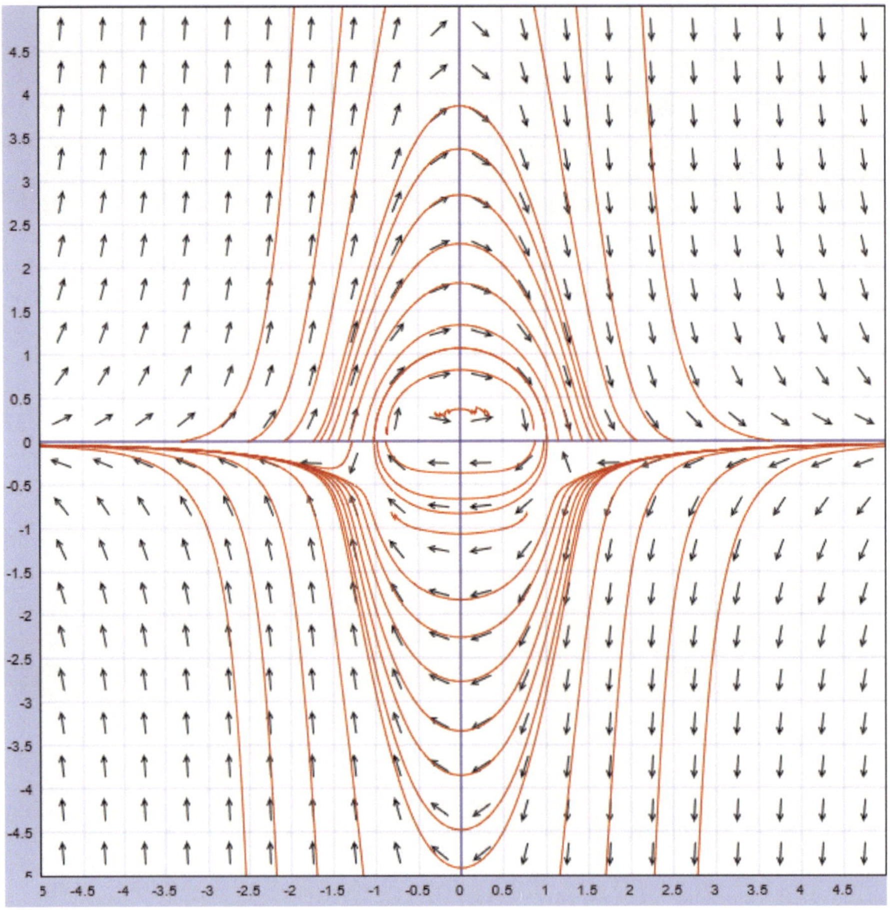

Figure 6.
Phase portrait and vector field of Eq. (35) for $n = 2$.

Remark 4. *Eq. (43) is a class of $(n+1)$ nonlinear $(m+1)$th order autonomous systems that can exhibit isochronous oscillations. It is interesting to note that the constant φ can be determined by using two initial conditions*

$$x(t=0) = x_0, \quad \dot{x}(t=0) = v_0 \tag{45}$$

whereas the Cauchy initial value problem requires q initial conditions for qth order systems of differential equations. Additionally, we can prove the following results.

Theorem 1.5: Let

$$\frac{d^m}{dt^m}\left\{\left[\dot{x}^2\sum_{\ell=0}^{n} x^{2\ell} + \dot{x}^3 + \dot{x}\left(x^2-1\right) + x^{2n+2} - 1\right]x\right\} = 0. \tag{46}$$

Then, Eq. (46) has the general and exact isochronous harmonic solution

$$x(t) = \cos(t+\varphi). \tag{47}$$

Proof. Eq. (46) can be rewritten in the form

$$\left[\dot{x}^2\sum_{\ell=0}^{n}x^{2\ell}+\dot{x}^3+\dot{x}(x^2-1)+x^{2n+2}-1\right]\frac{d^m x}{dt^m}$$
$$+x\frac{d^m}{dt^m}\left[\dot{x}^2\sum_{\ell=0}^{n}x^{2\ell}+\dot{x}^3+\dot{x}(x^2-1)+x^{2n+2}-1\right]=0. \tag{48}$$

Since $\dot{x}^2\sum_{\ell=0}^{n}x^{2\ell}+\dot{x}^3+\dot{x}(x^2-1)+x^{2n+2}-1=0$, when $x(t)=\cos(t+\varphi)$, the first term of Eq. (48) is zero. The second term is also zero under Theorem 1.3. Thus, Theorem 1.5 is proved.

Theorem 1.6: Let

$$\frac{d^m}{dt^m}\left\{\left[\dot{x}^2\sum_{\ell=0}^{n}x^{2\ell}+\dot{x}^3+\dot{x}(x^2-1)+x^{2n+2}-1\right]e^x\right\}=0. \tag{49}$$

Then, Eq. (49) admits the general and exact isochronous sinusoidal solution

$$x(t)=\cos(t+\varphi) \tag{50}$$

Proof. Writing Eq. (49) yields

$$\left[\dot{x}^2\sum_{\ell=0}^{n}x^{2\ell}+\dot{x}^3+\dot{x}(x^2-1)+x^{2n+2}-1\right]e^x$$
$$+e^x\frac{d^m}{dt^m}\left[\dot{x}^2\sum_{\ell=0}^{n}x^{2\ell}+\dot{x}^3+\dot{x}(x^2-1)+x^{2n+2}-1\right]=0. \tag{51}$$

allows us to note that the second term is zero under Theorem 1.3. Since $\dot{x}^2\sum_{\ell=0}^{n}x^{2\ell}+\dot{x}^3+\dot{x}(x^2-1)+x^{2n+2}-1=0$ for $x(t)=\cos(t+\varphi)$, the first term of Eq. (51) is equal to zero, so Theorem 1.6 is proved.

Theorem 1.7: Let

$$\frac{d^m}{dt^m}\left\{\left[\dot{x}^2\sum_{\ell=0}^{n}x^{2\ell}+\dot{x}^3\left(1+\sum_{\ell=0}^{n}x^{2\ell+2}\right)+\dot{x}(x^{2n+4}-1)+x^{2n+2}-1\right]x\right\}=0 \tag{52}$$

Then, Eq. (52) exhibits the general and exact isochronous harmonic solution

$$x(t)=\cos(t+\varphi). \tag{53}$$

Proof. Eq. (52) can take the form

$$\left[\dot{x}^2\sum_{\ell=0}^{n}x^{2\ell}+\dot{x}^3\left(1+\sum_{\ell=0}^{n}x^{2\ell+2}\right)+\dot{x}(x^{2n+4}-1)+x^{2n+2}-1\right]\frac{d^m x}{dt^m}$$
$$+x\frac{d^m}{dt^m}\left[\dot{x}^2\sum_{\ell=0}^{n}x^{2\ell}+\dot{x}^3\left(1+\sum_{\ell=0}^{n}x^{2\ell+2}\right)+\dot{x}(x^{2n+4}-1)+x^{2n+2}-1\right]=0 \tag{54}$$

Since $\dot{x}^2\sum_{\ell=0}^{n}x^{2\ell}+\dot{x}^3\left(1+\sum_{\ell=0}^{n}x^{2\ell+2}\right)+\dot{x}\left(x^{2n+4}-1\right)+x^{2n+2}-1=0$ when $x(t)=\cos(t+\varphi)$, the first term of Eq. (54) is equal to zero. The second term is equal to zero under Theorem 1.4. Therefore, Theorem 1.7 is proved.

Theorem 1.8: Let

$$\frac{d^m}{dt^m}\left\{\left[\dot{x}^2\sum_{\ell=0}^{n}x^{2\ell}+\dot{x}^3\left(1+\sum_{\ell=0}^{n}x^{2\ell+2}\right)+\dot{x}\left(x^{2n+4}-1\right)+x^{2n+2}-1\right]e^x\right\}=0. \quad (55)$$

Then, Eq. (55) admits the general and exact isochronous solution

$$x(t)=\cos(t+\varphi). \quad (56)$$

Proof. Applying the rule of differentiation of a product of two functions, Eq. (55) can be written as

$$\begin{aligned}&\left[\dot{x}^2\sum_{\ell=0}^{n}x^{2\ell}+\dot{x}^3\left(1+\sum_{\ell=0}^{n}x^{2\ell+2}\right)+\dot{x}\left(x^{2n+4}-1\right)+x^{2n+2}-1\right]e^x+\\&e^x\frac{d^m}{dt^m}\left[\dot{x}^2\sum_{\ell=0}^{n}x^{2\ell}+\dot{x}^3\left(1+\sum_{\ell=0}^{n}x^{2\ell+2}\right)+\dot{x}\left(x^{2n+4}-1\right)+x^{2n+2}-1\right]=0.\end{aligned} \quad (57)$$

Under Theorem 1.4, the second term of Eq. (57) is equal to zero when $x(t)=\cos(t+\varphi)$. The first term is also zero when $x(t)=\cos(t+\varphi)$ since

$$\dot{x}^2\sum_{\ell=0}^{n}x^{2\ell}+\dot{x}^3\left(1+\sum_{\ell=0}^{n}x^{2\ell+2}\right)+\dot{x}\left(x^{2n+4}-1\right)+x^{2n+2}-1=0.$$

This completes the proof of Theorem 1.8.

Remark 5. *If* $H(x,\dot{x})=b$, *when* $x=\cos(t+\varphi)$, *then*

$$\frac{d^m}{dt^m}\{[H(x,\ \dot{x})-b]Q(x,\ \dot{x})\}=0, \quad (58)$$

has the exact solution $\cos t$, where $Q(x,\dot{x})\neq 0$ is a function of its arguments. Now, we can investigate higher-order nonautonomous nonlinear systems.

4. Nonautonomous nonlinear systems

In recent decades, nonautonomous systems have been the subject of intensive investigation in the literature, given their applications in physics and applied mathematics [33–35]. In particular, these systems have been used to describe time-varying parameter processes in many areas of physical and life sciences [33, 35]. Nonautonomous systems are generally investigated within the framework of the qualitative theory of differential equations. The Lyapunov method is often used to study the stability, boundedness, and conditions for the existence of periodic solutions of these systems. However, the recent literature shows that the classical existence theorems are not sufficient to predict the behavior of nonlinear dynamic systems. Additionally, qualitative results are not sufficient for engineering and industrial applications [23]. By definition [34, 35], a nonautonomous dynamic system is distinguished from an autonomous system by the fact that the solution of the associated initial value problem depends not only on the elapsed time $t - t_0$ but also on the initial time t_0. In this part, we prove the existence of nonautonomous dynamic systems whose solution to the initial value problem does not depend on the initial time t_0. To that end, we have the following result.

Theorem 1.9: Consider the Hamiltonian $H(x, \dot{x})$ such that

$$H(x, \dot{x}) = b, \tag{59}$$

when $x = \cos(t + \varphi)$, where b and φ are constants. Then, the nonautonomous equation

$$\frac{d^m}{dt^m}[H(x,\ \dot{x}) - b]Q(t) = 0, \tag{60}$$

has the general and exact isochronous sinusoidal solution

$$x(t) = \cos(t + \varphi), \tag{61}$$

where $Q(t) \neq 0$ is a function of t.

Proof. Using the rule of differentiation of a product of two functions, we can rewrite Eq. (60) in the form

$$(H - b)\frac{d^m Q(t)}{dt^m} + Q(t)\frac{d^m}{dt^m}(H - b) = 0. \tag{62}$$

From Eq. (59), the first term of Eq. (62) is equal to zero for $x(t) = \cos(t + \varphi)$. Now $\frac{d^m}{dt^m}(H - b) = \frac{d^{m-1}}{dt^{m-1}}\left[\frac{d}{dt}(H - b)\right]$ so that $\left[\frac{d}{dt}(H - b)\right] = \frac{dH}{dt} = 0$, using Eq. (59) when $x(t) = \cos(t + \varphi)$. This completes the proof of Theorem 1.9.

4.1 Examples of illustration

4.1.1 Example 1

Let us consider Eq. (20), where $b = 1$. Then, the nonlinear $(m + 1)th$ order nonautonomous equation

$$\frac{d^m}{dt^m}\left\{\left[\dot{x}^2\sum_{\ell=0}^{n}x^{2\ell} + \dot{x}^3 + \dot{x}(x^2 - 1) + x^{2n+2} - 1\right]\cos t\right\} = 0, \tag{63}$$

exhibits isochronous oscillations corresponding to the general and exact sinusoidal solution

$$x(t) = \cos(t + \varphi). \tag{64}$$

Solution (64) is also the general and exact isochronous sinusoidal solution of the harmonic oscillator

$$\ddot{x} + x = 0. \tag{65}$$

Thus, the constant φ can be determined using two initial conditions

$$x(t=0)=x_0, \quad \dot{x}(t=0)=v_0 \tag{66}$$

such that, as is well-known, solution (64) does not depend on the initial time t_0.

4.1.2 Example 2

Consider Eq. (31) where $b=1$. Then, we have the nonlinear $(m+1)th$ order nonautonomous equation

$$\frac{d^m}{dt^m}\left\{\left[\dot{x}^2\sum_{\ell=0}^{n}x^{2\ell}+\dot{x}^3\left(1+\sum_{\ell=0}^{n}x^{2\ell+2}\right)+\dot{x}\left(x^{4n+2}-1\right)+x^{2n+2}-1\right]\cos t\right\}=0, \tag{67}$$

that can exhibit isochronous sinusoidal oscillations with the general and exact solution $\cos(t+\varphi)$. It is interesting to note that the classes of Eqs. (63) and (67) contain $(n+1)$ nonlinear $(m+1)th$ order nonautonomous systems that reproduce in an exact way the isochronous harmonic oscillations of the harmonic oscillator. In this context, we can present a conclusion for the chapter.

5. Conclusion

In this chapter, we explicitly proved some results concerning isochronous sinusoidal oscillations of nonlinear systems. These results contribute to recent developments and major advances in the field of second-order and higher-order autonomous and nonautonomous nonlinear dynamic system theory.

Author details

Jean Akande†, Kolawolé Kêgnidé Damien Adjaï†, Marcellin Nonti†
and Marc Delphin Monsia*†
Department of Physics, University of Abomey-Calavi, Cotonou, Benin

*Address all correspondence to: monsiadelphin@yahoo.fr

† These authors contributed equally.

References

[1] Lilbre J, Mereu AC, Teixeira MA. Limit cycles of the generalized polynomial Liénard differential equations. In: Mathematical Proceedings of the Cambridge Philosophical Society. 2010;**148**(2):363-383. DOI: 10.1017/S0305004109990193

[2] Chavarriga J, Garcia B, Llibre J. Polynomial first integrals of quadratic vector fields. Journal of Differential Equations. 2006;**230**(2): 393-421. DOI: 10.1016/j.jde.2006.07.022

[3] Gine J, Grau MM, Llibre J. Polynomial and rational first integrals for planar homogeneous polynomial differential systems. Publ.Mat. 2014:255-278. DOI: 10.5565/PUBLMAT_Extra14_14

[4] Christopher C, Llibre J, Pantazi C, Walcher S. On planar polynomial vector fields with elementary first integrals. Journal of Differential Equations. 2019; **267**(8):4572-4588. DOI: 10.1016/j.jde.2019.05.007

[5] Jordan DW, Smith P. Nonlinear Ordinary Differential Equations: An Introduction for Scientists and Engineers. Fourth ed. New York: Oxford University Press; 2007

[6] Mickens RE. Oscillations in Planar Dynamic Systems, Series on Advances in Mathematics for Applied Sciences. World Scientific; 1996

[7] Mickens RE. Truly Nonlinear Oscillators. Singapore: World Scientific; 2010

[8] Adjaï KKD, Koudahoun LH, Akande J, Kpomahou YJF, Monsia MD. Solutions of the Duffing and Painlevé-Gambier equations by generalized Sundman transformation. Journal of Mathematics and Statistics. 2018;**14**(1):241-252. DOI: 10.3844jmssp.2018.241.252

[9] Monsia MD, Kpomahou YJF. Simulating nonlinear oscillations of visco-elastically damped mechanical systems. Engineering, Technology & Applied Science Research. 2014;**4**(6):714-723

[10] Kpomahou YJF, Monsia MD. Asymptotic perturbation analysis for nonlinear oscillations in viscoelastic systems with hardening exponent, Int. J. Adv. Appl. Math. and Mech. 2015;**3**(1): 49-56

[11] Sabatini M. On the period function of Lienard systems. Journal of Differential Equations. 1999;**152**(1):467-487

[12] Sabatini M. On the period function of $\ddot{x} + f(x)\dot{x}^2 + g(x) = 0$. Journal of Differential Equations. 2004;**196**(1): 151-168

[13] Sabatini M. Characterizing isochronous centres by Lie brackets. Differential Equations and Dynamical Systems. 1997;**5**(1):91-99

[14] Christopher C, Devlin J. On the classification of Lienard systems with amplitude-independent periods. Journal of Differential Equations. 2004;**200**(1): 1-17

[15] Christopher C, Devlin J. Isochronous centers in planar polynomial systems. Siam J. Math. Anal. 1997;**28**(1):162-177. DOI : 10.1137/50036141093259245

[16] Guha P, Choudhury GA. The Jacobi last multiplier and isochronicity of Lienard type systems. Reviews in Mathematical Physics. 2013;**25**:(6): 1330009-1-1330009-31

[17] Kovacic I, Rand R. About a class of nonlinear oscillators with amplitude-independent frequency. Nonlinear Dynamics. 2013;**74**(1):455-465

[18] Akande J, Adjaï KKD, Yehossou AVR, Monsia MD. Exact and sinusoidal periodic solutions of Lienard equation without restoring force. Int. J. Anal. Appl. 2022;**20**(4):1-6

[19] Kpomahou YJF, Nonti M, Adjaï KKD, Monsia MD. On the linear harmonic oscillator solution for a quadratic Lienard type equation. Mathematical Physics. 2021. Available online: https://viXra.org/abs/2101.0010v1 (preprint)

[20] Yessoufou AB, Adjaï KKD, Akande J, Monsia MD. Modified Emden type Oscillator Equations with Exact Harmonic solutions. Int. J. Anal. Appl. 2022;**20**(39):1-17

[21] Akplogan ARO, Adjaï KKD, Akande J, Avossevou GYH, Monsia MD. Modified Van der Pol-Helmohltz oscillator equation with exact harmonic solutions. 2021. DOI: 10.21203/rs.3.rs-1229125v1 (preprint)

[22] Adjaï KKD, Nonti M, Akande J, Monsia MD. Unusual non-polynomial Van der Pol oscillator equations with exact harmonic and isochronous solutions. 2021. DOI: 10.13140/RG.2.2.17308.41606 (preprint)

[23] Adjaï KKD, Akande J, Yehossou AVR, Monsia MD. Periodic solutions and limit cycles of mixed Lienard-type differential equations. AIMS Mathematics. 2022;7(8): 15195-15211

[24] Akande J, Adjaï KKD, Nonti M, Monsia MD. Counter-examples to the existence theorems of limit cycles of differential equations. 2021. DOI: 10.13140/RG.2.2.15940.76167 (preprint)

[25] Monsia MD. On the exact periodic solution of a truly nonlinear oscillator equation. 2020. Available online: https://viXra.org/pdf/2009.0057v2 (preprint)

[26] Monsia MD. The non-periodic solution of a truly nonlinear oscillator with power nonlinearity. 2020. Available online: https://viXra.org/pdf/2009.0174v1 (preprint)

[27] Doutetien EA, Yehossou AR, Mallick P, Rath B, Monsia MD. On the general solutions of a nonlinear pseudo-oscillator equation and related quadratic lienard systems. Proceedings of the Indian National Science. 2020;**86**(4):1361-1365

[28] Françoise JP. Isochronous systems and perturbation theory. Journal of Nonlinear Mathematical Physics. 2005; **12**:315-326

[29] Akande J, Adjaï KKD, Yehossou AVR, Monsia MD. On unusual first integrals. 2022. DOI: 10.13140/RG.2.2.16734.72006 (preprint)

[30] Adjaï KKD, Akande J, Monsia MD. On certain first integrals. 2022. DOI: 10.13140/RG.2.2.35928.57601v1 (preprint)

[31] Adjaï KKD, Akande J, Monsia MD. Higher-order nonautonomous isochronous dynamical systems. 2022. DOI: /10.13140/RG.2.2.18766.33606 (preprint)

[32] Adjaï KKD, Akande J, Monsia MD. Limit cycles and isochronous systems via first integrals. 2022. DOI: 10.13140/RG.2.2.14252.54409 (preprint)

[33] Akande J, Adjaï KKD, Monsia MD. On damped Mathieu and periodic Lienard type equations. 2021. DOI: 10.6084/m9.figshare.14547102.V1 (preprint)

[34] Kloeden PE, Rasmussen M. Nonautonomous Dynamical Systems. AMS, Mathematical Surveys and Monographs. New York: American Mathematical Society; 2011

[35] Kloeden PE, Pötzsche C. Nonautonomous Dynamical Systems in the Life Sciences. London: Springer; 2013

Chapter 9

Control Configuration Selection for Nonlinear Systems

Sujatha Vijayaraghavan

Abstract

Very popular research in the field of Modern control Engineering is design of controllers for nonlinear systems. It is obvious that all the real-world systems are multivariable and nonlinear in nature which is highly challenging to control these nonlinear systems as it exhibits complexity. In addition to these, all the real systems exhibit uncertainty due to slow or sudden changes in process parameters. Hence, the design of robust nonlinear controller should have an ability to handle these uncertainties. The design of controllers for nonlinear system needs proper selection of appropriate input–output pairing. This book chapter focus on the conventional and proposed method of control configuration selection for nonlinear systems.

Keywords: input–output pairing, closed loop undesired responses, benchmark nonlinear systems, linearization, control configuration, nonlinear controller

1. Introduction

Research in nonlinear systems is developing rapidly and it is observed that useful results tend to appear. The classical approach in nonlinear system is linearization and thereby design of linear controllers. This classical approach is recommended when the nonlinearities are mild. Also, this approach is not recommended when the nonlinearities are pronounced more. For such case, variable transformation techniques are adopted and hence effective controllers are designed. The philosophy of nonlinear controller design is indicated using following four schemes: i. Local linearization, ii. Local linearization with adaptation, iii. Linearization using variable transformations and iv. Special purpose procedure.

Most of the nonlinear systems exhibit sustained oscillations for wide range of operating point. Investigation of stability for nonlinear systems is based on linearization of nonlinear system around the steady state. If the linearized system is stable at the vicinity, then it is concluded that the corresponding nonlinear system will be stable in the vicinity of the point.

There are four methods involved to analyse the dynamic behaviour of nonlinear systems. Rigorous analytical approach is used to characterise qualitatively the dynamics of nonlinear system. Analytical approach will evolve to the state of the nonlinear system. Exact linearization by variable transformation technique will first carry out the transformation, then dynamic analysis on linear transformed version and finally transforming back to original variables. Numerical analysis will processes the numerical values at specific points in time. Finally in approximate linearization method nonlinear system will be approximated to linear system.

Many researchers from 'drive by wire' cars to 'fly by air' flight control systems have shown interest in analysis and design of nonlinear control strategies. This growing interest in design of nonlinear control is due to improvement in linear control system and hard nonlinearities analysis. Hence researcher need to deal with model uncertainties and robust design.

Modern technology requires high speed and accurate robots. Inverted pendulum is an example of nonlinear system which finds application in positioning of robots and control of manipulators. As the nonlinear systems exhibit limit cycle, it is not ease to use Kalman test for checking controllability and observability. Also, stability is not simply location of poles as the system is having multiple stable/unstable equilibrium points.

In eighteenth century, nonlinear control was introduced to control steam engine by using centrifugal flyball governor reviewed by Jamshed Iqbal et al. [1]. Lyapunov [2] in 1892 proposed the stability for nonlinear system by finding stability of linear approximation of nonlinear system at equilibrium point is equivalent to analysing stability for nonlinear system at vicinity of equilibrium point. Two benchmark nonlinear system on Duffing's research [3] in 1918 on nonlinear vibrations and van der pol findings [4] in 1926 on electronic oscillations representing nonlinear control systems. The various phenomenon in nonlinear systems is jump resonance, limit cycle, subharmonic oscillations and frequency entrainment. Control Engineering in 1930 [5], Poincare approximated servomechanisms to second order system using phase-plane method. During second world war, research in nonlinear control led to control of guided vehicles for defence. In the year 1940–1960, nonlinear systems are represented analytically using describing function and phase plane method [6]. Modern era for nonlinear control was developed in the year 1960. The key applications of nonlinear control are defence sector and industrial arena. As nonlinear real-world systems are multivariable, high dimensional, poorly modelled which is outside the boundary of classical control theory. Thus, nonlinear control falls under the modern control Engineering in which digital controllers are used Fuller 1979 [7, 8]. In 1970, scientist proposed dynamic system can be viewed as energy transformation mechanism. Sontag and Wang [9] in 1995 proposed input–output stability of nonlinear system using elementary subsystems. By the introduction of geometric control theory by Isidori [10] in 2013 to analyse the stability of nonlinear system. 1990s is considered as decade of 'activation process' in nonlinear control systems.

The aim of this chapter is to analysis the nonlinear system and nonlinear control. The rest of the chapter is carried out as follows: Section 2 discusses benchmark nonlinear system. Conventional method of input–output pairing is explained in Section 3. The proposed method of control structure selection and determination of input/output pairs are given in Section 4. At the end, conclusion is drawn.

2. Process description

A nonlinear system does not obey the principle of superposition. In this system, the response to sum of inputs is not equal to sum of individual responses. Also, response to step input of magnitude A is not equal to A times magnitude of step, step-down response is not equal to mirror image of step-up response. A sinusoidal input to nonlinear system will not lead to perfect sinusoidal response.

Most of the chemical processes are nonlinear. Some of the examples of nonlinear chemical process [11] are as follows:

1. The blending process

2. Stirred mixing tank process

3. Nonisothermal CSTR process - mildly nonlinear system [12]

4. pH neutralisation process–moderate to high nonlinear system and

5. Distillation column–highly non-linear system

2.1 The blending process

In the blending process shown in **Figure 1**, it is required to blend the pure material A and pure material B whose respective flowrates F_A and F_B. The objective of blending process is to control the flow rate and composition.

Representing the mathematical modelling of blending process:

Total mass balance:

$$F = F_A + F_B \tag{1}$$

Component A mass balance:

$$x = F_A/F_A + F_B \tag{2}$$

On linearizing, the process transfer function is obtained as:

$$G_{11}(s) = \frac{R'(s)}{W_1'(s)} = 10^{-3} \tag{3}$$

$$G_{12}(s) = \frac{R'(s)}{W_2'(s)} = 10^{-3} \tag{4}$$

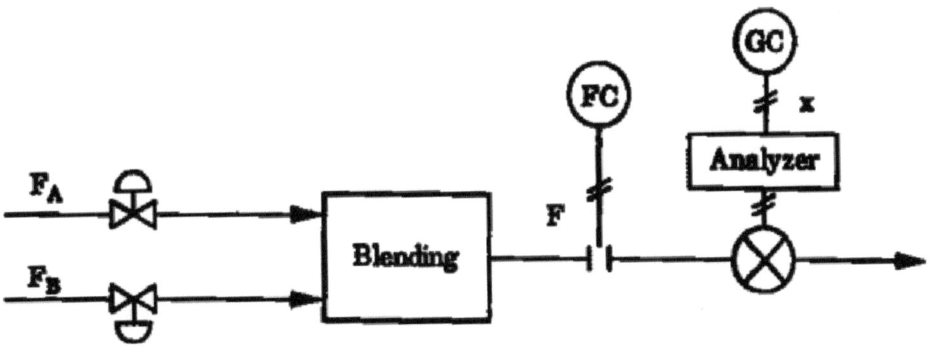

Figure 1.
Blending process.

$$G_{21}(s) = \frac{X'(s)}{W_1'(s)} = \frac{-2.5 * 10^{-6}}{s + 4 * 10^{-4}} \tag{5}$$

$$G_{22}(s) = \frac{X'(s)}{W_2'(s)} = \frac{-2.5 * 10^{-6}}{s + 4 * 10^{-4}} \tag{6}$$

2.2 Stirred mixing tank reactor

The stirred mixing tank reactor is having two input variables: hot and cold stream flowrate and two output variables: liquid level and temperature of the liquid in the tank.

For this stirred mixing tank reactor in **Figure 2**, assume cross sectional area is uniform, liquid physical properties are constant. The mathematical model of the tank reactor is

$$A_c \frac{dh}{dt} = F_H + F_C + F_D - k\sqrt{h} \tag{7}$$

$$\rho C_p A_c \frac{d(hT)}{dt} = \rho C_p \tag{8}$$

This process is considered as nonlinear because of square root term and product functions of h and T.

Process transfer function for stirred mixing tank reactor is:

$$G(s) = \begin{bmatrix} \frac{0.7}{1+9s} & 0 \\ \frac{2}{1+8s} & \frac{0.4}{1+9s} \end{bmatrix} \tag{9}$$

2.3 CSTR process

One of the nonlinear chemical processes is nonisothermal CSTR shown in **Figure 3**.

In this process, irreversible chemical reaction takes place, where the feed material having composition C moles/volume enters the reactor at the temperature, T_f.

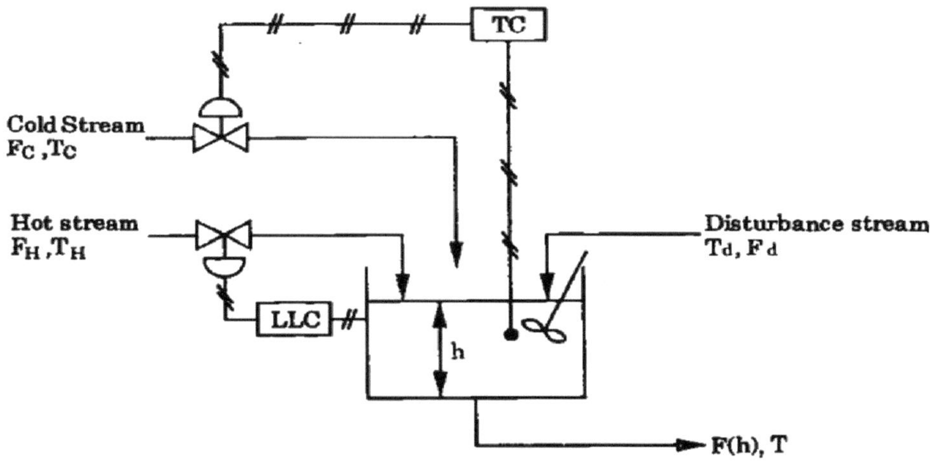

Figure 2.
Stirred mixing tank reactor.

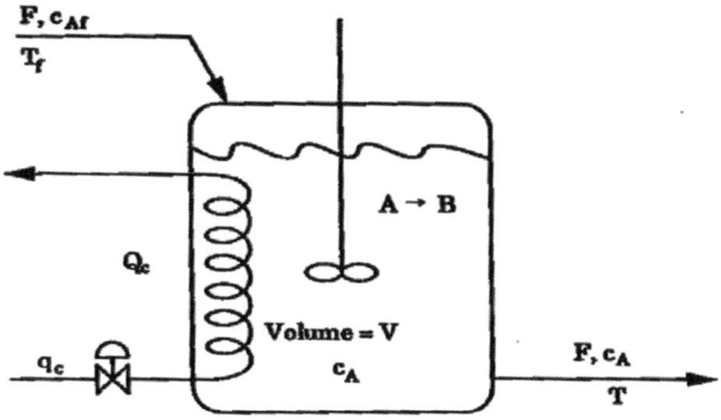

Figure 3.
The nonisothermal CSTR.

Assuming the concentration and temperature are uniform throughout, exit temperature and composition are also same as within the reactor.

The mathematical model for this process is:

$$\frac{dC_A}{dt} = \frac{-1}{\theta} c_A - k_0 e^{-\left(\frac{E}{RT}\right)} c_A + \frac{1}{\theta} c_{Af} \tag{10}$$

$$\frac{dT}{dt} = \frac{-1}{\theta} T + \beta k_0 e^{-\left(\frac{E}{RT}\right)} c_A + \frac{1}{\theta} T_f - \chi \tag{11}$$

The CSTR process is having two manipulated variables and two controlled variables (reaction temperature and concentration). Based on mathematical model of CSTR process, it is observed that the nonlinearities are due to nonlinear function of temperature – involving exponential of temperature and products of concentration and function of temperature.

The process transfer function matrix of CSTR process is presented in Eq. (12):

$$\begin{pmatrix} C_A \\ T \end{pmatrix} = \begin{pmatrix} \frac{0.022e^{-0.33s}}{15s+1} & \frac{5e^{-0.33s}}{21s+1} \\ \frac{0.0056e^{-6s}}{21s+1} & \frac{5.9e^{-0.33s}}{21s+1} \end{pmatrix} \begin{pmatrix} F \\ F_j \end{pmatrix} \tag{12}$$

2.4 pH neutralisation process

Most of the pH processes exhibit nonlinear behaviour to degree of either mild or high. The pH neutralisation process is shown in **Figure 4**.

The dynamic model of pH neutralisation system is derived using conservation and equilibrium relations. Assuming perfect mixing, constant density and complete solubility of the ions involved.

The following differential equations for the effluent reaction invariants can be derived:

$$A_1 h_1 \frac{dW_{a4}}{dt} = q_{1e}(W_{a1} - W_{a4}) + q_2(W_{a2} - W_{a4}) + q_3(W_{a3} - W_{a4}) \tag{13}$$

$$A_1 h_1 \frac{dW_{b4}}{dt} = q_{1e}(W_{b1} - W_{b4}) + q_2(W_{b2} - W_{b4}) + q_3(W_{b3} - W_{b4}) \tag{14}$$

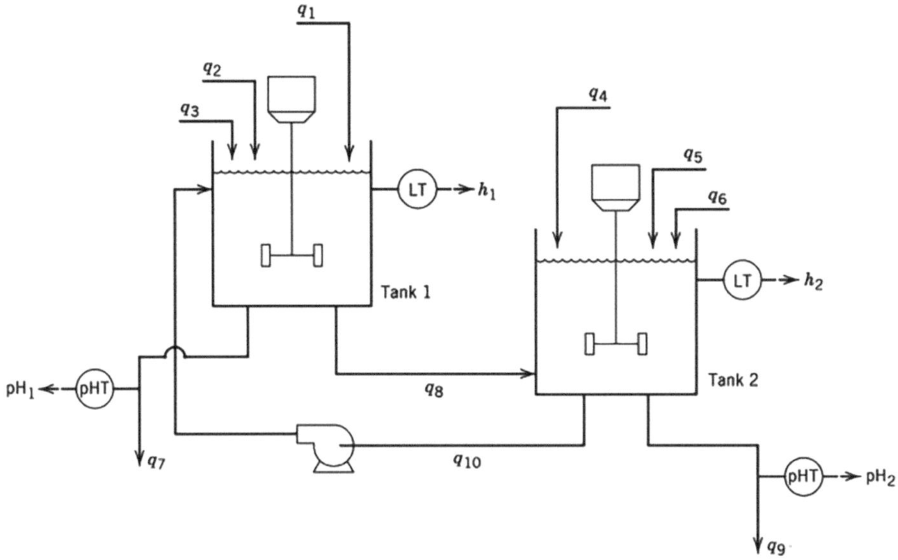

Figure 4.
pH neutralisation process.

The pH and level transmitters are modelled as first order transfer functions. The desired flow rates q_1 and q_3 serve as setpoints.

The process is treated as square MIMO systems where pH_2 and h_2 are to be controlled variables using Q_4 and Q_6 as the manipulated variables with Q_1 and Q_3 held constant.

The resulting process transfer function matrix is represented in the following Eq. (15).

$$\begin{pmatrix} pH_2 \\ h_2 \end{pmatrix} = \begin{pmatrix} \frac{-0.32e^{-0.8s}}{2.36s+1} & \frac{0.32e^{-0.4s}}{2.03s+1} \\ \frac{0.42e^{-0.4s}}{3.32s+1} & \frac{0.41e^{-0.1s}}{2.07s+1} \end{pmatrix} \begin{pmatrix} Q_4 \\ Q_6 \end{pmatrix} \tag{15}$$

2.5 Distillation column

The design of binary distillation process shown in **Figure 5** is a highly nonlinear chemical process.

The potential manipulated variables of the distillation process are reflux (FR) and reboiler flow rate (FV) whereas distillate (XD) and bottom composition (XB) are the controlled variables here. This nonlinear system is having 3 input variables and 2 output variables.

The transfer functions are obtained for the distillation column is described by the following equation.

$$X_D(s) = \frac{0.0747e^{-3s}}{12s+1} F_R(s) - \frac{0.0667e^{-2s}}{15s+1} F_V(s) + \frac{0.7e^{-5s}}{14.4s+1} X_F(s) \tag{16}$$

$$X_B(s) = \frac{0.1173e^{-3.3s}}{11.7s+1} F_R(s) - \frac{0.1253e^{-2s}}{10.2s+1} F_V(s) + \frac{1.3e^{-5s}}{12s+1} X_F(s) \tag{17}$$

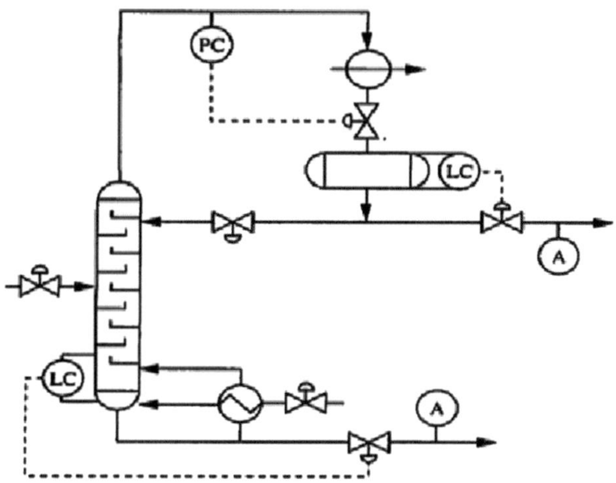

Figure 5.
Binary distillation tower.

whose transfer function matrix Eq. (18) is described as:

$$\begin{pmatrix} X_D(s) \\ X_B(s) \end{pmatrix} = \begin{pmatrix} \frac{0.0747e^{-3s}}{12s+1} & \frac{-0.0667e^{-2s}}{15s+1} \\ \frac{0.1173e^{-3.3s}}{11.7s+1} & \frac{-0.1253e^{-2s}}{10.2s+1} \end{pmatrix} \begin{pmatrix} F_R(s) \\ F_V(s) \end{pmatrix} \tag{18}$$

3. Conventional method of loop pairing for nonlinear systems

For all the processes namely SISO and MIMO it is required to pair the input and output variables before designing the controller. Later the controller can be designed according to any one of such input–output pair. Further any one of this configuration will lead to better overall system performance.

For the linear systems, Relative Gain Array (RGA) is obtained based on transfer function models [11]. Interaction analysis using RGA is based on steady-state information. But for nonlinear systems, by assuming that process model is available, two approaches are used to obtain RGA.

1. Using steady state version of nonlinear model from first principles, it is possible to obtain the analytical expressions.

2. By linearizing the nonlinear model around a steady state using approximate K matrix.

3.1 RGA based loop pairing for blending process

RGA is computed using the steady state version of nonlinear model for this process.

For the blending process, two input variables are u_1 and u_2 and the two output variables are F and x.

$$F = u_1 + u_2 - \text{Linear} \tag{19}$$

$$X = u_1/u_1 + u_2 - \text{Nonlinear} \tag{20}$$

For this 2x2 MIMO process, the elements of RGA is given by:

$$\lambda = \frac{\left(\frac{dF}{du_1}\right)_{bothloopsopen}}{\left(\frac{dF}{du_1}\right)_{secondloopclosed}} \tag{21}$$

From Eq. (19), when both loops open, $\left(\frac{dF}{du_1}\right)_{bothloopsopen} = 1.$

Upon closing the second loop, what value u_2 will take in order for any change in u_1, x is restored to desired steady state value x^*. Solving for u_2 in terms of u_1 and x^*.

$$u_2 = \frac{u_1}{x} - x \tag{22}$$

Thus, when the second loop is closed, subs Eq. (22) in Eq. (19)

$$F = u_{1+} \frac{u_1}{x} - x$$

$$F = \frac{u_1}{x} \tag{23}$$

Differentiate F w.r.t u_1, we get $\left(\frac{dF}{du_1}\right)_{secondloopclosed} = \frac{1}{x}$

Finally, $\lambda = \frac{1}{\frac{1}{x}}$

$$\lambda = x \tag{24}$$

Therefore, RGA for blending system is

$$\lambda = \begin{pmatrix} x & 1-x \\ 1-x & x \end{pmatrix} \tag{25}$$

where x is the mole fraction of species A in the blend.

RGA depends on only one steady state operating point, x whose value lies between 0 and 1.

Loop pairing for blending process:

1. When x^* is closer to 1, recommended pairing is F-u_1 and m-u_2.

2. When product composition is closer to 1, recommended pairing is F-u_1 and m-u_2.

3. When product composition is closer to 0, recommended pairing is F-u_2 and m-u_1.

4. When x^* = 0.5, which input variable is used to control which output variable.

3.2 RGA based loop pairing for stirred mixing tank reactor

Using the technique of approximating the nonlinear model around the steady state value, RGA is obtained for stirred mixing tank reactor.

Steady state gain matrix for the reactor is

$$K = G(0) = \frac{1}{k}\begin{bmatrix} 2\sqrt{h_s} & 2\sqrt{h_s} \\ \frac{(T_H - T_s)}{\sqrt{h_s}} & \frac{(T_C - T_s)}{\sqrt{h_s}} \end{bmatrix} \tag{26}$$

Two output variables are y_1 – liquid level and y_2 – temperature
Two input variables are u_1 – hot stream flowrate and u_2 – cold stream flowrate
RGA for this system at steady state operating point T_S,

$$\Lambda = \begin{bmatrix} \frac{T_C - T_s}{T_C - T_H} & \frac{-(T||H - T_s)}{T_C - T_H} & \frac{-(T||H - T_s)}{T_C - T_H} & \frac{T_C - T_s}{T_C - T_H} \end{bmatrix} \tag{27}$$

For illustrations, numerical values of $T_C = 15°C$ and $T_H=65°C$
Condition 1: $T_S > 40°C$, (T_S = 55°C)

$$\Lambda = \begin{bmatrix} 0.8 & 0.2 \\ 0.2 & 0.8 \end{bmatrix} \tag{28}$$

RGA recommends (u_1-y_1) and (u_2-y_2) pairing.
Condition 2: $T_S < 40°C$, (T_S = 25°C)

$$\Lambda = \begin{bmatrix} 0.2 & 0.8 \\ 0.8 & 0.2 \end{bmatrix} \tag{29}$$

RGA recommends (u_2-y_1) and (u_1-y_2) pairing.
Condition 3: T_S = 40°C,

$$\Lambda = \begin{bmatrix} 0.5 & 0.5 \\ 0.5 & 0.5 \end{bmatrix} \tag{30}$$

Here either pairing is equally bad.
Condition 4: $T_S = T_H$

$$\Lambda = \begin{bmatrix} 1 & 0 \\ 0 & 1 \end{bmatrix} \tag{31}$$

Here we can achieve the perfect control.

3.3 RGA based loop pairing for mild, mild to high and high nonlinear process

RGA for mild, mild to high and high nonlinear processes are given in **Table 1** based on steady state value of process transfer function matrix.

It is clear that from RGA matrix for all the benchmark process, the desirable input–output pair is (y_1-m_1); (y_2-m_2).

4. Proposed loop pairing for the nonlinear process

The proposed method is based on finding the area under the closed loop undesired response and choosing the pair based on minimum area under the response [13, 14]. The controllers are designed using the method proposed by Panda [15] **Figures 10** and **11**.

S.No	Process	Relative Gain Array
1	CSTR	$\begin{pmatrix} 1.2750 & -0.2750 \\ -0.2750 & 1.2750 \end{pmatrix}$
2	pH	$\begin{pmatrix} 0.4940 & 0.5060 \\ 0.5060 & 0.4940 \end{pmatrix}$
3	Distillation column	$\begin{pmatrix} 6.0937 & -5.0937 \\ -5.0937 & 6.0937 \end{pmatrix}$

Table 1.
Conventional method of loop pairing for nonlinear chemical process.

4.1 Loop pairing for blending process

Case 1: Diagonal pairing

Based on the comparing the area tabulated in **Table 2** under undesired response for both diagonal and off diagonal shown in **Figures 6–9**, the recommended input output pairing is 1–1/2–2 using the proposed method.

Blending Process			
(y_1-u_1); (y_2-u_2) pairing		**(y_2-u_1); (y_1-u_2) pairing**	
y_2	1.0393e-05	y_2	−3.8772e-24
y_1	6.6693e-05	y_1	−9.0875e-06

Table 2.
Comparison of area under undesired response of blending process.

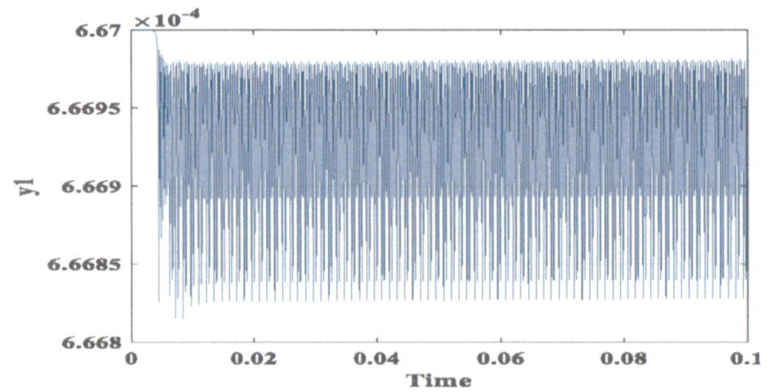

Figure 6.
Closed loop undesired response (y_1) for diagonal pairing.

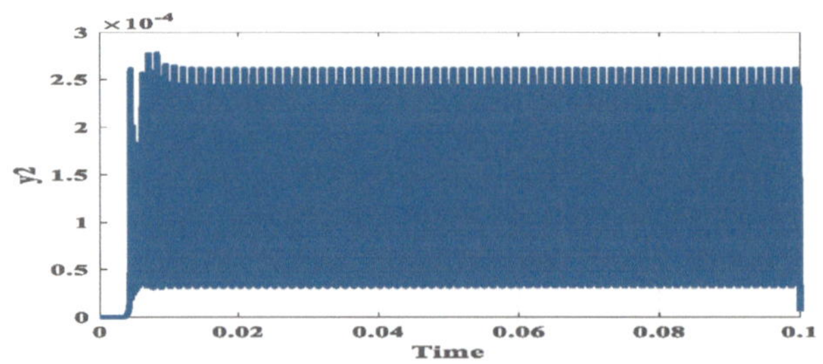

Figure 7.
Closed loop undesired response (y_2) for diagonal pairing.

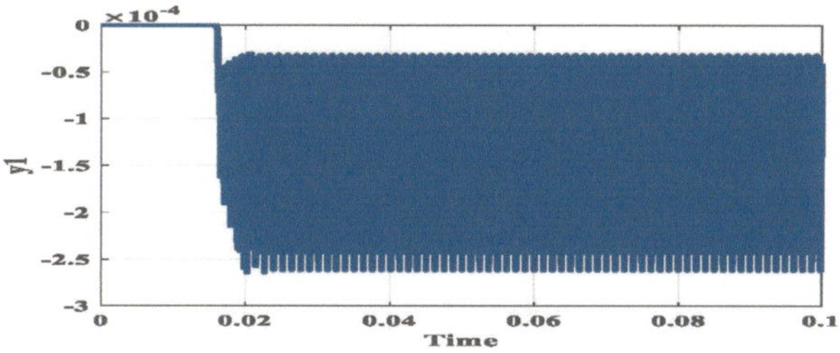

Figure 8.
Closed loop undesired response (y_1) for off diagonal pairing.

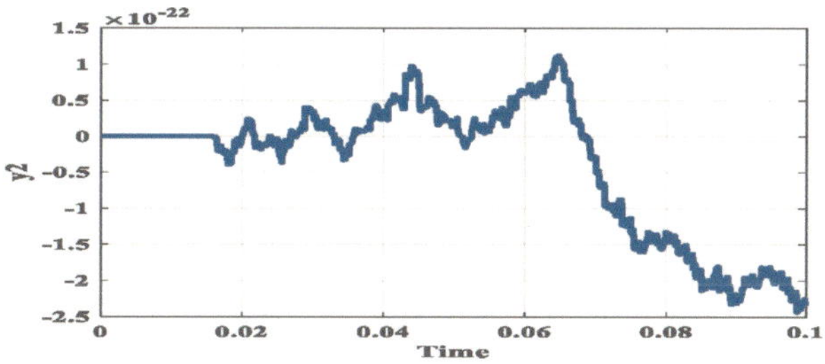

Figure 9.
Closed loop undesired response (y_2) for off diagonal pairing.

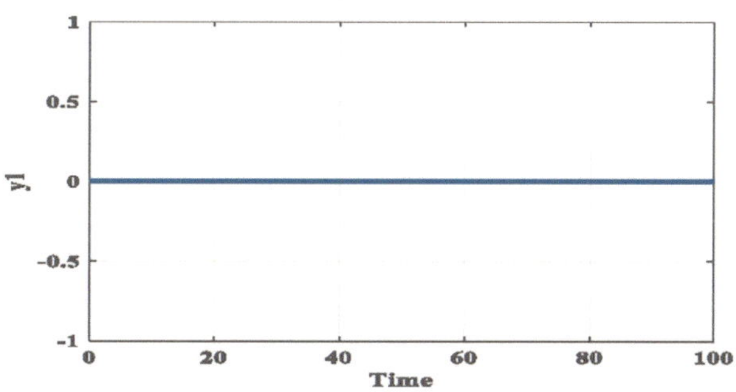

Figure 10.
Stirred mixing tank undesired response (y_1) for diagonal pairing.

4.2 Loop pairing for stirred mixing tank process

Case1: Diagonal pairing: Figures 10 and **11**
Case 2: off-diagonal pairing: Figures 12 and **13**
Closed loop undesired response is shown in **Figures 9–12** and area is compared in **Table 3** conclude that off diagonal pairing is the recommended input output pairing.

4.3 Loop pairing of CSTR process

By assuming both the diagonal and off diagonal pairing, responses are obtained and its area under the responses are compared to find the desired input output pairing.

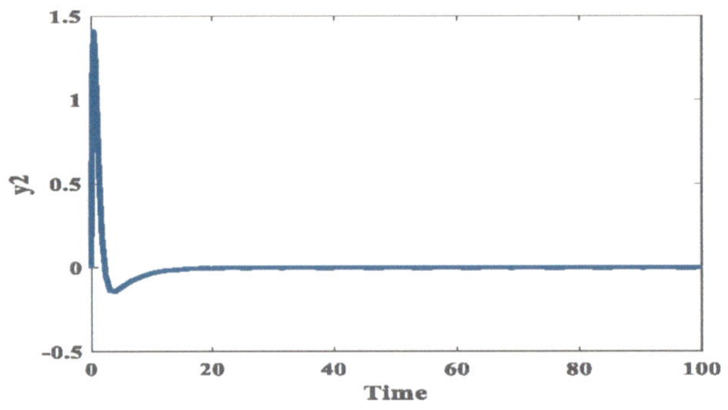

Figure 11.
Stirred mixing tank undesired response (y_2) for diagonal pairing.

Mixing stirred tank reactor			
(y_1-u_1); (y_2-u_2) pairing		**(y_2-u_1); (y_1-u_2) pairing**	
y_2	0.6568	y_2	0
y_1	0	y_1	0.6325

Table 3.
Area under the undesired response.

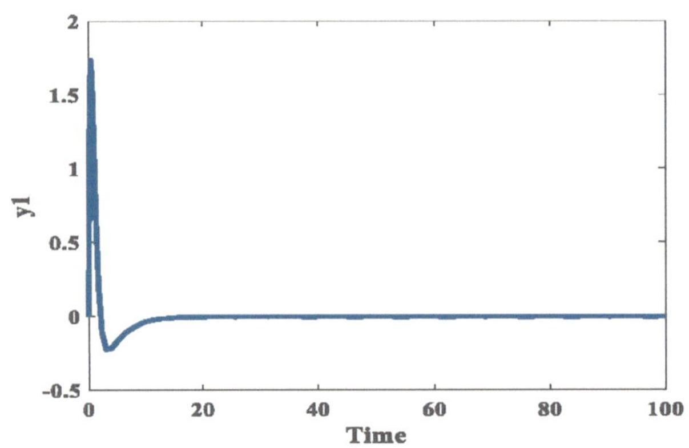

Figure 12.
Stirred mixing tank undesired response (y_1) for off diagonal pairing.

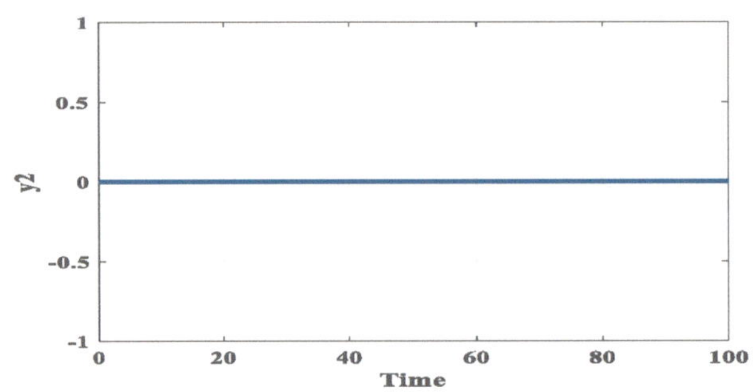

Figure 13.
Stirred mixing tank undesired response (y_2) for off diagonal pairing.

Case 1: Diagonal pairing: (y_1-m_1); (y_2-m_2) pairing

Figure 14 represents the closed loop undesired response, y_2 of CSTR process by assuming the diagonal pairing when there is change in input m_1 while m_2 = 0.

For CSTR process, the closed-loop undesired response, y_1 when there is change in m_2 for diagonal pairing of CSTR process is represented in **Figure 15**.

Case 2: Off-diagonal pairing: (y_2-m_1); (y_1-m_2) pairing

Figure 16 shows that for CSTR process, the closed loop undesired response y_2 for the change in m_2 by assuming off-diagonal pairing.

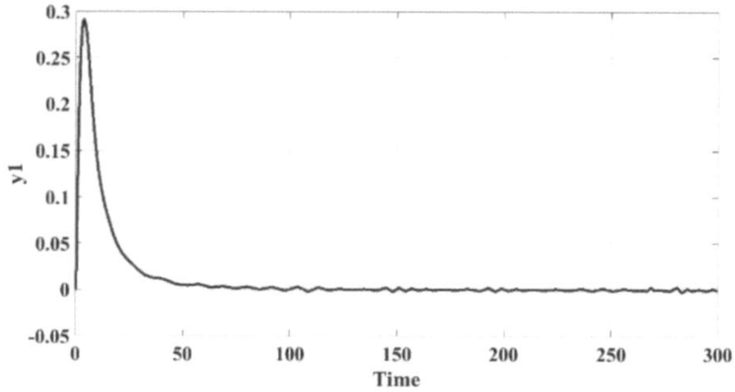

Figure 14.
Closed loop undesired response for diagonal pairing (y_2/m_1).

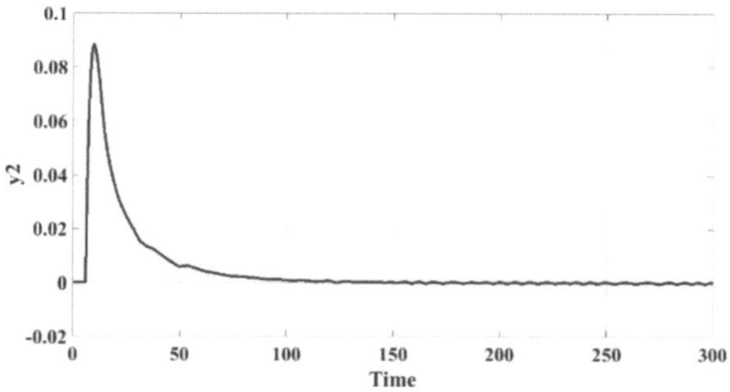

Figure 15.
Closed loop undesired response for diagonal pairing (y_1/m_2).

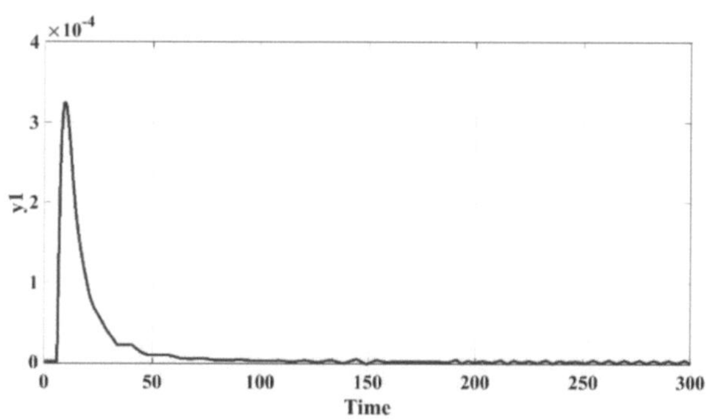

Figure 16.
Closed loop undesired response for off diagonal pairing (y_2/m_2).

The closed loop undesired response y_1 for the change in m_1 for off diagonal pairing in CSTR process is represented in **Figure 17**.

4.4 Loop pairing of pH process

For the pH process, closed loop undesired responses are obtained for both diagonal and off-diagonal pairing as shown in **Figures 18–21**. The areas obtained under these curves are given in **Table 4** and the same is compared to obtain the desired input–output pair based on minimum value.

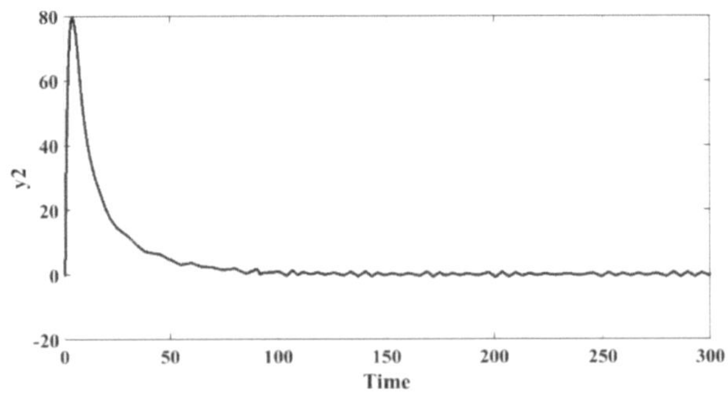

Figure 17.
Closed loop undesired response for off diagonal pairing (y_1/m_1).

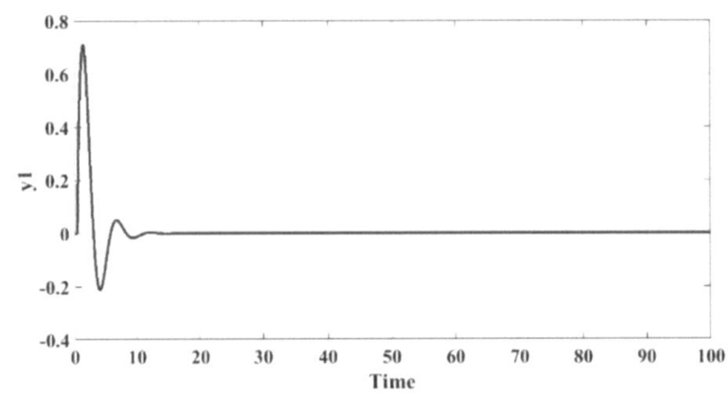

Figure 18.
Closed loop undesired response for diagonal pairing (y_2/m_1).

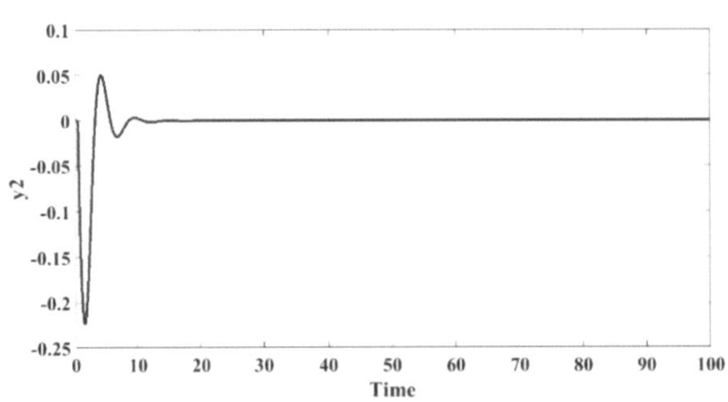

Figure 19.
Closed loop undesired response for diagonal pairing (y_1/m_2).

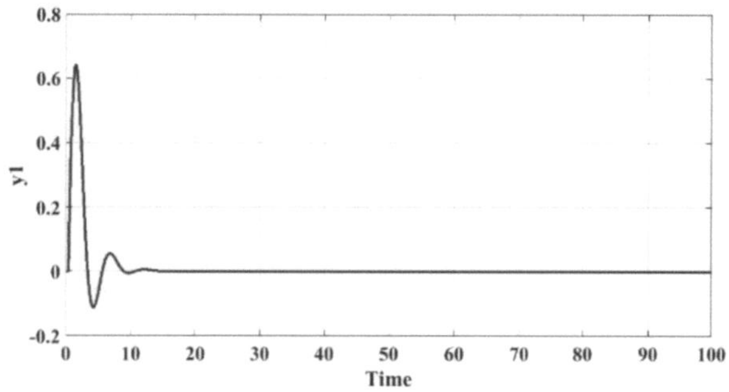

Figure 20.
Closed loop undesired response for off diagonal pairing (y_2/m_2).

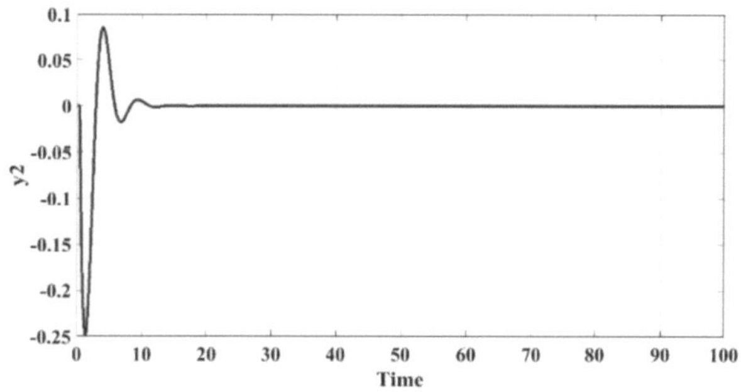

Figure 21.
Closed loop undesired response for off diagonal pairing (y_1/m_1).

CSTR Process			
(y_1-m_1); (y_2-m_2) pairing		**(y_2-m_1);(y_1-m_2) pairing**	
$(y_2 m_1)$	0.0036	$(y_2 m_2)$	0.0637
$(y_1 m_2)$	15.7309	$(y_1 m_1)$	0.8989
pH Process			
$(y_2 m_1)$	0.3333	$(y_2 m_2)$	0.2539
$(y_1 m_2)$	0.8328	$(y_1 m_1)$	1.0930
Distillation column			
$(y_2 m_1)$	51.0823	$(y_2 m_2)$	29.0406
$(y_1 m_2)$	26.0268	$(y_1 m_1)$	45.7813

Table 4.
Proposed loop pairing - Comparison of areas under the load responses.

Case 1: **Diagonal pairing: (y_1-m_1); (y_2-m_2) pairing**

Figure 18 represents the closed loop undesired response y_2 for the change in m_1 for diagonal pairing of pH process.

For diagonal pairing in pH process, closed loop undesired response y_1 for change in m_2 is shown in **Figure 19**.

Case 2: Off diagonal pairing (y_2-m_1); (y_1-m_2) pairing

For the pH process, the closed loop undesired response y_2 for change in m_2 is represented in **Figure 20** for off-diagonal pairing.

For off diagonal pairing in pH process, **Figure 21** represents the closed-loop undesired response y_1 for the change in m_1.

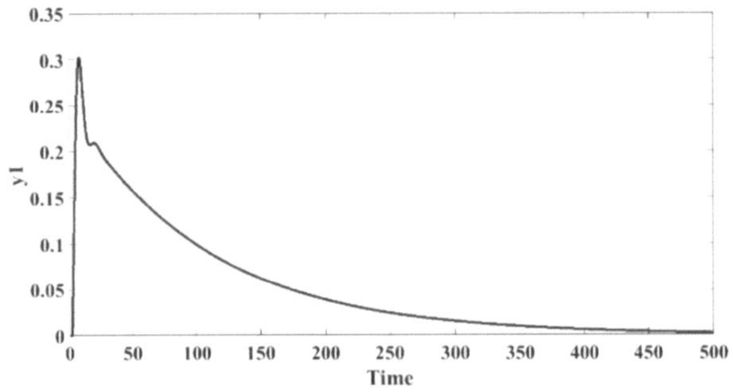

Figure 22.
Closed loop undesired response for diagonal pairing (y_2/m_1).

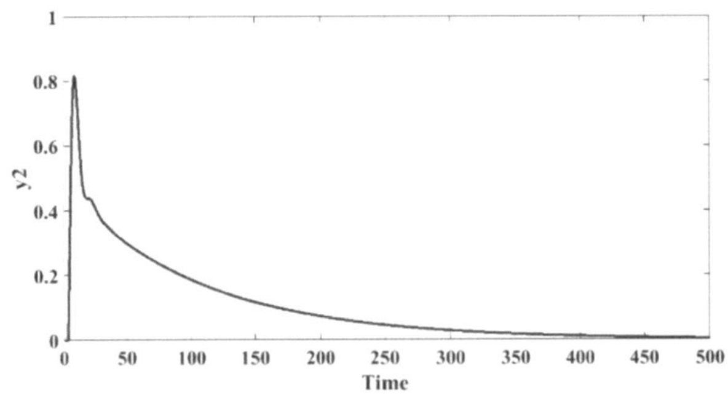

Figure 23.
Closed loop undesired response for diagonal pairing (y_1/m_2).

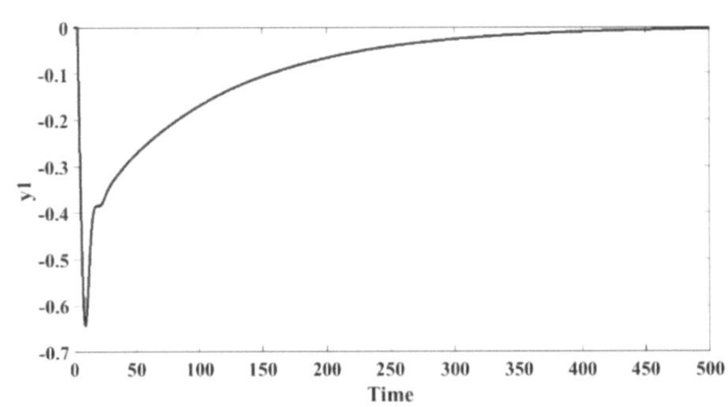

Figure 24.
Closed loop undesired response for off diagonal pairing (y_2/m_2).

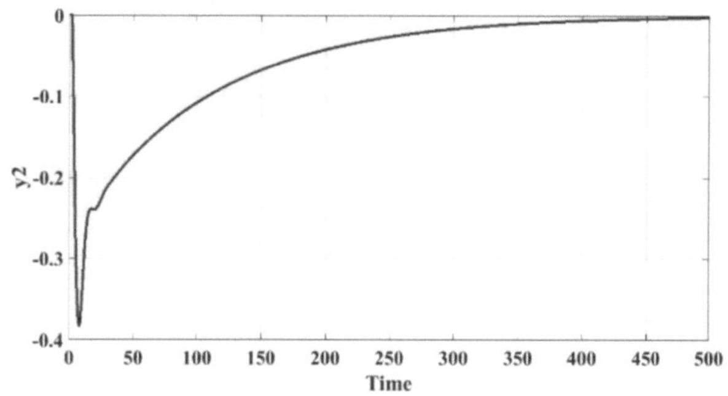

Figure 25.
Closed loop undesired response for off diagonal pairing (y_1/m_1).

4.5 Loop pairing of distillation column

Similar to CSTR and pH process, the pairing for distillation column is also carried out to choose the correct input–output pairing.

Case 1: Diagonal pairing:

Closed loop undesired response for diagonal pairing of distillation column is represented in **Figure 22.**

Figure 23 shows that closed loop response of distillation column for diagonal pairing.

Case 2: Off- diagonal pairing

For the distillation column, closed loop undesired response y_2 for the change in m_2 for off diagonal pairing is shown in **Figure 24.**

Figure 25 represents that the closed loop undesired response y_1 for off diagonal pairing of distillation column

Similarly, **Figures 22–25** represent undesired responses for distillation column and its area is given in **Table 4.**

From the **Table 4,** for the CSTR, pH and distillation column benchmark nonlinear chemical processes, it is clear that the minimum area is obtained only for (y_1-m_1); (y_2-m_2) pairing. Hence, desirable pairing for all these processes is **(y_1-m_1); (y_2-m_2)**.

5. Conclusion

As real-world physical systems are nonlinear, it is required to control these nonlinear processes. In order to design the nonlinear controller, one needs to choose the proper input–output pair. This chapter discusses the conventional method of loop pairing for the class of benchmark nonlinear system. RGA is calculated based on steady state information for the nonlinear system. The proposed method of input–output pair is also calculated for nonlinear benchmark process. The proposed method of input–output pair is validated with the conventional method. The proposed control configuration selection is based on closed loop response whereas the conventional method of pairing is based on gain. Thus, using the proposed method of control configuration selection one can design the good nonlinear control.

Author details

Sujatha Vijayaraghavan
Department of Mechatronics Engineering, SRM Institute of Science and Technology, Chennai, India

*Address all correspondence to: dr.vijaysuji@gmail.com

References

[1] Iqbal J, Ullah M, Khan SG, Khelifa B, Ćuković S. Nonlinear control systems – A brief overview of historical and recent advances. Nonlinear Engineering. 2017; **6**(4):301-312

[2] Lyapunov AM. The general problem of motion stability. Annals of Mathematics Studies. 1892;**17**

[3] Duffing G. Erzwungene Schwingungen bei veränderlicher Eigenfrequenz und ihre technische Bedeutung: R. Vieweg & Sohn; 1918

[4] Van der Pol B. On "relaxation-oscillations". The London, Ed-inburgh, and Dublin Philosophical Magazine and Journal of Science. 1926;**2**: 978-992

[5] Bennett S. A history of Control Engineering 1800–1930. UK: Peter Peregrinus, Stevenage; 1979

[6] Kalman R, Bertram J. Control system analysis and design via the second method of Lyapunov. Transactions of the American Society of Mechanical Engineers. 1960;**1**:394-400

[7] Fuller AT. The early development of control theory. Transactions of the AMSE Journal of Dynamic Systems, Measurement, and Control. 1976a;**98**: 109-118

[8] Fuller A. The early development of control theory II. Transactions of the AMSE Journal of Dynamic Systems, Measurement, and Control. 1976b;**98**: 224-235

[9] Sontag ED, Wang Y. On characterizations of the input-to-state stability property. Systems & Control Letters. 1995;**24**:351-359

[10] Isidori A. Nonlinear Control Systems. Springer Science & Business Media; 2013

[11] Ogunnaike BA, Ray WH. Process Dynamics, Modeling and Control. Oxford/New York: Oxford University Press; 1994

[12] Sujatha V, Panda RC. Time domain modeling and control of complex non-linear chemical processes using relay feedback test. Transactions of Institute of Measurement and Control. 2020; **42**(15):2885-2907

[13] Panda RC, Sujatha V. Introduction to PID Controllers – Theory, Tuning and Application to Frontier Areas. Croatia: London, UK.InTech Publisher; 2012. p. 105

[14] Sujatha V, Panda RC. Control configuration selection for multi input multi output processes. Journal of Process Control. 2013;**23**:1567-1574

[15] Panda RC. Synthesis of PID controller using desired closed loop criteria. Industrial and Engineering Chemistry Research. 2009;**47**(22): 1684-1692

Chapter 10

Feedback Linearization Control of Interleaved Boost Converter Fed by PV Array

Erdal Şehirli

Abstract

One of the powerful methods of nonlinear control is the feedback linearization technique. This technique consists of input state and input-output linearization methods. In this chapter, the feedback linearization technique, including input state and input-output linearization methods, is described. Then, input-output linearization method is used for output voltage control of interleaved boost converter. Firstly, mathematical model of the interleaved boost converter is derived after that the method is applied. Besides, the interleaved boost converter is fed by a PV array under irradiation level and ambient temperature change. As a result of the simulation study, output voltage control of interleaved boost converter under reference voltage change is realized as desired.

Keywords: feedback linearization, interleaved, boost, PV

1. Introduction

In nature, most of the system is nonlinear. However, the analysis and design of a nonlinear controller require complex mathematical procedures. On the other hand, linear methods provide easy analysis and design of control systems. Nonetheless conducting the control of a wide range and with parameters changes, linear control and analysis methods are not so powerful, especially in the power electronics system. Power electronics systems have a highly nonlinear nature because of the switching states of the power switch. So, for designing the proper controller for such systems, the usage of nonlinear control methods is required.

In literature, [1–3] give the fundamental analysis, design, and methods of nonlinear systems. Ramirez and Ortega [4] applies nonlinear control methods to the basic power electronics converters. Kazimierczuk [5] describes the design analysis and operation of power electronics converters, including buck, boost, and buck-boost converters. Feedback linearization technique classified under nonlinear control is applied to fundamental power electronics converter, including boost converter in [6, 7], buck converter in [8], buck-boost converter in [9, 10]. Sira-Ramirez et al. [11, 12] present another nonlinear control method that is a sliding mode controller for boost converter, [13] buck converter, and [14] buck-boost converter. Furthermore, adaptive-based nonlinear controller is presented for boost converter [15], buck converter [16], and buck-boost converter [17]. Robust nonlinear controller is designed for buck in [18], boost in [19], and buck-boost converter in [20].

In this chapter, firstly feedback linearization technique, one of the most useful nonlinear methods, is described. Then, input-output linearization method classified under the feedback linearization technique is applied to the interleaved boost converter. Besides, as a power supply of interleaved boost converter, PV array is used with solar irradiation and ambient temperature changes. Furthermore, a nonlinear controller is designed to control the output voltage of the interleaved boost converter. After designing the nonlinear controller of interleaved boost converter, it is compared to a linear controller. As a result of the simulation study, it is concluded that a nonlinear controller for the output voltage of interleaved boost converter gives better results than a linear type controller.

2. Feedback linearization

Feedback linearization techniques have become very popular in recent years because of providing the linear equivalent systems of nonlinear systems by exact linearization. Feedback linearization techniques provides the transformations of nonlinear systems into fully or partly linear systems, algebraically so that linear control techniques can be used. In feedback linearization, linearization is realized by exact state transformation and feedback, making these techniques different from conventional linearization aiming linear approximation of the system.

Feedback linearization technique can be classified into two methods that are input-state linearization and input–output linearization.

2.1 Input-state linearization

In input-state linearization method, it is aimed to linearize state Eq. (1) completely. In order to cancel the nonlinearities in the original system, state transformation and input transformation are used. After applying the proper transformation, nonlinear system is transformed into the linear system [2].

If there is a nonlinear system given with the form of Eq. (1).

$$\dot{x} = f(x, u) \tag{1}$$

There should be a state transformation given in Eq. (2) and an input transformation in Eq. (3) in order to apply input-state linearization.

$$z = z(x) \tag{2}$$

$$u = u(x, v) \tag{3}$$

There are some points to bear in mind about applying input-state control, which are as follows:

- Even though the results obtained by input-state linearization control is valid in a large region, they may not be global. There may also occur some singularity points, while the initial state is at those points, controller may not bring the system to the equilibrium point.

- To apply control law, state transformation in Eq. (2) should be available. If state components are not physically meaningful or could not be measured, original states should be used to compute state components.

- If there is uncertainty in the model or in any parameters of the system, it causes an error in the calculation of new state **z** and control input **u**.

2.2 Input-output linearization

Another feedback linearization method is input-output linearization method. In this method, the main process is to generate a linear differential relation between the system output and new control input. This method can be summarized in three stages, as follows [1, 2]:

1. Differentiate the system output till control input appears,

2. Select new control input in order to guarantee tracking convergence and cancel nonlinearities,

3. Examine the stability of internal dynamics.

In order to explain input-output linearization method, think about a system given in Eqs. (4) and (5).

$$\dot{x} = f(x) + g(x)u \tag{4}$$

$$y = h(x) \tag{5}$$

To have input-output relation, output should be differentiated till input appears. After differentiating Eq. (5), Eq. (6) is acquired as in refs. [1, 2, 10].

$$\dot{y} = \frac{\partial h}{\partial x}[f(x) + g(x)u] = L_f h(x) + L_g h(x)u \tag{6}$$

$L_f h$ and $L_g h$ in Eq. (6) are described as Lie derivatives of $f(x)$ and $h(x)$ and given in Eq. (7).

$$L_f h(x) = \frac{\partial h}{\partial x} f(x),\ L_g h(x) = \frac{\partial h}{\partial x} g(x) \tag{7}$$

After r_j times derivation of the output, considering the condition in Eq. (8), Eq. (9) is acquired.

$$L_{gi} L_f^{r-1} h_i(x) \neq 0 \tag{8}$$

$$y_i^{r_i} = L_f^{r_i} h_i + \sum_i^n \left(L_{gi} L_f^{r_i-1} h_i\right) u_i \tag{9}$$

If there is a multi-input, multi-output system, considering to apply Eq. (9) to all outputs, Eq. (10) is obtained.

$$\begin{bmatrix} y_1^{r_1} \\ \cdots \\ \cdots \\ y_n^{r_n} \end{bmatrix} = \begin{bmatrix} L_f^{r_1} h_1(x) \\ \cdots \\ \cdots \\ L_f^{r_n} h_n(x) \end{bmatrix} + \begin{bmatrix} L_{g1} L_f^{r_1-1} h_1 & \cdots & L_{gn} L_f^{r_n-1} h_1 \\ \vdots & \ddots & \vdots \\ L_{g1} L_f^{r_n-1} h_1 & \cdots & L_{gn} L_f^{r_n-1} h_n \end{bmatrix} \begin{bmatrix} u_1 \\ \cdots \\ \cdots \\ u_n \end{bmatrix} = \alpha(x) + E(x)u \tag{10}$$

After selecting new control variable, input–output linearization is acquired as in Eq. (11).

$$\begin{bmatrix} u_1 \\ \cdots \\ \cdots \\ u_n \end{bmatrix} = -E^{-1} \begin{bmatrix} L_f^{r_1} h_1(x) \\ \cdots \\ \cdots \\ L_f^{r_n} h_n(x) \end{bmatrix} + E^{-1} \begin{bmatrix} v_1 \\ \cdots \\ \cdots \\ v_n \end{bmatrix} \tag{11}$$

The relation between system output y and new control input v is given in Eq. (12). In Eq. (12), k is constant to be chosen ensuring the stability of the system.

$$\begin{bmatrix} y_1^{r_1} \\ \cdots \\ \cdots \\ y_n^{r_n} \end{bmatrix} = \begin{bmatrix} v_1 \\ \cdots \\ \cdots \\ v_n \end{bmatrix}, \begin{bmatrix} v_1 \\ \cdots \\ \cdots \\ v_n \end{bmatrix} = \begin{bmatrix} -k_{1(r-1)} y^{r-1} \ldots - k_{11(r-1)} y^1 - k_{10}\left(y_1 - y_1^*\right) \\ \cdots \\ \cdots \\ -k_{n(r-1)} y^{r-1} \ldots - k_{21(r-1)} y^1 - k_{20}\left(y_n - y_n^*\right) \end{bmatrix} \tag{12}$$

Also Eq. (13) gives the closed loop error dynamics relating to system output, reference values, and k constants as in Ref. [21].

$$\begin{bmatrix} e \\ \cdots \\ \cdots \\ e^r \end{bmatrix} = \begin{bmatrix} y - y^* \\ \cdots \\ y^r - y^{*^r} \end{bmatrix}, \begin{bmatrix} e_1^r + k_{1(r-2)} e_1^{r-1} + \ldots + k_{11} e_1^1 + k_{10} e_1 \\ \cdots \\ e_n^r + k_{n(r-1} e_n^{r-1} + \ldots + k_{21} e_n^1 + k_{20} e_n \end{bmatrix} = \begin{bmatrix} 0 \\ \cdots \\ \cdots \\ 0 \end{bmatrix} \tag{13}$$

2.3 Interleaved boost DC-DC converter

Circuit structure of interleaved boost DC-DC converter is shown in **Figure 1**. It is seen that two separate boost converters are connected to the DC bus. The difference of interleaved boost converter from boost converter is that both switches are conducted with time delay in order to have input current having less ripple content.

Operation of the interleaved boost converters can be summarized as follows: When S_1 is in switch-on position, S_2 is turned off, L_1 current increases linearly, L_2 current decreases, and D_2 conducts. When S_1 is switched off, S_2 is switched on and L_2 current increases linearly, L_1 current decreases, and D_1 conducts. While the inductor's current decreases, inductors transfer their energy to the load. Passive components of the interleaved boost converter can be chosen by using Eqs. (14) and (15) as in Ref. [5].

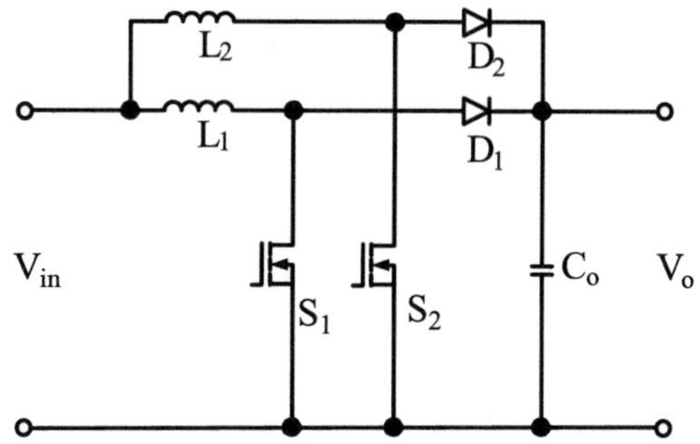

Figure 1.
Interleaved boost DC-DC converter.

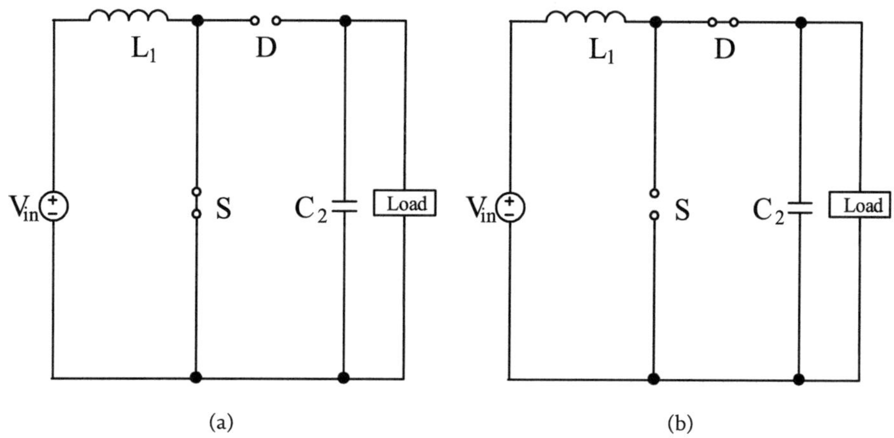

Figure 2.
(a) Switch-on, (b) switch-off position of the boost converter.

$$L = \frac{RD(1-D)^2}{2f_s} \tag{14}$$

$$C = \frac{DV_o}{f_s R \Delta V_C} \tag{15}$$

While deriving a mathematical model of interleaved boost DC-DC converter, model of boost DC-DC converter can be used. A mathematical model of the boost converter is obtained by switching on and off positions, shown in **Figure 2**. By applying Kirchhoff voltage and current laws for both circuits, a mathematical model of the boost converter is written.

At switch-on interval, after applying Kirchhoff voltage and current law, Eqs. (16) and (17) are obtained. The model for switch-on interval can be written in Eq. (19) with the form Eq. (18) of state-space representation.

$$\frac{di_L}{dt} = \frac{V_{in}}{L_1} \tag{16}$$

$$\frac{dV_o}{dt} = -\frac{V_o}{RC} \tag{17}$$

$$\dot{x} = Ax + Bu \tag{18}$$

$$\begin{bmatrix} \dot{i_L} \\ V_o \end{bmatrix} = \begin{bmatrix} 0 & 0 \\ 0 & -1/_{RC} \end{bmatrix} \begin{bmatrix} i_L \\ V_o \end{bmatrix} + \begin{bmatrix} 1/_L \\ 0 \end{bmatrix} V_{in} \tag{19}$$

At the switch-off interval, Kirchhoff voltage and current law are applied to **Figure 2b** and Eqs. (20) and (21) are obtained. It is also written in the form of Eq. (22).

$$\frac{di_L}{dt} = \frac{V_{in}}{L_1} - \frac{V_o}{L_1} \tag{20}$$

$$\frac{dV_o}{dt} = -\frac{i_L}{C} - \frac{V_o}{RC} \tag{21}$$

$$\begin{bmatrix} \dot{i_L} \\ V_o \end{bmatrix} = \begin{bmatrix} 0 & -1/_L \\ 1/_C & -1/_{RC} \end{bmatrix} \begin{bmatrix} i_L \\ V_o \end{bmatrix} + \begin{bmatrix} 1/_L \\ 0 \end{bmatrix} V_{in} \tag{22}$$

Mathematical model of the boost converter can be derived in Eq. (24) by using state-space average technique given in Eq. (23).

$$A = dA_1 + (1 - d)A_2, \ B = dB_1 + (1 - d)B_2 \tag{23}$$

$$\begin{bmatrix} \dot{i_L} \\ V_o \end{bmatrix} = \begin{bmatrix} 0 & {}^{-1+d}/_L \\ {}^{1-d}/_C & -1/_{RC} \end{bmatrix} \begin{bmatrix} i_L \\ V_o \end{bmatrix} + \begin{bmatrix} 1/_L \\ 0 \end{bmatrix} V_{in} \tag{24}$$

State-space mathematical mode in Eq. (24) can be reordered as in Eq. (25) to apply input–output linearization technique. As a system input in Eq. (25), d is chosen.

$$\begin{bmatrix} \dot{i_L} \\ V_o \end{bmatrix} = \begin{bmatrix} {}^{V_o}/_L + {}^{V_{in}}/_L \\ {}^{i_L}/_C - {}^{V_o}/_{RC} \end{bmatrix} + \begin{bmatrix} {}^{V_o}/_L \\ {}^{-i_L}/_C \end{bmatrix} d \tag{25}$$

The purpose of the control is to regulate output voltage V_o, so as an output variable V_o is chosen as in Eq. (26).

$$y = h(x) = V_o \tag{26}$$

The way of using input–output linearization techniques is to derive system output until system input is obtained in output. So, after derivation of system output V_o, Eq. (27) is obtained.

$$\dot{y} = \dot{V_o} = {}^{i_L}/_C - {}^{V_o}/_{RC} {}^{-i_L d}/_C \tag{27}$$

It is observed in (27) that at the first derivation, system input is found at system output. It means that the relative degree of the system is "1." Eq. (27) is rearranged in Eq. (28) with respect to system input.

$$d = \left(-\dot{V_o} + {}^{i_L}/_C - {}^{V_o}/_{RC}\right) \frac{C}{\dot{i_L}} \tag{28}$$

The next stage is to choose a new control input. The control input is chosen regarding to relative degree in Eq. (29).

$$y^1 = V_1 \tag{29}$$

After choosing new control input as in Eq. (30), and replacing it in Eq. (28), Eq. (31) is obtained. Eq. (31) is the system input, nonlinear controller is operated with respect to it.

$$V_1 = k_1\left(V_o - V_o^*\right) \tag{30}$$

$$d = \left(-k_1\left(V_o - V_o^*\right) + {}^{i_L}\!/_{C} - {}^{V_o}\!/_{RC}\right)\frac{C}{i_L} \tag{31}$$

In order to provide the operation of the interleaved boost converter, S_1 is switched by using the duty cycle calculated in Eq. (31), and S_2 is switched by using the same duty cycle having 90° delay.

3. Simulations

Interleaved boost DC-DC converter fed by PV array is controlled by input–output linearization technique by means of the simulation. Simulation study is realized by Matlab/Simulink software. The circuit diagram of the study is shown in **Figure 3**. It is seen in the figure that there is a nonlinear controller that is based on the Eq. (31) in chapter 2.

It is seen that interleaved boost converter is connected to the output of the PV array. Because of the interleaved nature of the converter, input current has a lower

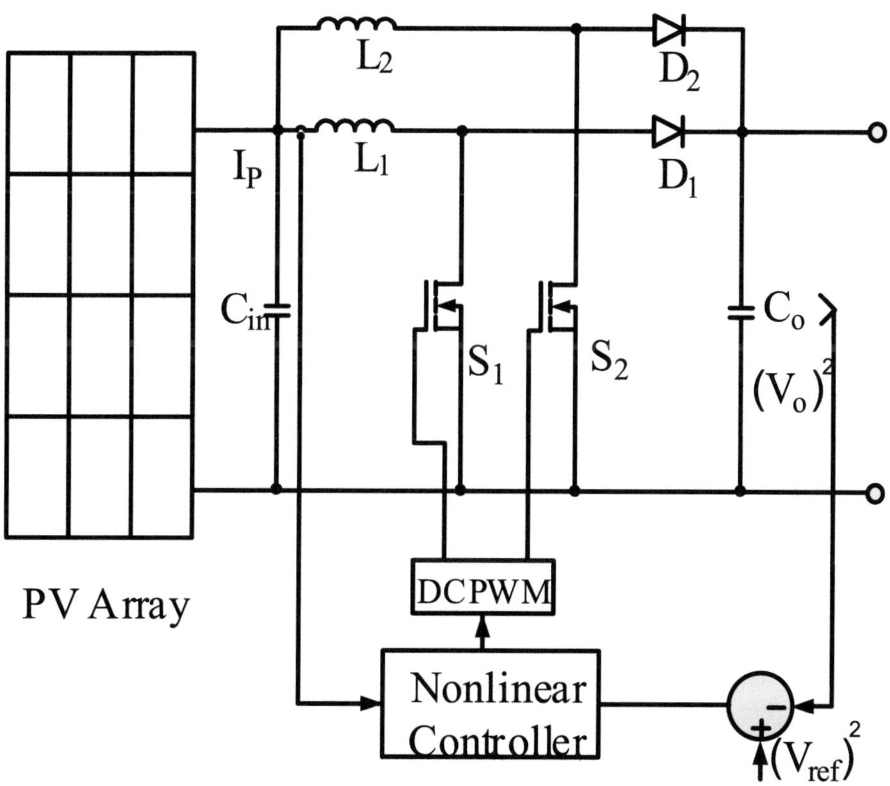

Figure 3.
PV-fed interleaved boost DC-DC converter with nonlinear control.

ripple than the classical boost converter. The simulation diagram of the circuit is shown in **Figure 4**.

Parameters used in the simulation are given in **Table 1**.

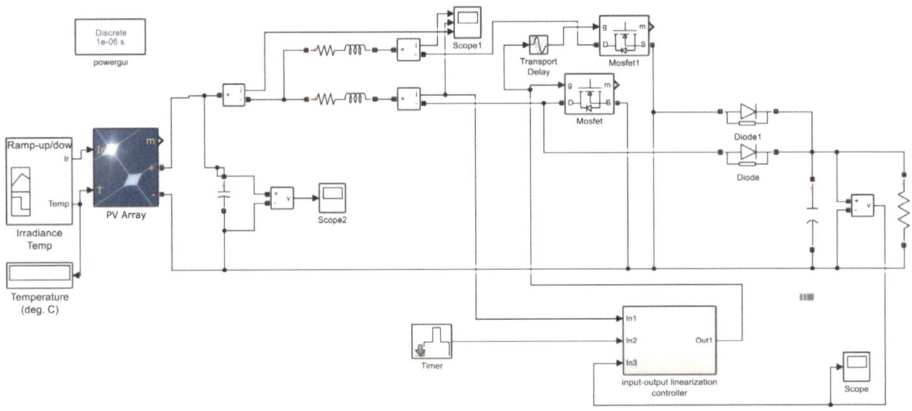

Figure 4.
PV-fed interleaved boost DC-DC converter simulation diagram.

L1, L2	C	R	f_{sw}	PV Array	PV Array
600 μH	1000 μH	250 Ω	69 kHz	40 prl, 2 srs	305 W Pmax, 64.2 Voc

Table 1.
Parameter values used in the study.

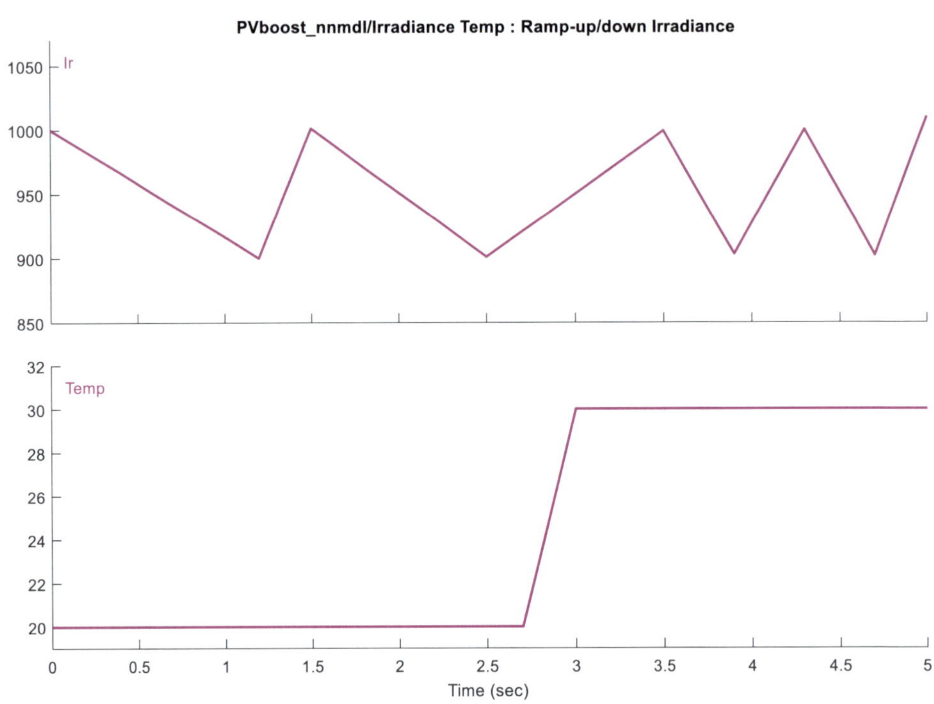

Figure 5.
Irradiation level and ambient temperature change of PV array.

Simulation studies are realized under irradiation and ambient temperature change of the PV array; these changes are sketched in **Figure 5**.

Output voltage under reference change is obtained as in **Figure 6** by the nonlinear controller. Reference voltage is changed from 150 V to 200 V at 1 s, from 200 V to 250 V at 2 s, from 250 V to 200 V at 3 s, and from 200 V to 150 V at 4 s. Under the reference changes, output voltage is achieved as desired with −0.2 V at 150 V reference, −0.5 V at 200 V reference, −1.1 V at 250 V reference, steady-state error. Also, steady-state error is obtained as 0.015 s at 150 V reference, 0.005 s at 200 V

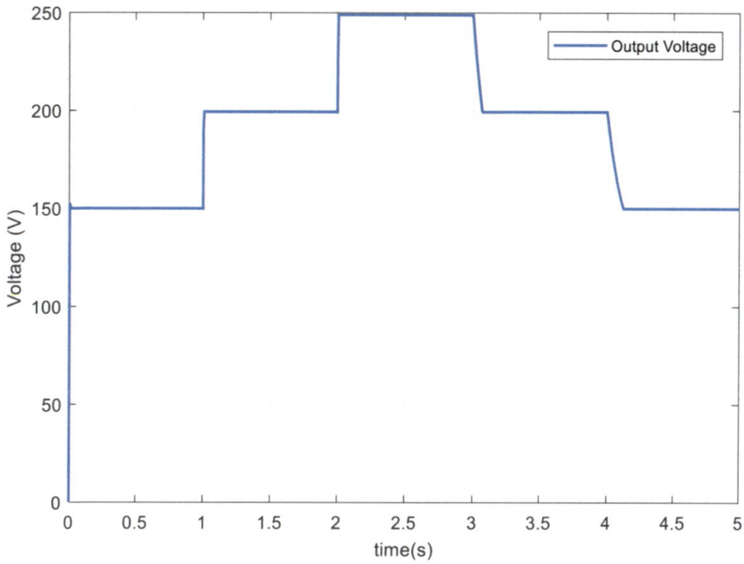

Figure 6.
Output voltage of interleaved boost DC-DC converter.

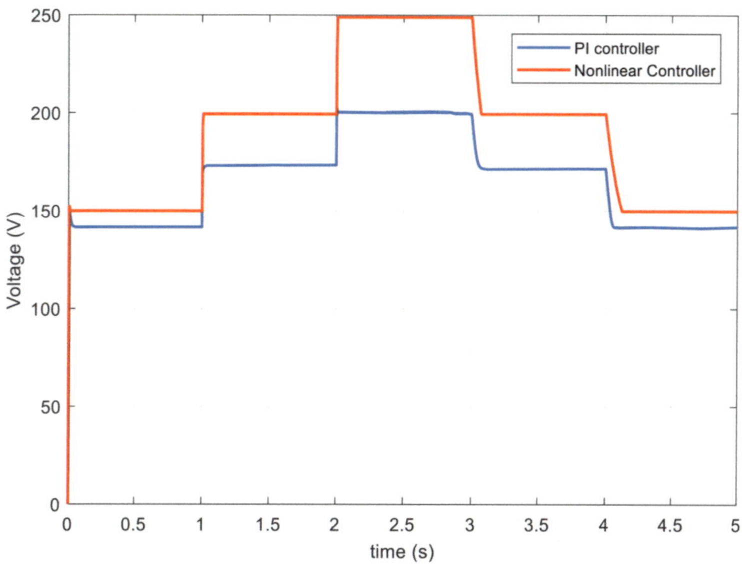

Figure 7.
Output voltage of interleaved boost DC-DC converter with PI and nonlinear controller.

reference, 0.008 s at 250 V reference, and 0.07 s at second 200 V reference, 0.125 s at second 150 V.

In order to compare the performance of the nonlinear controller, the same system is controlled by the PI controller. In **Figure 7**, the results obtained by both controllers are sketched.

Figure 7 shows that by PI controller desired reference voltages are not acquired, however, nonlinear controller provides desired reference voltages.

4. Conclusions

In this chapter, firstly feedback linearization techniques, including input-state and input–output linearization, are described. Then input–output linearization technique is applied to interleaved boost converter that is connected to the output of the PV array. Besides, solar irradiation and ambient temperature of PV array are changed during the simulation study.

The result obtained by the nonlinear controller is compared to the linear PI controller. It is determined by the study that the nonlinear controller ensures the desired output voltage with a maximum 1.1 V steady-state error and 0.125 s settling time, whereas the linear PI controller could not provide the reference voltage as it desired.

In future work, the implementation of the study is targeted to be carried out.

Conflict of interest

The authors declare no conflict of interest.

Author details

Erdal Şehirli
Kastamonu University, Kastamonu, Turkey

*Address all correspondence to: esehirli@kastamonu.edu.tr

References

[1] Khalil H. Nonlinear Systems. 3rd ed. New Jersey: Pearson Education; 2000. p. 505

[2] Slotine JJE, Weiping L. Applied Nonlinear Control. 1st ed. New Jersey: Prentice Hall; 1990. p. 207

[3] Isıdori A. Nonlinear Control Systems: An Introduction. 2nd ed. Heidelberg: Springer; 1995. p. 178. DOI: 10.1007/BFb0006368

[4] Ramirez HS, Ortega R. Control Design Techniques in Power Electronics Devices. 1st ed. Germany: Springer; 2010. p. 235

[5] Kazimierczuk MK. Pulse-Width Modulated DC-DC Power Converters. 1st ed. Singapore: Wiley; 2008. p. 23

[6] Zheng H, Shuai D. Nonlinear control of boost converter by state feedback exact linearization. In: Proceedings of the IEEE 24th Chinese Control and Decision Conference (CCDC). Taiyuan, China: IEEE; 2012. pp. 3502-3506

[7] Arora S, Balsara P, Bhatia D. Input–output linearization of a boost converter with mixed load (constant voltage load and constant power load). IEEE Transactions on Power Electronics. 2019; **34**:815-825. DOI: 10.1109/TPEL.2018.2813324

[8] Salimi M, Siami S. Closed-loop control of DC-DC buck converters based on exact feedback linearization. In: Proceedings of the IEEE 4th International Conference on Electric Power and Energy Conversion Systems (EPECS). Sharjah: IEEE; 2015. p. 1

[9] Ding-xin S. State feedback exact linearization control of buck-boost converter. In: Proceedings of the IEEE International Power Electronics and Application Conference and Exposition. Shanghai, China: IEEE; 2014. p. 1

[10] Sehirli E, Altınay M. Input-output linearization control of single-phase buck-boost power factor corrector. In: Proceedings of the IEEE 47th International Universities Power Engineering Conference (UPEC); 2012; UK: IEEE; 2012. p. 1-6

[11] Sira-Ramirez H, Rios-Bolivar M. Sliding mode control of dc-to-dc power converters via extended linearization. IEEE Transactions on Circuits and Systems I: Fundamental Theory and Applications. 1994;**41**:652-661. DOI: 10.1109/81.329725

[12] Oucheriah S, Guo R. PWM-based adaptive sliding-mode control for boost DC–DC converters. IEEE Transactions on Industrial Electronics. 2013;**60**: 3291-3294. DOI: 10.1109/TIE.2012.2203769

[13] Wang Z, Li S, Li Q. Discrete-time fast terminal sliding mode control design for DC–DC Buck converters with mismatched disturbances. IEEE Transactions on Industrial Informatics. 2020;**16**:1204-1213. DOI: 10.1109/TII.2019.2937878

[14] Lineas-Flores J, Juarez-Abad A, Hernandez-Mendez A, Castro-Heredia O, Guerrero-Castellanos JF, Heredia-Barba R, et al. Sliding mode control based on linear extended state observer for DC-to-DC Buck–Boost power converter system with mismatched disturbances. IEEE Transactions on Industry Applications. 2022;**58**:940-950. DOI: 10.1109/TIA.2021.3130017

[15] Bellinaso L, Figueira H, Basquera MF, Vieira RP, Grundling HA,

Michels L. Cascade control with adaptive voltage controller applied to photovoltaic boost converters. IEEE Transactions on Industry Applications. 2019;**55**:1903-1912. DOI: 10.1109/TIA.2018.2884904

[16] Wei Z, Bao-bin L. Analysis and design of DC-DC buck converter with nonlinear adaptive control. In: Proceedings of the IEEE 7th International Conference on Computer Science & Education (ICCSE). Melbourne, Australia: IEEE; 2012. pp. 1036-1038

[17] Soriano-Rangel C, He W, Mancilla-David F, Ortega R. Voltage regulation in Buck–Boost converters feeding an unknown constant power load: An adaptive passivity-based control. IEEE Transactions on Control Systems Technology. 2021;**29**:395-402. DOI: 10.1109/TCST.2019.2959535

[18] Song P, Cui C, Bai Y. Robust output voltage regulation for DC-DC Buck converters under load variations via sampled-data Sensorless control. IEEE Access. 2018;**6**:10688-10698. DOI: 10.1109/ACCESS.2018.2794458

[19] Mummamdi V, Mohan BK. Robust digital voltage-mode controller for fifth-order boost converter. IEEE Transactions on Industrial Electronics. 2011;**58**:263-277. DOI: 10.1109/TIE.2010.2044130

[20] Buso S. Design of a Robust Voltage Controller for a Buck-boost converter using μ-synthesis. IEEE Transactions on Control Systems Technology. 1999;**7**:222-229. DOI: 10.1109/87.748148

[21] Lee T. Input output linearization and zero-dynamics control of three-phase AC/DC voltage-source converters. IEEE Transactions on Power Electronics. 2003;**18**:11-22. DOI: 10.1109/TPEL.2002.807145

Chapter 11

Nonlinear Intelligent Predictive Control for the Yaw System of Large-Scale Wind Turbines

Dongran Song, Ziqun Li, Jian Yang, Mi Dong, Xiaojiao Chen and Liansheng Huang

Abstract

This chapter presents a nonlinear intelligent predictive control using multi-step prediction model for the electrical motor-based yaw system of an industrial wind turbine. The proposed method introduces a finite control set under constraints for the demanded yaw rate, predicts the multi-step yaw error using the control set element and the prediction wind directions, and employs an exhaustive search method to search the control output candidate giving the minimal value of the objective function. As the objective function is designed for a joint power and actuator usage optimization, the weighting factor in the objective function is optimally determined by the fuzzy regulator that is optimized by an intelligent algorithm. Finally, the proposed method is demonstrated by simulation tests using real wind direction data.

Keywords: nonlinear model predictive, intelligent algorithm, yaw control, wind turbine

1. Introduction

With the increase in social demand, the scale of wind power generation continues to expand. The total installed capacity of global wind power in 2020 has reached 743GW, which means that wind turbines (WTs) are moving towards large-scale and high-capacity. Typical WT controls include pitch, torque, and yaw control, of which about 80% of the research is on the first two, while yaw control has received limited attention [1]. Meanwhile, with the large-scale development of WTs [2], problems such as power reduction and load increase caused by the yaw misalignment can no longer be ignored. According to the investigation result, the potential energy loss due to yaw misalignment is about 2.7% and the failure rate of yaw system accounts for approximately 12.5% of the total failure rate of WTs [3]. Therefore, there is an urgent need to improve the yaw control performance of WT.

Yaw control changes the direction of blade rotating surface by turning the nacelle horizontally. Traditional yaw control methods include Logic Control [4], PID, fuzzy PID [5], and so on. Yet, wind direction sensor suffers from the disturbance of rotor rotation, which makes it difficult to accurately measure the incoming wind direction. In order to

IntechOpen

avoid measurement error, some wind direction estimation methods are proposed, including the hill-climbing search algorithm [6] and so on. However, these improved methods have limited effect on large-scale WTs, and they are rarely applied in industry. In short, with the development of WT towards large-scale, traditional yaw control methods generally have shortcomings, which promote the development of new methods.

As the development of advanced prediction technologies like LiDAR [7], more recent research has concentrated on the advanced predictive controls [8]. Model predictive control (MPC) is a typical representative of predictive control, which has been proposed for the torque and yaw control of WTs, and has achieved good control performance [9–11]. The MPC for yaw system involves performance indicators such as energy capture efficiency and yaw actuator usage [12]. By adding weight coefficients, each performance indicator can be combined into a single objective function. Obviously, the setting of weight coefficient could influence the control performance. In order to find the connotative knowledge, potential regulations and methods, the Pareto optimization theory is used in [13] to explore and gives the suggestion that weight coefficients should be regulated according to the wind characteristics. However, how to effectively adjust the weight coefficients in real time remains unsolved.

Fuzzy logic (FL) is a potential solution to regulate the weight coefficient for model-predictive yaw control. FL is an abstraction of the approximate reasoning characteristics of human decision-making, which has been applied in many fields [14]. Yet, the excessive dependence of FL on expert experience leads to artificially set membership functions (MFs) and fuzzy rules that could limit the control performance. To this end, the optimization of FL is proposed to enhance its effect [15]. In summary, the advantages of FL and the potential room for optimization make it possible to effectively regulate the weight coefficient of MPC.

Motivated by above observations, in this study, the nonlinear MPC (NMPC) method with multi-step prediction models for the yaw control system is proposed. Specifically, an "ideal" NMPC controller that employs perfect previewed wind directions into the prediction model is used in this study and the NMPC problem is solved by using an exhaustive search method based on the sequential diagram. Further, a novel method of using the mind of FL to dynamically regulate the weight coefficient of the NMPC is proposed, which is called as fuzzy inference weight coefficient regulator (FIWR). Specifically, the fuzzy rules and MFs of FIWR are simultaneously optimized by an advanced intelligent algorithm, so as to fully exert the advantage of FIWR. By doing so, it is achieved the deep optimization of NMPC performance for yaw system.

2. Model and methodology

This study aims at proposing and studying the NMPC method for the yaw control system on a horizontal-axis WT. The yaw model for WTs will be introduced first, followed by the design of finite-set NMPC. On this basis, a weight coefficient regulator based on fuzzy inference is proposed, and the multi-objective optimization problem of fuzzy inference is summarized. Then a proposed solution strategy for this problem is introduced, and a multi-objective intelligent optimization algorithm is improved to solve it.

2.1 Yaw system modeling

The yaw system can be modeled according to the three types of yaw dynamics: rigid yaw, flexible yaw, and controlled yaw torque. Since the yaw rate of large WTs is very slow, the yaw rate is set to a fixed value of 0.5 deg./s. There are three yaw situations for WTs: no deflection (0 deg./s), clockwise deflection (0.5 deg./s) and counterclockwise deflection (−0.5 deg./s). In realistic operation, considering the safety requirement of the yaw actuator, the yaw rate at current moment is affected by the yaw rate at previous moment and meets following constraint:

$$\dot{\theta}_{np}(k+1) = \begin{cases} 0.5\text{deg/s} & \exists\dot{\theta}_{np}(k) \in \{0, 0.5\text{deg/s}\} \\ 0\text{deg/s} & \exists\dot{\theta}_{np}(k) \in \{0, 0.5, -0.5\text{deg/s}\} \\ -0.5\text{deg/s} & \exists\dot{\theta}_{np}(k) \in \{0, -0.5\text{deg/s}\} \end{cases} \quad (1)$$

where $\dot{\theta}_{np}(k)$ is the yaw rate in k-th control period, and $\dot{\theta}_{np}(k+1)$ is the yaw rate in $(k + 1)$-th control period. According to Eq.(1), in a certain state at current moment, the control action of the system at next moment is an element in a finite set. This yaw action mode provides the basis for the finite-set NMPC.

2.2 Nonlinear model predictive control

The NMPC for the yaw system aims at maximizing energy capture through tracking the wind direction while avoiding over-usage of the yaw actuator. Accordingly, the yaw error and the yaw actuator usage are used to form the overall objective function. In the following, the proposed NMPC method will be specified in terms of the prediction model, the objective function, and the finite-set NMPC solver.

2.2.1 Multi-step prediction model

The primary control objective of the yaw control system is to minimize the yaw error. Thus, the yaw error θ_{ye} is selected as the state variable, and its one-step prediction model in the form of the discrete equation can be given as follows:

$$\theta_{ye}(k+1|k) = \theta_{wd}(k+1|k) - \theta_{np}(k+1|k) \quad (2)$$

where k is the k-th control period; $\theta_{ye}(k+1|k)$, $\theta_{wd}(k+1|k)$ and $\theta_{np}(k+1|k)$ are the next-step prediction values of yaw error, wind direction, and nacelle position, respectively.

Since the nacelle is rotated by the yaw control system at a certain yaw rate, the predicted nacelle position $\theta_{np}(k+1|k)$ can be obtained by:

$$\theta_{np}(k+1|k) = \theta_{np}(k) + \dot{\theta}_{np}(k+1) \cdot T_s \quad (3)$$

where $\theta_{np}(k)$ is the nacelle position at the k-th control period, T_s is the control period, and $\dot{\theta}_{np}(k+1)$ is the yaw rate at the $(k + 1)$-th control period.

By using Eq. (2) and (3), the m-step prediction model of the yaw error $\theta_{ye}(k+m|k)$ can be obtained by:

$$\theta_{ye}(k+m|k) = \theta_{wd}(k+m|k) - \theta_{np}(k) - \sum_{i=1}^{m} \dot{\theta}_{np}(k+i) \cdot T_s \quad (4)$$

where $\theta_{wd}(k+m|k)$ is the variable that needs to be predicted, which can be predicted by LiDAR or some prediction methods.

2.2.2 Objective function

The first goal of yaw control is to improve the energy capture of the WT, and the second goal is to reduce the yaw actuator usage time. Considering that the energy capture of the WT has a cosine-squared relation to the yaw error, the two objectives can be express as:

$$E_{cap} = \sum_{i=k+1}^{k+m} \frac{1}{2}\rho A_r C_p V_0^3 \cos^2(\theta_{ye}(i)) \tag{5}$$

$$t_{yaw} = \sum_{i=k+1}^{k+m} \left(|\dot{\theta}_{np}(i)| > 0\right) \cdot T_s \tag{6}$$

where ρ is air density, A_r is rotor area, C_p is aerodynamic power coefficient, V_0 is the effective wind speed. Considering the dimensional difference, the two control objectives after normalization can be expressed as follows:

$$\xi = \frac{(E_{ideal} - E_{cap})}{E_{ideal}} = 1 - \left(\sum_{n=k+1}^{k+m} \cos^2(\theta_{ye}(n))\right)/m \tag{7}$$

$$\zeta = \frac{t_{yaw}}{t_{tol}} = \left(\sum_{n=k+1}^{k+m} \left(|\dot{\theta}_{np}(n)| > 0\right)\right)/m \tag{8}$$

where ξ is the energy capture loss ratio caused by the yaw error, E_{cap} is the energy capture considering yaw error, E_{ideal} is the energy capture in an ideal state; ζ is the yaw actuator usage ratio, t_{yaw} is the yaw time, t_{tol} is the running time of the WT.

By adding a weight coefficient between ξ and ζ, the objective function QF of the NMPC can be written as:

$$\begin{aligned} QF &= (1-\omega)\cdot\xi + \omega\cdot\zeta \\ &= (1-\omega)\cdot\left(1 - \left(\sum_{n=k+1}^{k+m} \cos^2(\theta_{ye}(n))\right)/m\right) \\ &\quad + \omega\cdot\left(\left(\sum_{n=k+1}^{k+m} \left(|\dot{\theta}_{np}(n)| > 0\right)\right)/m\right) \end{aligned} \tag{9}$$

where ω is the weight coefficient, which is used to balance energy capture loss ratio ξ and yaw actuator usage ratio ζ.

2.2.3 Finite-set NMPC solver

So far, the NMPC problem for the yaw system has been formulated by Eq. (1)–(9), which is a nonlinear optimization issue under constraints. To facilitate the problem solver, the control horizon is selected to be equal to the prediction horizon. According to Eq. (1), under the finite prediction horizon, the control law of NMPC always belongs to a finite set. Thus, the designed optimal problem can be effectively solved by using an exhaustive search (ES) method.

Figure 1 illustrates the case of the NMPC with the three-step prediction model, from which the ES method is explained as follows:

During the initialization period ($m = 0$), the yaw control system is inactivated, and thus the total yaw state is 1.

- For $m = 1$, because the current yaw rate is zero. Accordingly, there are three potential solutions and the total yaw states are 3.
- For $m = 2$, the situation is different. Constrained by Eq. (1), there are only two candidate values in the case of $\dot{\theta}_{np}(m = 1) = \pm 0.5$ deg/s, while there are three candidate values in the case of $\dot{\theta}_{np}(m = 1) = 0$. Consequently, along the sequential order, there are seven potential solutions and the total yaw states are 7.
- For $m = 3$, the situation becomes slightly complicated but similar to the two-step prediction model. There are seventeen solutions for the three-step prediction model and the total yaw states are 17.

2.3 Intelligent fuzzy inference weight coefficient regulator

There is a contradiction between increasing energy capture and reducing yaw actuator usage. In Eq. (9), ω is used to balance ξ and ζ in QF, so the choice of weight coefficient affects the performance of NMPC to a large extent. Even if an optimal constant ω is selected according to the Pareto theory, it cannot ensure that the best control performance is always provided. Therefore, the FIWR is proposed to dynamically adjust ω according to the predicted wind direction in each control period. Moreover, to better play the effect of FIWR and avoid the subjectivity of manual tuning, the intelligent optimization of FIWR is also necessary.

2.3.1 Design of FIWR

The proposed FIWR scheme is shown in **Figure 2**. This is a fuzzy inference system with two inputs and one output. Inputs WD_{av} and WD_{std} have three and five linguistic

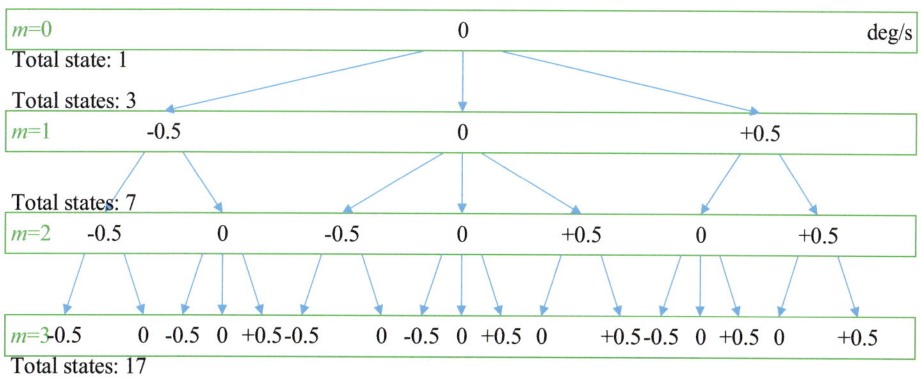

Figure 1.
Sequential diagram of the ES for a three-step prediction model.

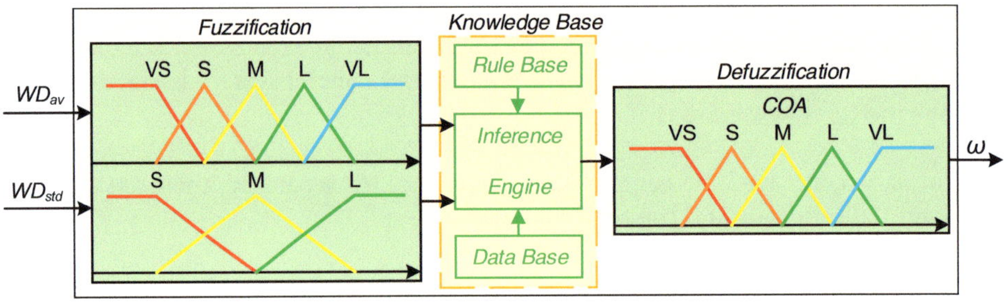

Figure 2.
The scheme of FIWR.

values, respectively, and output ω has five linguistic values. The initial membership function (IMF) adopts the equally divided triangular MF. Fuzzy inference adopts Mamdani-type algorithm and the center of area (COA) is utilized in defuzzification process.

2.3.1.1 Design of input/output

The yaw error is the core part of the input. Which directly determines the action of the yaw system. Therefore, the adjustment of ω takes the yaw error as a reference. Because the future yaw action is the control law to be solved, which is unknown, the predicted yaw error here refers to the difference between the predicted average wind direction of the $(k + i)$-th control period and the first sampled value of the nacelle position of the current k-th control period, expressed by $\theta_{ye}(k+i \rightarrow k)$, which denotes the difference between the future wind direction obtained by LiDAR and the nacelle position at the current moment.

Different from the error and error derivative method used by ordinary two-input fuzzy inference, the designed input of the FIWR is related to the statistical characteristics of the predicted yaw error. The two inputs are designed as the weighted average and standard deviation of yaw error in prediction horizon m respectively, named WD_{av} and WD_{std}. The calculation of WD_{av} is:

$$WD_{av} = \sum_{i=1}^{m}(m+1-i)\cdot\theta_{ye}(k+i \rightarrow k)/\sum_{i=1}^{m} i \tag{10}$$

where m is the prediction step, and $\theta_{ye}(k+i \rightarrow k)$ can be calculated by setting the yaw rate as zero in Eq. (4). Based on WD_{av}, WD_{std} is calculated by:

$$WD_{std} = \sqrt{\sum\nolimits_{i=1}^{m}\left(\theta_{ye}(k+i \rightarrow k) - WD_{av}\right)^2/m} \tag{11}$$

In practice, the yaw error might be affected by some subtle factors, so moving average filter is presented to process the wind direction data. In this study, the filter value of each sample is the mean value of the N sample values in the sliding window. For wind direction, N usually takes 12.

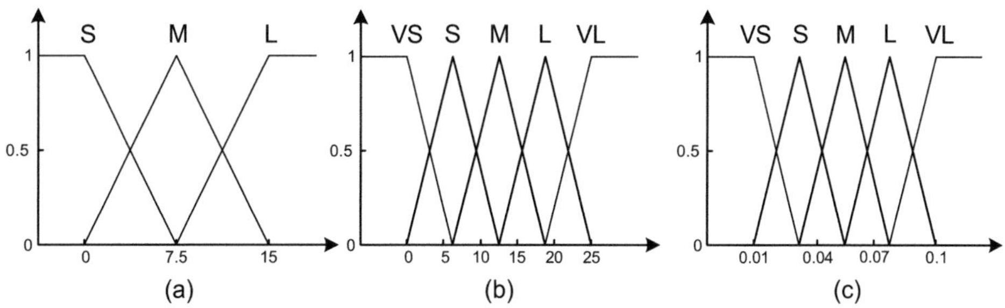

Figure 3.
Initial membership functions.

WD_{std}		VS	S	M	L	VL
WD_{av}	S	VL	L	M	M	S
	M	S	S	M	L	L
	L	VS	VS	S	L	VL

Table 1.
Initial fuzzy rules.

2.3.1.2 Design of MFs and fuzzy rules

The IMFs corresponding to input WD_{av}, WD_{std} and output ω are illustrated in **Figure 3**, respectively. The types of IMFs are all selected as sensitive and simple triangular MFs, with the bottom edge equal and overlapping with adjacent IMFs by 50%. The universe of discourse (UOD) of WD_{av} and WD_{std} is defined as [0 deg., 15 deg] and [0 deg., 25 deg] respectively. The best value range of ω is [0, 0.1], so the UOD of output is defined on [0, 0.1]. The linguistic values VS, S, M, L, and VL represent very small, small, medium, large, and very large, respectively. WD_{av} and WD_{std} are mapped to 3 and 5 linguistic values, respectively, so the fuzzy rule table will contain 15 different rules. The initial fuzzy rules of the proposed FIWR are listed in **Table 1**. The derivation of the fuzzy rules is based on the expert experience, that is, a larger yaw error and a smaller standard deviation will lead to the yaw action towards improving energy capture.

2.3.2 Intelligent optimization of FIWR

The advantage of fuzzy inference is that it can fully incorporate expert experience. However, when the expert experience is insufficient or wrong, the result of fuzzy inference will no longer be reliable; and the fuzzy relationship under complex input sometimes cannot be directly given by the expert experience. Therefore, the optimization of FIWR is proposed.

2.3.2.1 Fuzzy optimization problem formulation

The goal of the proposed FIWR is to reduce the yaw actuator usage and the energy capture loss, so the optimization problem can be expressed as:

$$\min F(x_{membership}, x_{rule}) = \left(\frac{\int \xi dt}{\int dt}, \frac{\int \zeta dt}{\int dt} \right)$$
$$s.t. \begin{cases} x_{membership} \in \Omega_m \\ x_{rule} \in \Omega_r \end{cases} \tag{12}$$

where $x_{membership}$ and x_{rule} is the optimization vector of MFs and fuzzy rules, respectively; Ω_m and Ω_r is the feasible regions of the two optimization vectors, respectively.

For Ω_m, there are three kinds of constraints: the number constraint of MF, the type constraint of MF, and the position constraint of MF. As for this study, in order to simplify the optimization problem, the optimization variables corresponding to the first two constraints are fixed, that is, the number and type of MF do not need to be optimized. Assuming that the position of each MF is uniquely determined by a certain vertex of the triangle, the optimization dimension of MF is further reduced. Obviously, the optimization of position is subject to constraints, that is, a small linguistic value cannot exceed a large linguistic value and each MF must be changed within the UOD.

For Ω_r, it is affected by the number of inputs and outputs. For the fuzzy inference using Mamdani model, the consequence of the fuzzy rule is a certain fuzzy set of output. If there are s inputs and h outputs in fuzzy inference, the feasible region of the fuzzy rule can be expressed as:

$$\Omega_r = rule\left(\prod_{i=1}^{h} num_i^{\prod_{j=1}^{s} num_j} \right) \tag{13}$$

where num_j is the number of linguistic values of j-th input, and num_i is the number of linguistic values of i-th output.

2.3.2.2 Fuzzy optimization problem simplification

Although Eq. (12) has been simplified, it is still difficult to solve directly. For this complex problem, it is necessary to simplify the problem as much as possible on the premise of ensuring a certain solution accuracy. Therefore, a solution strategy is proposed to simplified the complex optimization problem with the purpose of quickly and reliably solving FIWR optimization parameters.

As mentioned earlier, the MFs are subject to order constraints that the MF with smaller linguistic value must be front of the MF with larger linguistic value. This constraint is to avoid repeated searches and ensure the logical accuracy of fuzzy inference. However, if the output MF is no longer constrained, the current linguistic value sequence of output can be used as a reference for the optimization of fuzzy rule. Therefore, the fuzzy rule is associated with MF, which can greatly reduce the complexity of optimization problem and ensure the search ability.

Specifically, the output linguistic value sequence after the sequence change can be expressed as:

$$B = \boldsymbol{A} \cdot \boldsymbol{S} \tag{14}$$

where $\boldsymbol{A}$ is the original sequence, and $\boldsymbol{S}$ is the identity matrix after elementary transformation, called the transformation matrix.

For example, in a certain change, the output linguistic value sequence is transformed into [S M L VS VL], then it can be expressed as:

$$[\mathrm{SMLVSVL}] = [\mathrm{VSSMLVL}]\begin{bmatrix} 0 & 0 & 0 & 1 & 0 \\ 1 & 0 & 0 & 0 & 0 \\ 0 & 1 & 0 & 0 & 0 \\ 0 & 0 & 1 & 0 & 0 \\ 0 & 0 & 0 & 0 & 1 \end{bmatrix} \tag{15}$$

After each iteration, $\boldsymbol{A}$ and $\boldsymbol{B}$ are known, so $\boldsymbol{S}$ can be calculated according to Eq. (14). Then the updated fuzzy rule table can be calculated according to $\boldsymbol{S}$:

$$R_{new}[3 \times 5] = R_{old}[3 \times 5] \cdot S[5 \times 5] \tag{16}$$

where $R_{new}[3 \times 5]$ and $R_{old}[3 \times 5]$ are the fuzzy rule tables after and before the update, respectively, with a size of 3×5. In order to ensure the exploitation of the solution process, each linguistic value in the fuzzy rule table is transformed correspondingly with a certain probability. For example, the S linguistic value is changed to VS with a small probability p. This probabilistic processing procedure can improve the search ability that find the optimal MF under the current fuzzy rule.

2.3.2.3 Improved AGA-MOPSO solver

Although Eq. (12) is simplified by the solution strategy, its objective function is complex, which makes it difficult to be solved by ordinary multi-objective optimization algorithms. Therefore, an improved multi-objective particle swarm optimization (MOPSO) algorithm based on adaptive grid algorithm MOPSO (AGA-MOPSO) is designed. AGA-MOSPO is an efficient variant of PSO for multi-objective problem. By combining with adaptive grid algorithm (AGA), it achieves nice balance between exploration and exploitation [16].

For a two-dimensional multi-objective problem, AGA-MOPSO first calculates the search ranges $\left[\min f_1^k, \max f_1^k\right]$ and $\left[\min f_2^k, \max f_2^k\right]$ of the objective space after k-th iteration, then calculates the grid number of the i-th particle according to the following equation:

$$\left(x_1^i, x_2^i\right) = \left(Int\left(\frac{f_1^i - \min f_1^k}{\left(\max f_1^k - \min f_1^k\right)/M}\right) + 1, Int\left(\frac{f_2^i - \min f_2^k}{\left(\max f_2^k - \min f_2^k\right)/M}\right) + 1\right) \tag{17}$$

where x_1^i and x_2^i is the grid numbers of the particles, M is the number of grids, f_1^i and f_2^i are the fitness values of the two targets respectively, and Int is rounding.

M will be adaptively increased with the iteration to balance the computational cost and accuracy. The density information of each grid can be obtained according to the grid number. According to the density information, the global optimal particles are selected and the Archive set is clipped.

Considering the complexity of the optimization problem, the following improvements are proposed for the selection of the global optimal particle and the truncation of Archive set in AGA-MOPSO:

(1) First is to use two methods to determine the corresponding g_{best} for each particle: 1) Select g_{best} with the smallest grid density. This method focuses on the exploration of the search space to improve the ductility of Pareto front (PF). 2) Select the global optimal particle as g_{best} according to the technique for order preference by similarity to an ideal solution (TOPSIS):

$$d = \frac{\sqrt{\left(f_1 - f_1^{\min}\right)^2 + \left(f_2 - f_2^{\min}\right)^2}}{\sqrt{\left(f_1 - f_1^{\min}\right)^2 + \left(f_2 - f_2^{\min}\right)^2} + \sqrt{\left(f_1 - f_1^{\max}\right)^2 + \left(f_2 - f_2^{\max}\right)^2}} \tag{18}$$

where d is the deviation between a certain point and the ideal point. The smaller the d, the smaller the deviation from the ideal point. This method focuses on the exploitation of the search space to make PF closer to the real optimal solution. In the early stage of algorithm, the probability of choosing g_{best} by the first method is greater, so as to find as many non-dominated solutions as possible; in the later stage, the probability of choosing the second method is greater to approximate the true solution.

(2) Second is to truncate the Archive set by adaptive dynamic threshold. It is calculated using:

$$Th \cdot M = C \tag{19}$$

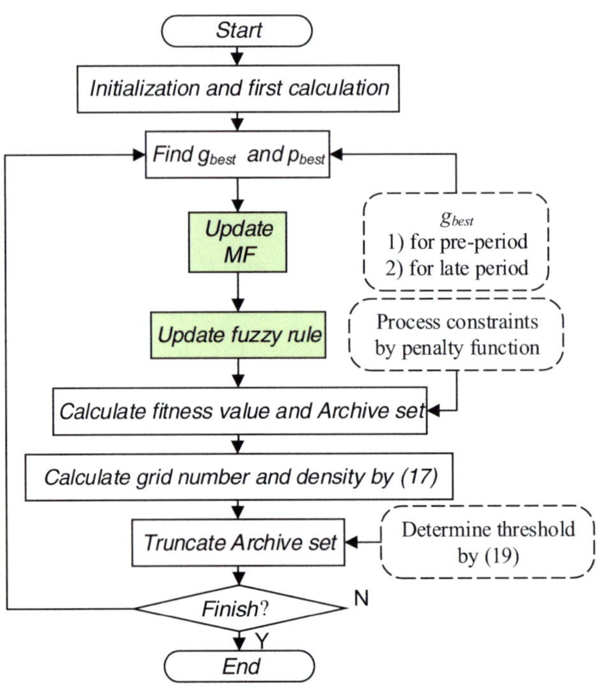

Figure 4.
The flowchart of the improved AGA-MOPSO.

where Th is the threshold, and C is a constant. When the number of particles in the grid exceeds Th, the grid is truncated; Th is reduced following an increased M. This ensuring that the number of particles on the PF is relatively stable.

The main procedure of the improved AGA-MOPSO solver is shown in **Figure 4**.

3. Validation and discussion

The experiment is based on MATLAB/SIMULINK. First, the optimized parameters of FIWR are obtained by the solution strategy that run on MATLAB. Then, the proposed FIWR-NMPC controller is simulated in SIMULINK. The common experimental parameters in **Table 2** are set to the same value. The UOD is set to a uniform value [0, 10] to facilitate the handling of the constraints of optimization variables, so the universe conversion scale coefficients corresponding to WD_{av}, WD_{std}, and ω are 1.5, 2.5, and 0.01, respectively.

The wind direction data used in the experiment is from the actual wind direction data of a wind farm in operation for one day with a sampling period of 1 s, as shown in **Figure 5(a)**. The wind direction after sliding average filtering is shown in **Figure 5(b)**.

3.1 Optimization results

The optimization results are analyzed and discussed first. Taking the case of prediction step m = 6 as an example, the optimization results of AGA-MOPSO are shown in **Figure 6**, where the horizontal axis is the yaw actuator usage ratio, and the vertical axis is the energy capture loss ratio. After 50 iterations, the particles finally converge to PF. The particles are evenly distributed in the search space near the PF, which indicate that this associated idea will not lead to the coupling optimization of PF and fuzzy rule in the search process.

The optimization results of the design variables are shown in **Table A** in Appendix. Considering space reasons, only ten points on PF are randomly selected to discuss. The simplification solution strategy associates fuzzy rule with MF to optimize, so as to reduce optimization complexity. The MF of ω does not strictly follow the linguistic value order. This random information is utilized to adjust the fuzzy rule, thereby simultaneously optimizing the fuzzy rule and the MF.

The prediction step m in the proposed FIWR-NMPC is variable and has a greater impact on the performance of controller. Therefore, the influence of m on the optimization effect of FDWE is discussed. **Figure 7** shows the optimization result under m = 1–6. In Figure 11, the PF when m =1 is significantly different from the PF when m = 2–6; as m increases, the yaw actuator usage ratio and energy capture loss ratio

AGA-MOPSO	**Population size**	**Number of iterations**	**Number of grids**
	100	**50**	**[10 20]**
FIWR	Deduction method	Defuzzification	UOD
	COA	Mamdani-type	[0 10]
NMPC	Sampling period(s)	Control period(s)	Prediction step
	1	30	[1 6]

Table 2.
Common parameters in the experiment.

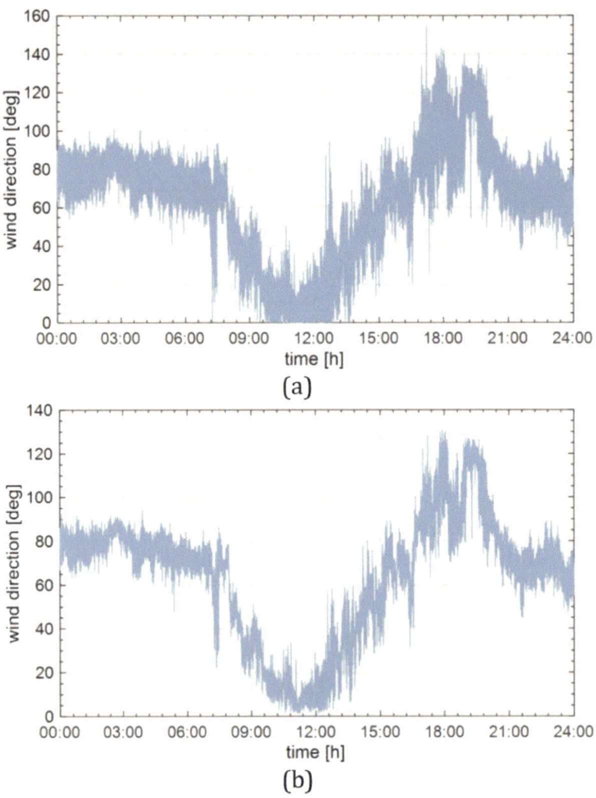

Figure 5.
24-hour wind direction used in experiment: (a) actual wind direction; (b) filtering wind direction.

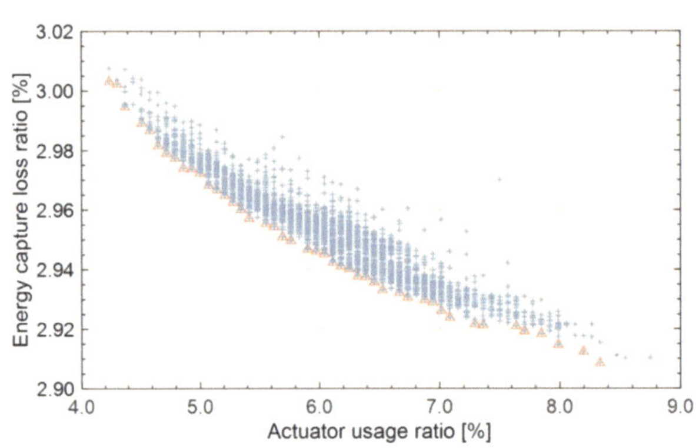

Figure 6.
Iterative results.

show a downward trend. This is because m =1 is very short, and the predicted wind information is very few. But the actual wind direction greatly fluctuates due to the existence of turbulence, and the yaw system has a large time lag. So, it is difficult for the yaw system to track the change of the wind direction. When m increases, the predicted wind direction information increases, and the controller can start yawing several control periods in advance to compensate for the time lag.

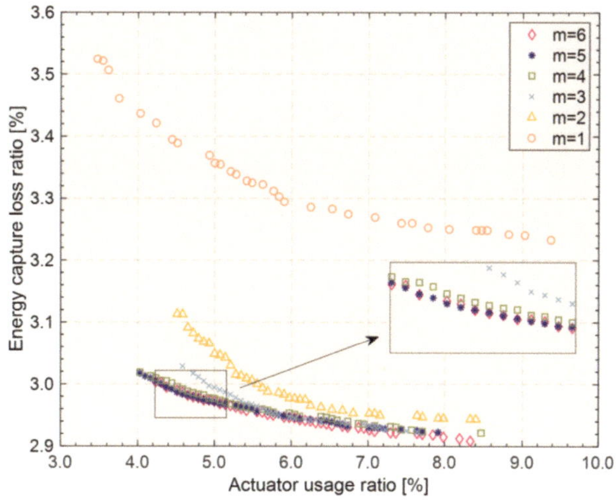

Figure 7.
Optimization results under different m.

Furthermore, in the enlarged part, the PF of m =5 and m =6 is very close, and the increase of m gradually reduces the improvement effect of the FIWR-NMPC control performance. These results show that the increase in m can provide better performance for FIWR-NMPC, but there is a limit to the performance improvement. Among the six FIWR-NMPCs with different m, the controller with m =6 provides the best performance, which is very close to the ultimate performance.

3.2 Simulation results

A specific FIWR-NMPC controller and a baseline NMPC controller is designed based on the foregoing discussion. The MFs and fuzzy rules in FIWR are shown in **Figure 8** and **Table 3**. The MFs and fuzzy rules are derived from the optimal solution obtained through TOPSIS. The remaining parameters of FIWR-NMPC and baseline NMPC are shown in **Table 2**, where m =6.

The simulation results are shown in **Figure 9**. **Figure 9(a)** and **(b)** respectively represent the energy capture loss ratio and the yaw actuator usage ratio of the WT within 24 hours. Compare with the baseline NMPC, the proposed FIWR-MPC

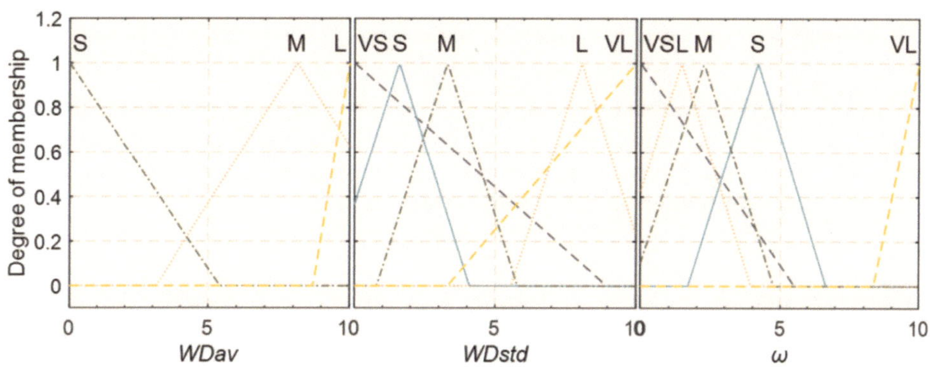

Figure 8.
MFs used by FIWR.

WD_{std}		VS	S	M	L	VL
WD_{av}	S	VL	L	M	M	VL
	M	M	M	L	S	S
	L	VS	VS	S	S	VS

Table 3.
Fuzzy rules used by FIWR.

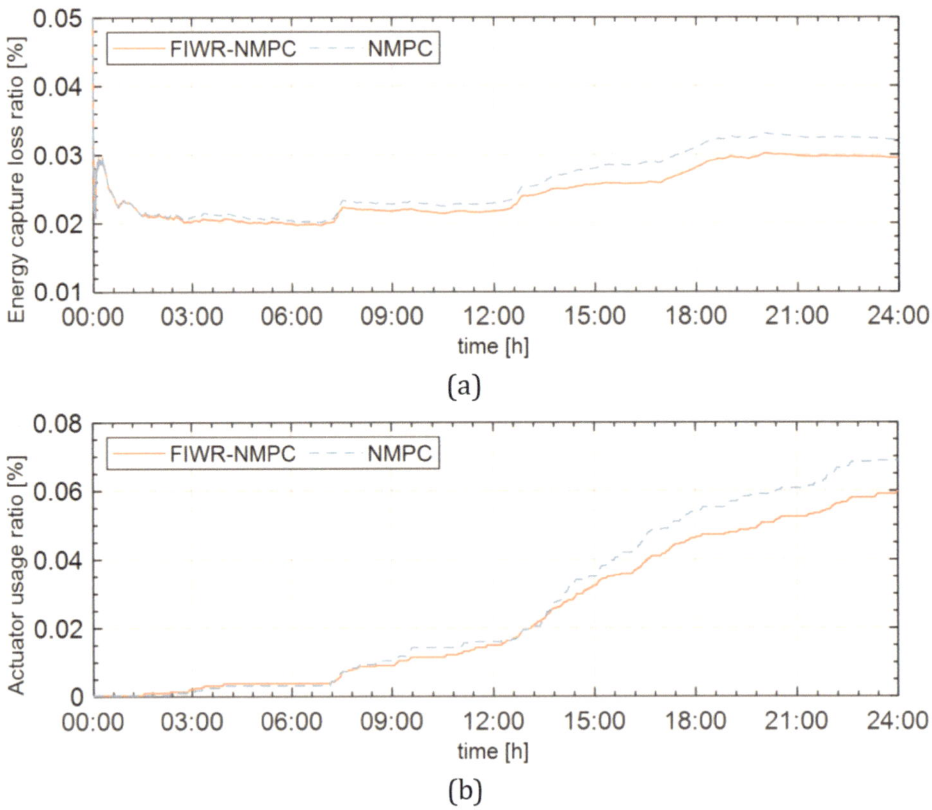

Figure 9.
Main performance of the two controllers: (a) energy capture loss ratio; (b) yaw actuator usage ratio.

increases the energy capture by about 0.3% while reducing the yaw actuator usage ratio by about 1%. This improvement benefits from the dynamically adjusted weight coefficient, that is, dynamically weighing the two control objectives based on the predicted wind information to determine the yaw action.

4. Conclusions

In this study, an advanced nonlinear model predictive control solution including multi-step prediction models has been proposed and investigated for the yaw control system of a horizontal-axis wind turbine. The noticeable feature of the proposed solution is to use a finite control set under constraints for the possible demanded yaw rate, and thus the optimal control demand for the yaw system has been conveniently solved using an exhaustive search method based on the sequential diagram. On the other hand, the weighting coefficient in the objective function of NMPC has been dynamically tuned by employing the fuzzy inference regulator, as the proposed solution is designed for a joint energy capture and actuator usage optimization which is basically a two objective tradeoff that depends on the selection of the weighting factor. To give full play to the ability of the regulator, its parameter tuning is refined into an optimization problem and a solution strategy is designed to simplify it, and then the optimal fuzzy rules and membership functions are solved by the improved AGA-MOPSO algorithm. The final optimized FIWR-NMPC achieves deep optimization of wind turbine yaw performance. The important investigation findings include:

- Both fixed-weight NMPC and FIWR-NMPC can achieve higher energy capture with lower yaw time. NMPC can achieve 96.875% energy capture efficiency at 6.875% yaw time ratio, while FIWR-NMPC can achieve 97.052% energy capture efficiency at 5.792% yaw time ratio. FIWR-NMPC further improves the yaw performance by dynamically adjusting ω according to wind direction information.

- The fuzzy optimization problem simplified by the solution strategy can be solved reliably by AGA-MOPSO. The optimization method is promising to provide guidance for the design of fuzzy inference problem.

- Along with the extended prediction horizon of the NMPC, the energy capture performance is enhanced while maintaining the same yaw actuator usage, while the enhanced performance achieved by the NMPC is limited. The simulation results show that when the predicted horizon $m = 6$, its influence on the control effect tends to be stable.

- Future work can be carried out from the following aspects:

- Verify algorithm performance in professional simulation software like Bladed to further improve the possibility of practical application.

- Another research focus can be focused on adaptive tuning of NMPC control parameters to further improve yaw performance.

- FIWR provides a promising direction for NMPC research. In addition to the yaw system, control systems with multiple contradictory targets can use this method for performance optimization. The application in other WT control systems could be considered in the future.

Acknowledgements

This work is supported by the National Natural Science Foundation of China under Grant 52177204; the Natural Science Foundation of Hunan Province (No. 2020JJ4744); the Innovation-Driven Project of Central South University (No. 2020CX031).

Conflict of interest

The authors declare no conflict of interest.

A. Appendix

(a) Optimization results of MFs.

	WD_{av}				WD_{stt}					ω		
S	M	L	VS	S	M	L	VL	VS	S	M	L	VL
7.8700	6.6634	2.9170	5.6587	2.9202	3.6664	7.4053	3.2371	3.9158	6.0601	7.5499	0.5157	8.4095
6.0657	7.8755	8.2295	8.9000	4.6107	5.4806	9.7508	3.6173	4.3579	3.0437	3.2677	0.0000	8.0467
6.1503	5.7286	8.3112	7.8680	1.0916	3.3298	5.8968	1.3362	6.4390	4.3686	3.7788	4.4940	5.8137
7.7184	9.6993	7.4724	8.5182	2.2270	5.5094	8.2835	4.7723	4.5060	5.1648	4.5060	1.4068	7.9818
5.0738	6.9428	9.3414	9.1123	2.4585	4.4399	8.0474	2.5071	5.5342	3.1609	3.2971	1.5578	6.9253
5.9604	7.1217	8.0914	8.2626	2.6369	4.7863	7.5243	2.8854	5.0664	3.3863	3.6186	1.9524	6.7971
5.6626	6.3262	7.9887	8.0156	1.9404	3.6519	6.5687	2.2624	5.3948	5.3088	3.5967	2.3074	6.4714
5.1838	7.6663	7.3318	7.9048	4.1286	6.1454	8.4669	3.0868	4.3406	2.4830	3.0608	0.7845	7.3532
6.9129	7.6576	7.2776	9.3193	2.6147	4.9671	8.1121	4.1930	3.9820	2.7621	4.6776	1.6381	7.6662
6.2102	7.5423	7.7496	8.9233	2.9329	5.1472	8.4352	3.7087	4.4784	2.7644	4.3071	1.1391	6.9978

(b) Optimization results of fuzzy rules.

WD_{av}	S	M	L	S	M	L	S	M	L	S	M	L	S	M	L
WD_{std}	VS	VS	VS	S	S	S	M	M	M	L	L	L	VL	VL	VL
1	VL	M	VS	L	S	VS	S	L	S	M	S	S	VS	S	VL
2	VS	L	VS	M	M	VL	M	L	S	L	M	S	VL	L	VS
3	VL	L	VL	L	L	VL	L	S	L	M	L	L	VS	L	VL
4	VL	S	VS	M	M	VL	S	M	M	M	M	M	VL	S	VS
5	VL	L	VS	S	L	VL	S	M	S	S	S	L	VS	S	VL
6	VL	S	VS	S	L	VL	L	S	M	S	M	S	VS	S	VL
7	VL	L	VL	M	M	VS	M	M	M	L	M	M	VS	S	VL
8	VS	L	VS	L	M	VL	S	S	S	L	L	L	VL	L	VS
9	VL	S	VS	S	M	VL	S	S	M	S	L	L	VS	L	VL
10	VL	L	VL	S	S	VS	M	L	L	L	M	L	VS	S	VL

Table A.
Design variable optimization results.

Author details

Dongran Song[1]*, Ziqun Li[1], Jian Yang[1], Mi Dong[1], Xiaojiao Chen[2]
and Liansheng Huang[2]

1 School of Automation, Central South University, Changsha, China

2 Institute of Plasma Physics, Chinese Academy of Sciences, Hefei, China

*Address all correspondence to: humble_szy@163.com

References

[1] Yang J, Fang L, Song D, et al. Review of control strategy of large horizontal-axis wind turbines yaw system. Wind Energy. 2020;**24**(2):97-115. DOI: 10.1002/we.2564

[2] Song D, Shanmin X, Huang L, et al. Multi-site and multi-objective optimization for wind turbines based on the design of virtual representative wind farm. Energy. 2022;**252**:123995. DOI: 10.1016/j.energy.2022.123995

[3] Perez JMP, Marquez FPG, Tobias A, et al. Wind turbine reliability analysis. Renewable & Sustainable Energy Reviews. 2013;**23**:463-472. DOI: 10.1016/j.rser.2013.03.018

[4] Song D, Fan X, Yang J, et al. Power extraction efficiency optimization of horizontal-axis wind turbines through optimizing control parameters of yaw control systems using an intelligent method. Applied Energy. 2018;**224**: 267-279. DOI: 10.1016/j.apenergy. 2018.04.114

[5] Campos-Mercado E, Cerecero-Natale LF, Garcia-Salazar O, et al. Mathematical modeling and fuzzy proportional-integral-derivative scheme to control the yaw motion of a wind turbine. Wind Energy. 2020;**24**(4):379-401. DOI: 10.1002/we.2579

[6] Guo F, Jiang W, Shao H, et al. Research on the wind turbine yaw system based on PLC. In: Presented at the 2017 29th Chinese Control and Decision Conference. IEEE; 2017

[7] Han Zhao, Lawu Zhou, Yu Liang et al. "A 2.5MW wind turbine TL-EMPC yaw strategy based on ideal wind measurement by LiDAR". IEEE Access. 2021;**9**:89866-89877. DOI: 10.1109/ access.2021.3089513

[8] Song D, Yang Y, Zheng S, et al. New perspectives on maximum wind energy extraction of variable-speed wind turbines using previewed wind speeds. Energy Conversion and Management. 2020;**206**:112496. DOI: 10.1016/j. enconman.2020.112496

[9] Song D, Li Z, Wang L, et al. Energy capture efficiency enhancement of wind turbines via stochastic model predictive yaw control based on intelligent scenarios generation. Applied Energy. 2022;**312**:118773. DOI: 10.1016/j. apenergy.2022.118773

[10] Song D, Yanping T, Wang L, et al. Coordinated optimization on energy capture and torque fluctuation of wind turbines via variable weight NMPC with fuzzy regulator. Applied Energy. 2022; **312**:118821. DOI: 10.1016/j.apenergy. 2022.118821

[11] Song D, Liu J, Yang Y, et al. Maximum wind energy extraction of large-scale wind turbines using nonlinear model predictive control via yin-Yang grey wolf optimization algorithm. Energy. 2021;**221**:119866. DOI: 10.1016/j.energy.2021.119866

[12] Song D, Chang Q, Zheng S, et al. Adaptive model predictive control for yaw system of variable-speed wind turbines. Journal of Modern Power Systems and Clean Energy. 2021;**9**(1): 219-224. DOI: 10.35833/mpce.2019. 000467

[13] Song D, Li Q, Cai Z, et al. Model predictive control using multi-step prediction model for electrical yaw system of horizontal-Axis wind turbines. IEEE Transactions on Sustainable Energy. 2019;**10**(4):2084-2093. DOI: 10.1109/tste.2018.2878624

[14] Masero E, Francisco M, Maestre JM, et al. Hierarchical distributed model predictive control based on fuzzy negotiation. Expert Systems with Applications. 2021;**176**:1-13. DOI: 10.1016/j.eswa.2021.114836

[15] Castillo O, Martínez-Marroquín R, Melin P, et al. Comparative study of bio-inspired algorithms applied to the optimization of type-1 and type-2 fuzzy controllers for an autonomous mobile robot. Information Sciences. 2012;**192**: 19-38. DOI: 10.1016/j.ins.2010.02.022

[16] Yang J, Zhou J, Fang R, et al. Multi-objective particle swarm optimization based on adaptive grid algorithms. Journal of System Simulation. 2008;**21**:5843-5847. DOI: 10.16182/j.cnki.joss.2008.21.041